Linear Algebra and Its Applications in Programming

Editors-in-chief:

Guo Shu-Li

School of Automation

Beijing Institute of Technology

Han Li-Na

Department of Cardiovascular Internal Medicine

Nanlou Branch of Chinese PLA General Hospital

北京理工大学出版社

BEIJING INSTITUTE OF TECHNOLOGY PRESS

线性代数及其在规划中的应用

主编：
郭树理
北京理工大学自动化学院

韩丽娜
中国人民解放军总医院南楼心血管内科

北京理工大学出版社
BEIJING INSTITUTE OF TECHNOLOGY PRESS

Top-Quality Course Construction of Beijing Institute of Technology

Linear Algebra and Its Applications in Programming

Editors-in-chief:

Guo Shu-Li Han Li-Na

Editors:

Lu Bei-Bei Guo Yang Hao Xiao-Ting

Wang Chun-Xi Si Quan-Jin

北京理工大学出版社

BEIJING INSTITUTE OF TECHNOLOGY PRESS

北京理工大学精品课程建设

线性代数及其在规划中的应用

主编：

郭树理　韩丽娜

编委：

卢贝贝　郭　阳　郝晓亭

王春喜　司全金

北京理工大学出版社

BEIJING INSTITUTE OF TECHNOLOGY PRESS

Introduction

In 1991, I first learned an undergraduate course on linear algebra. In 1995, I learned advanced linear algebra as a master degree course and in 1998, I learned linear algebra theory as a necessary Ph. D degree course. It was in 2009 that I first taught a graduate course in advanced linear algebra and its applications. Until 2013, hundreds of students already attended this course and benefited from it in their later control engineer careers. Over these years, I tried to choose different references and arranged a set of lecture notes for Engineering and Science students.

Prof. Han Li-Na first learned linear programming theory in 1998 and since then she focused on 3D cardiovascular system modelling and its medical analysis by using programming theory.

It is very necessary to edit a top-quality book about linear algebra and its applications in programming to meet our graduate students' needs now.

Here were some of my motives to write this book.

1) To have something as short and inexpensive as possible.

2) To organize the material in the most simple-minded, straightforward manner.

3) To order the material linearly.

4) To order many topics from fundamental linear algebra theory to matrix theory and its applications in linear and nonlinear programming. This is both a foundational course and a topic course.

5) To offer an alternative for control science majors to the linear and nonlinear programming courses.

This book may be divided into three parts.

Part I is a survey of abstract algebra with emphasis on linear algebra. Chapter 1 provides the background such that the basic concepts are products of sets, partial orderings, equivalence relations, functions, and the integers, especially that there are the properties of surjective, injective, bijective, and the notion of a solution of an equation. Chapter 2 is the most difficult part because groups are the central objects of algebra. In later chapters we will define rings and modules and see that they are special cases of groups. Also ring homomorphism and module homomorphism are special cases of group homomorphism. This chapter and the next two chapters are restricted to the most basic topics. The approach is to do quickly the fundamentals of groups, rings, and matrices, and to push forward to the chapter on linear algebra. In Chapter 3, rings are additive abelian groups with a second operation called multiplication. The connection between the two operations is provided by the distributive law. For example, ideals are also normal subgroups and ring homomorphisms are also group homomorphisms. In Chapter 4, many topics, such as invertible matrices, transpose, elementary matrices, systems of equations, and determinant, are all classical. Its highlights are these theorems, such that a square matrix is a unit in the matrix ring if its determinant is a unit in the ring, that similar matrices have the same determinant, trace, and characteristic polynomial, and that an endomorphism on a finitely generated vector space has a well-defined determinant, trace, and characteristic polynomial.

Part II presents the basic concepts of vector and tensor analysis, vector spaces, linear transformations, determinants and matrices, tensor algebra. Our intention is to present to Engineering and Science students a modern introduction to vectors and tensors and the systematic development of concepts. This part is intended as a text covering the central concepts of practical techniques on vector spaces, linear transformations, determinants and matrices, tensor algebra. It is designed for either undergraduates or graduates who have technical backgrounds in mathematics, engineering, or science. And this part should be useful to system analysts, operations researchers, numerical analysts, management scientists, and other practical specialists.

Part III provides several features as follows. Chapter 10 is devoted to a presentation of the theory and methods of polynomial-time algorithms for linear programming. These methods include, especially, interior point methods that have revolutionized linear programming. Chapter 11 includes an expanded treatment of necessary conditions, manifested by not only first- and second-order necessary conditions for optimality, but also by zeroth-order conditions that use no derivative information. This part continues to present the important descent methods for unconstrained problems, but there is new material on convergence analysis and on Newton's methods for both linear and nonlinear programming. Chapter 12 now includes the global theory of necessary conditions for constrained problems, expressed as zeroth-order conditions. Also interior point methods for general nonlinear programming are explicitly discussed within the sections on penalty and barrier methods. A significant addition to Chapter 12 is an expanded presentation of duality from both the global and local perspectives.

It is well known that mathematics theory and its applications are difficult and heavy subjects. Our styles are to make this book a little lighter. This book works best when viewed lightly and read as poems and songs, and I hope all readers enjoy it. Every effort has been extended to make every subject move rapidly and to make the sealing-in from one topic to the next as seamless as possible. The goal is to stay focused and go forward, because mathematics is learned in hindsight. We would have made the subjects short enough, every topic shorter, but I did not have any more time.

We wish to thank the many students who over the years have given us comments concerning this book and those who encouraged us to carry it out. It is difficult to do anything in life without help from friends, and many of my friends have contributed much to this text. My sincere gratitude goes especially to prof. Wu Qing-He, Prof. Wang Jun-Zheng and Prof. Zhang Bai-Hai.

We wish to thank Graduate School, Beijing Institute of Technology for its support through Top-Quality Course Construction Funds (YJPK2014 - A14, YJPK2014 - A16), and Teaching Team Building Funds (YJXTD - 2014 - A05).

We also wish to take this opportunity to thank Prof. Xia Yuan-Qing for critically checking the entire manuscript and offering improvements on many points.

Last, I am indebted to Ms. Mo Li, Ms. Liang Tong-Hua for working with me for three months and for taking care of the typing and preparation of this book's manuscript.

Guo Shu-Li
School of Automation
Beijing Institute of Technology

Han Li-Na
Department of Cardiovascular Internal Medicine
Nanlou Branch of Chinese PLA General Hospital

Contents

Part Ⅰ

Part Ⅱ

Part Ⅲ

Part I

Chapter 1

Background and Fundamentals of Mathematics

The first chapter of this book shows the fundamental knowledge for both algebra and mathematics, which is very important for the future study. The basic concepts are set, relations, functions, integers and so on. We define some symbols firstly.

$\exists$ means "there exists."

$\exists\,!$ means "there exists a unique."

$\forall$ means "for each."

$\Rightarrow$ means "implies."

1.1 Basic Concepts

Any set called an index set is assumed to be non-void. Suppose T is an index set and for each $t \in T, A_t$ is a set.

$$\bigcup_{t \in T} A_t = \{x : \exists\ t \in T \text{ with } x \in A_t\}$$

$$\bigcap_{t \in T} A_t = \{x : \text{if } t \in T, x \in A_t\} = \{x : \forall\, t \in T, x \in A_t\}$$

Let $\varnothing$ be the null set. If $A \cap B = \varnothing$, then A and B are said to be disjoint.

Cartesian products: If X and Y are sets, $X \times Y = \{(x, y) : x \in X \text{ and } y \in Y\}$. In other words, the Cartesian product of X and Y is defined to be the set of all ordered pairs whose first term is in X and whose second term is in Y.

Example 1.1.1: $R \times R = R^2$ = the plane.

Definition: If each of $X_1, \ldots, X_n$ is a set, $X_1 \times \ldots \times X_n = \{(x_1, \ldots, x_n): x_i \in X_i \text{ for } 1 \leqslant i \leqslant n\}$ = the set of all ordered n-tuples whose ith term is in X_i.

Example 1.1.2: $R \times \cdots \times R = R^n$ = real n-space.

1.2 Relations

If A is a non-void set, a non-void subset $R \subset A \times A$ is called a relation on A. If $(a, b) \in R$ we say that a is related to b, and we write this fact by the expression $a \sim b$. Here are several properties which a relation may possess.

1) If $a \in A$, then $a \sim a$. (reflexive)

2) If $a \sim b$, then $b \sim a$. (symmetric)

3) If $a \sim b$ and $b \sim a$, then $a = b$. (anti-symmetric)

4) If $a \sim b$ and $b \sim c$, then $a \sim c$. (transitive)

Definition: A relation which satisfies 1), 3), and 4) is called a partial ordering. In this case we write $a \sim b$ as $a \leqslant b$. Then

1) If $a \in A$, then $a \leqslant a$.

2) If $a \leqslant b$ and $b \leqslant a$, then $a = b$.

3) If $a \leqslant b$ and $b \leqslant c$, then $a \leqslant c$.

Definition: A linear ordering is a partial ordering with the additional property that, if $a, b \in A$, then $a \leqslant b$ or $b \leqslant a$.

Example 1.2.1: $A = R$ with the ordinary ordering, is a linear ordering.

Hausdorff maximality principle (HMP): Suppose S is a non-void subset of A and $\sim$ is a relation on A. This defines a relation on S. If the relation satisfies any of the properties 1), 2), 3), or 4) on A, the relation also satisfies these properties when restricted to S. In particular, a partial ordering on A defines a partial ordering on S. However the ordering may be linear on S but not linear on A. HMP is that any linearly ordered subset of a partially ordered set is contained in a maximal linearly ordered subset.

Monotonic: A collection of sets is said to be monotonic if, given any two sets of the collections, one is contained in the other.

Corollary to HMP: Suppose X is a non-void set and A is some non-void collection of subsets of X, and S is a subcollection of A which is monotonic. Then $\exists$ a maximal monotonic subcollection of A which contains S.

Proof: Define a partial ordering on A by $V \leqslant W$, and apply HMP.

Equivalence relations: A relation satisfying properties 1), 2), and 4) is called an equivalence relation.

Definition: If $\sim$ is an equivalence relation on A and $a \in A$, we define the equivalence class containing a by $cl(a) = \{x \in A : a \sim x\}$.

Theorem 1.2.1:

1) If $b \in cl(a)$, then $cl(b) = cl(a)$. Thus we may speak of a subset of A being an equivalence class with no mention of any element contained in it.

2) If each of $U, V \subset A$ is an equivalence class and $U \cap V \neq \varnothing$, then $U = V$.

3) Each element of A is an element of one and only one equivalence class.

Definition: A partition of A is a collection of disjoint non-void subsets whose union is A. Note that if A has an equivalence relation, the equivalence classes form a partition of A.

Theorem 1.2.2: Suppose A is a non-void set with a partition. Define a relation on A by $a \sim b$ if a and b belong to the same subset of the partition. Then $\sim$ is an equivalence relation, and the equivalence classes are just the subsets of the partition.

1.3 Functions

There are two ways of defining a function. One is the "intuitive" definition, and the other is the "graph" or "ordered pairs" definition. Because the first one is more order, so in this book we use the "intuitive" to define a function.

Definition: If X and Y are non-void sets, a function or mapping or map with domain X and range Y, is an ordered triple (X, Y, f) where f assigns to each $x \in X$ a well defined element $f(x) \in Y$. The statement that (X, Y, f) is a function is written as $f: X \to Y$ or $X \xrightarrow{f} Y$.

Graph: The graph of a function (X, Y, f) is the subset $\Gamma \subset X \times Y$ defined by $\Gamma = \{(x, f(x)) : x \in X\}$.

Theorem 1.3.1: If $f: X \to Y$, then the graph $\Gamma \subset X \times Y$ has the property that each $x \in X$ is the first term of one and only one ordered pair in Γ. Conversely, if Γ is a subset of $X \times Y$ with the property that each $x \in X$ is the first term of one and only ordered pair in Γ, then $\exists ! f: X \to Y$ whose graph is Γ. The function is defined by "$f(x)$ is the second term of the ordered pair in Γ whose first term is x."

Example 1.3.1: Here $X = Y$ and $f: X \to Y$ is defined by $f(x) = x$ for all $x \in X$. The identity on X is denoted by I_X or just $I: X \to X$.

Example 1.3.2: Suppose $y_0 \in Y$. Define $f: X \to Y$ by $f(x) = y_0$ for all $x \in X$.

Restriction: Given $f: X \to Y$ and a non-void subset S of X, define $f|S: S \to Y$ by $(f|S)(s) = f(s)$ for all $s \in S$.

Inclusion: If S is a non-void subset of X, define the inclusion $i: S \to X$ by $i(s) = s$ for all $s \in S$. Note that inclusion is a restriction of the identity.

Composition: Given $W \xrightarrow{f} X \xrightarrow{g} Y$, define $g \circ f = W \to Y$ by $(g \circ f)(x) = g(f(x))$.

Theorem 1.3.2: (The associative law of composition) If $V \xrightarrow{f} W \xrightarrow{g} X \xrightarrow{h} Y$, then $h \circ (g \circ f) = (h \circ g) \circ f$. This may be written as $h \circ g \circ f$.

Definition: Suppose $f: X \to Y$.

1) If $T \subset Y$, the inverse image of T is a subset of X, $f^{-1}(T) = \{x \in X: f(x) \in T\}$.

2) If $S \subset X$, the image *of* S is a subset of Y, $f(S) = \{f(s): s \in S\} = \{y \in Y: \exists s \in S \text{ with } f(s) = y\}$.

3) The image of f is the image of X, i.e. image $(f) = f(X) = \{f(x): x \in X\} = \{y \in Y: \exists x \in X \text{ with } f(x) = y\}$.

4) $f: X \to Y$ is surjective provided image $(f) = Y$, i.e. the image is the range, i.e. if $y \in Y$, $f^{-1}(y)$ is a non-void subset of X.

5) $f: X \to Y$ is injective or 1 - 1 provided $(x_1 \neq x_2) \Rightarrow f(x_1) \neq f(x_2)$, i.e. if x_1 and x_2 are distinct elements of X, then $f(x_1)$ and $f(x_2)$ are distinct elements of Y.

6) $f: X \to Y$ is bijective or is a 1 - 1 correspondence provided f is surjective and injective. In this case, there is function $f^{-1}: Y \to X$ with $f^{-1} \circ f = I_X: X \to X$ and $f \circ f^{-1} = I_Y: Y \to Y$. Note that $f^{-1}: Y \to X$ is also bijective and $(f^{-1})^{-1} = f$.

Example 1.3.3:

1) $f: R \to R$ defined by $f(x) = \sin(x)$ is neither surjective nor injective.

2) $R \to [-1, 1]$ defined by $f(x) = \sin(x)$ is surjective but not injective.

3) $f:[0,\pi/2]\to R$ defined by $f(x) = \sin(x)$ is injective but not surjective.

4) $f:[0,\pi/2]\to[0,1]$ defined by $f(x) = \sin(x)$ is bijective. ($f^{-1}(x)$ is written as $\arcsin(x)$ or $\sin^{-1}(x)$.)

5) $f:R\to(0,\infty)$ defined by $f(x) = e^x$ is bijective. ($f^{-1}(x)$ is written as $\ln(x)$).

Note: A function is not defined unless the domain and range are specified.

Pigeonhole principle: Suppose X is a finite set with n elements, Y is a finite set with m elements, and $f:X\to Y$ is a function.

1) If $n=m$, then f is injective iff f is surjective and iff f is bijective.

2) If $n>m$, then f is not injective.

3) If $n<m$, then f is not surjective.

Strips: If $x_0\in X$, $\{(x_0,y):y\in Y\}=(x_0,Y)$ is called a vertical strip.

If $y_0\in Y$, $\{(x,y_0):x\in X\}=(X,y_0)$ is called a horizontal strip.

Theorem 1.3.3: Suppose $S\subset X\times Y$. The subset S is the graph of a function with domain X and range Y if each vertical strip intersects S in exactly one point.

Solutions of equations: Now we restate these properties in terms of solutions of equations. Suppose $f:X\to Y$ and $y_0\in Y$. Consider the equation $f(x)=y_0$. Here y_0 is given and x is considered to be a "variable." A solution to this equation is any $x_0\in X$ with $f(x_0)=y_0$. Note that the set of all solutions to $f(x)=y_0$ is $f^{-1}(y_0)$. Also $f(x)=y_0$ has a solution if $y_0\in$ image (f) and if $f^{-1}(y_0)$ is non-void.

Note: Suppose $X\xrightarrow{f}Y\xrightarrow{g}W$ are functions, there are some properties of function:

1) If $g\circ f$ is injective, then f is injective.

2) If $g\circ f$ is surjective, then g is surjective.

3) If $g\circ f$ is bijective, then f is injective and g is surjective.

Example 1.3.4: $X=W=\{p\}$, $Y=\{p,q\}$, $f(p)=p$, and $g(p)=g(q)=p$. Here $g\circ f$ is the identity, but f is not surjective and g is not injective.

Definition: Suppose $f:X\to Y$ is a function. A left inverse of f is a function $g: Y\to X$ such that $g\circ f=I_X:X\to X$. A right inverse of f is a function $h:Y\to X$ such that $f\circ h=I_Y:Y\to Y$.

Right and left inverse: Now we introduce the right and left inverse from properties of function, which is shown in the following theorem.

Theorem 1.3.4: Suppose $f: X \to Y$ is a function.

1) f has a right inverse if f is surjective. Any such right inverse must be injective.

2) f has a left inverse if f is injective. Any such left inverse must be surjective.

Corollary: Suppose each of X and Y is a non-void set. Then $\exists$ an injective $f: X \to Y$ if $\exists$ a surjective $g: Y \to X$. Also a function from X to Y is bijective if it has a left inverse and a right inverse.

The axiom of choice: If $f: X \to Y$ is surjective, then f has a right inverse h. That is, for each $y \in Y$, it is possible to choose an $x \in f^{-1}(y)$ and thus to define $h(y) = x$.

Projection: If X_1 and X_2 are non-void sets, we define the projection maps $\pi_1: X_1 \times X_2 \to X_1$ and $\pi_2: X_1 \times X_2 \to X_2$ by $\pi_i(x_1, x_2) = x_i$.

Theorem 1.3.5: If Y, X_1, and X_2 are non-void sets, there is a $1-1$ correspondence between {functions $f: Y \to X_1 \times X_2$} and {ordered pairs of functions (f_1, f_2) where $f_1: Y \to X_1$ and $f_2: Y \to X_2$}.

Proof: Given f, define $f_1 = \pi_1 \circ f$ and $f_2 = \pi_2 \circ f$. Given f_1 and f_2, define $f: Y \to X_1 \times X_2$ by $f(y) = (f_1(y), f_2(y))$. Thus a function from Y to $X_1 \times X_2$ is merely a pair of functions from Y to X_1 and Y to X_2. It is summarized by the equation $f = (f_1, f_2)$.

Definition: Suppose T is an index set and for each $t \in T$, X_t is a non-void set. Then the product $\prod_{t \in T} X_t = \prod X_t$ is the collection of all "sequences" $\{x_t\}_{t \in T} = \{x_t\}$ where $x_t \in X_t$. (Thus if $T = Z^+$, $\{x_t\} = \{x_1, x_2, \ldots\}$.) For each $s \in T$, the projection map $\pi_s: \prod X_t \to X_s$ is defined by $\pi_s(\{x_t\}) = x_s$.

Theorem 1.3.6: If Y is any non-void set, there is a $1-1$ correspondence between {functions $f: Y \to \prod X_t$} and {sequences of functions $\{f_t\}_{t \in T}$ where $f_t: Y \to X_t$}. Given f, the sequence $\{f_t\}$ is defined by $f_t = \pi_t \circ f$. Given $\{f_t\}$, f is defined by $f(y) = \{f_t(y)\}$.

1.4 The Integers

In this section, lower case letters a, b, c,... will represent integers, i. e. elements of Z. Here we will establish the following three basic properties of the integers.

1) If G is a subgroup of Z, then $\exists\ n \geqslant 0$ such that $G = nZ$.

2) If a and b are integers, not both zero, and G is the collection of all linear combinations of a and b, then G is a subgroup of Z, and its positive generator is the greatest common divisor of a and b.

3) If $n \geqslant 2$, then n factors uniquely as the product of primes.

All these will follow long division, which we now state formally.

1.4.1 Long Division

Euclidean algorithm: Given a, b with $b \neq 0$, $\exists\,!\ m$ and r with $0 \leqslant r < |b|$ and $a = bm + r$. In other words, b divides a "m times with a remainder of r." For example, if $a = -17$ and $b = 5$, then $m = -4$ and $r = 3$, $-17 = 5(-4) + 3$.

Definition: If $r = 0$, we say that b divides a and a is a multiple of b. This fact is written as $b \mid a$. Note that $b \mid a \Leftrightarrow$ the rational number a/b is an integer $\Leftrightarrow \exists\,!$ m such that $a = bm \Leftrightarrow a \in bZ$.

Note:

1) Anything (except 0) divides 0. 0 does not divide anything.

2) ± 1 divides anything. If $n \neq 0$, the set of integers which n divides is $nZ = \{nm : m \in Z\} = \{\ldots, -2n, -n, 0, n, 2n, \ldots\}$. Also n divides a and b with the same remainder if n divides $(a - b)$.

Definition: A non-void subset $G \subset Z$ is a subgroup provided ($g \in G \Rightarrow -g \in G$) and ($g_1, g_2 \in G \Rightarrow (g_1 + g_2) \in G$). We say that G is closed under negation and closed under addition.

Theorem 1.4.1: If $n \in Z$, then nZ is a subgroup. Thus if $n \neq 0$, the set of integers which n divides is a subgroup of Z.

Theorem 1.4.2: Suppose $G \subset Z$ is a subgroup, then

1) $0 \in G$.

2) If g_1 and $g_2 \in G$, then $(m_1 g_1 + m_2 g_2) \in G$ for all integers m_1, m_2.

3) $\exists$! non-negative integer n such that $G = nZ$. In fact, if $G \neq \{0\}$ and n is the smallest positive integer in G, then $G = nZ$.

Proof: Since G is non-void, $\exists$ $g \in G$. Now $(-g) \in G$ and thus $0 = g + (-g)$ belongs to G, and so 1) is true. Part 2) is straightforward, so consider 3). If $G \neq 0$, it must contain a positive element. Let n be the smallest positive integer in G. If $g \in G, g = nm + r$ where $0 \leqslant r < n$. Since $r \in G$, it must be 0, and $g \in nZ$.

1.4.2 Relatively Prime

Now suppose $a, b \in Z$ and at least one of a and b is non-zero.

Theorem 1.4.3: Let G be the set of all linear combinations of a and b, i. e. $G = \{ma + nb : m, n \in Z\}$, then

1) G contains a and b.

2) G is a subgroup. In fact, it is the smallest subgroup containing a and b. It is called the subgroup generated by a and b.

3) Denote by (a, b) the smallest positive integer in G. By the previous theorem, $G = (a, b)Z$, and thus $(a, b) \mid a$ and $(a, b) \mid b$. Also note that $\exists$ m, n such that $ma + nb = (a, b)$. The integer (a, b) is called the greatest common divisor of a and b.

4) If n is an integer which divides a and b, then n also divides (a, b).

Proof of 4): Suppose $n \mid a$ and $n \mid b$, i. e. suppose $a, b \in nZ$. Since G is the smallest subgroup containing a and b, $nZ \supset (a, b)Z$, and thus $n \mid (a, b)$.

Here we show three conditions, which are equivalent to each other.

1) a and b have no common divisors, i. e. $(n \mid a$ and $n \mid b) \Rightarrow n = \pm 1$.

2) $(a, b) = 1$, i. e. the subgroup generated by a and b is all of Z.

3) $\exists$ $m, n \in Z$ with $ma + nb = 1$.

Definition: If any one of these three conditions is satisfied, we say that a and b are relatively prime.

1.4.3 Prime

Theorem 1.4.4: If a and b are relatively prime with a not zero, then $a \mid bc \Rightarrow a \mid c$.

Proof: Suppose a and b are relatively prime, $c \in Z$ and $a \mid bc$. Then there exists m, n with $ma + nb = 1$, and thus $mac + nbc = c$. Now $a \mid mac$ and $a \mid nbc$. Thus $a \mid (mac + nbc)$ and so $a \mid c$.

Definition: A prime is an integer $p > 1$ which does not factor, i. e. if $p = ab$, then $a = \pm 1$ or $a = \pm p$. The first few primes are $2, 3, 5, 7, 11, 13, 17, \ldots$

Theorem 1.4.5: Suppose p is a prime.

1) If a is an integer which is not a multiple of p, then $(p, a) = 1$. In other words, if a is any integer, $(p, a) = p$ or $(p, a) = 1$.

2) If $p \mid ab$, then $p \mid a$ or $p \mid b$.

3) If $p \mid a_1 a_2 \ldots a_n$, then p divides some a_i. Thus if each a_i is a prime, then p is equal to some a_i.

Proof: Part 1) follows immediately the definition of prime. Now suppose $p \mid ab$. If p does not divide a, then by 1), $(p, a) = 1$ and by the previous theorem, p must divide b. Thus 2) is true. Part 3) follows 2) and induction on n.

1.4.4 The Unique Factorization Theorem

The unique factorization theorem: Suppose n is an integer which is not 0, 1, or -1, then n may be factored into the product of primes and, except for order, this factorization is unique. That is, $\exists$ a unique collection of distinct prime $p_1, \ldots, p_k$ and positive integers $s_1, s_2, \ldots, s_k$ such that $n = \pm p_1^{s_1} p_2^{s_2} \ldots p_k^{s_k}$.

Proof: Factorization into primes is obvious, and we can prove the uniqueness from 3) in the theorem 1.4.5. The power of this theorem is uniqueness, not existence.

Theorem 1.4.6 (summary):

1) Suppose $|a| > 1$ has prime factorization $a = \pm p_1^{s_1} \ldots p_k^{s_k}$, then the only divisors or a are of the form $\pm p_1^{t_1} \ldots p_k^{t_k}$ where $0 \leqslant t_i \leqslant s_i$ for $i = 1, \ldots, k$.

2) If $|a| > 1$ and $|b| > 1$, then $(a, b) = 1$ if there is no common prime in their factorizations. Thus if there is no common prime in their factorizations, $\exists$ m, n with $ma + nb = 1$.

3) Suppose $|a| > 1$ and $|b| > 1$. Let $\{p_1, \ldots, p_k\}$ be the union of the distinct primes of their factorizations. Thus $a = \pm p_1^{s_1} \ldots p_k^{s_k}$ where $0 \leqslant s_i$ and $b = \pm p_1^{t_1} \ldots p_k^{t_k}$ where $0 \leqslant t_i$. Let u_i be the minimum of s_i and t_i. Then $(a, b) = p_1^{u_1} \ldots p_k^{u_k}$. For example $(2^3 \cdot 5 \cdot 11, 2^2 \cdot 5^4 \cdot 7) = 2^2 \cdot 5$.

4) Let v_i be the maximum of s_i and t_i, then $c = p_1^{v_1} \ldots p_k^{v_k}$ is the least common multiple of a and b. Note that c is a multiple of a and b, and if n is a multiple of a and b, then n is a multiple of c. Finally, the least common multiple of a and b is

$c = ab/(a,b)$. In particular, if a and b are relatively prime, then their least common multiple is just their product.

5) There is an infinite number of primes. (Proof: Suppose there were only a finite number of primes $p_1, p_2, \dots, p_k$, then no prime would divide $(p_1 p_2 \cdots p_k + 1)$.)

6) $\sqrt{2}$ is irrational. (Proof: Suppose $\sqrt{2} = m/n$ where $(m,n) = 1$, then $2n^2 = m^2$ and if $n > 1$, n and m have a common prime factor. Since this is impossible, $n = 1$, and so $\sqrt{2}$ is an integer. This is a contradiction and therefore $\sqrt{2}$ is irrational.)

7) Suppose c is an integer greater than 1, then $\sqrt{c}$ is rational if $\sqrt{c}$ is an integer.

Chapter 2

Groups

Groups are the central and basic objects of algebra. In this chapter we will introduce some new concepts around this basic notion, such as, subgroups, normal subgroups, homomorphisms and so on. Now we will present these concepts in detail.

2.1 Groups

Definition: Suppose G is a non-void set and $\phi: G \times G \to G$ is a function. ϕ is called a binary operation, and we will write $\phi(a,b) = a \cdot b$ or $\phi(a,b) = a + b$. Consider the following properties.

1) If $a,b,c \in G$, then $a \cdot (b \cdot c) = (a \cdot b) \cdot c$ or $a + (b + c) = (a + b) + c$.

2) if $a \in G$, then $\exists\ e = e_G \in G$ satisfies $e \cdot a = a \cdot e = a$, or $\exists\, \underline{0} = \underline{0}_G \in G$ satisfies $\underline{0} + a = a + \underline{0} = a$.

3) If $a \in G$, $\exists\, b \in G$ with $a \cdot b = b \cdot a = e$ (b is written as $b = a^{-1}$), or with $a + b = b + a = \underline{0}$ (b is written as $b = -a$).

4) If $a,b \in G$, then $a \cdot b = b \cdot a$ or $a + b = b + a$.

Definition: If properties 1), 2), and 3) hold, (G,ϕ) is said to be a group. If we write $\phi(a,b) = a \cdot b$, we say it is a multiplicative group. If we write $\phi(a,b) = a + b$, we say it is an additive group. If in addition, property 4) holds, we say the group is abelian or commutative.

Theorem 2. 1. 1: In the above definition, let (G, ϕ) be a multiplicative group.

1) Suppose $a, b, \bar{c} \in G$, if $a \cdot c = a \cdot \bar{c} \Rightarrow c = \bar{c}$. Also $c \cdot a = \bar{c} \cdot a \Rightarrow c = \bar{c}$. In other words, if $f: G \to G$ is defined by $f(c) = a \cdot c$, then f is injective. Also f is injective with f^{-1} given $f^{-1}(c) = a^{-1} \cdot c$.

2) e is unique, i. e. if $\bar{e} \in G$ satisfies 2), then $e = \bar{e}$. In fact, if $a, b \in G$, then $(a \cdot b = a) \Rightarrow (b = e)$ and $(a \cdot b = b) \Rightarrow (a = e)$. Recall that b is an identity in G provided it is a right and left identity for any a in G. However, group structure is so rigid that if $\exists\, a \in G$ such that b is a right identity for a, then $b = e$. Of course, this is just a special case of the cancellation law in 1).

3) Every right and left inverse is an inverse, i. e. if $a \cdot b = e$, then $b = a^{-1}$, and if $b \cdot a = e$, then $b = a^{-1}$.

4) If $a \in G$, then $(a^{-1})^{-1} = a$.

5) If $a, b \in G$, $(a \cdot b)^{-1} = b^{-1} \cdot a^{-1}$. Also $(a_1 \cdot a_2 \ldots a_n)^{-1} = a_n^{-1} \cdot a_{n-1}^{-1} \ldots a_1^{-1}$.

6) The multiplication $a_1 \cdot a_2 \cdot a_3 = a_1 \cdot (a_2 \cdot a_3) = (a_1 \cdot a_2) \cdot a_3$ is well defined. In general, $a_1 \cdot a_2 \ldots a_n$ is well defined.

7) Suppose $a \in G$. Let $a^0 = e$ and if $n > 0$, $a^n = a \ldots a$ (n times) and $a^{-n} = a^{-1} \ldots a^{-1}$ (n times). If $n_1, n_2, \ldots, n_t \in Z$, then $a^{n_1} \cdot a^{n_2} \ldots a^{n_t} = a^{n_1 + \ldots + n_t}$. Also $(a^n)^m = a^{nm}$.

Finally: if G is abelian and $a, b \in G$, then $(a \cdot b)^n = a^n \cdot b^n$.

Theorem 2. 1. 2: Suppose G is an additive group. If $a \in G$, let $a0 = \underline{0}$ and if $n > 0$, let $an = (a + \ldots + a)$ where the sum is n times, and $a(-n) = (-a) + \ldots + (-a)$, which we write as $(-a \ldots -a)$, then the following properties hold in general, except the first requires that G be abelian.

1) $(a + b)n = an + bn$

2) $a(n + m) = an + am$

3) $a(nm) = (an)m$

4) $a1 = a$

Example 2. 1. 1: $G = \mathbf{R}$, $G = \mathbf{Q}$, or $G = \mathbf{Z}$ with $\phi(a, b) = a + b$ is an additive abelian group.

Example 2. 1. 2: $G = \mathbf{R} - 0$ or $G = \mathbf{Q} - 0$ with $\phi(a, b) = ab$ is a multiplicative abelian group.

2.2 Subgroups

Theorem 2.2.1: Suppose G is a multiplicative group and $H \subset G$ is a non-void subset satisfying

1) if $a, b \in H$, then $a \cdot b \in H$.

2) if $a \in H$, then $a^{-1} \in H$.

Then $e \in H$ and H is a group under multiplicative. H is called a subgroup of G.

Proof: Since H is non-void, $\exists\, a \in H$. By 2), $a^{-1} \in H$ and so by 1), $e \in H$. The associative law is immediate and so H is a group.

Example 2.2.1: G is a subgroup of G and e is a subgroup of G. These are called the improper subgroups of G.

Example 2.2.2: If $G = \mathbf{Z}$ under addition, and $n \in \mathbf{Z}$, then $H = n\mathbf{Z}$ is a subgroup of $\mathbf{Z}$. By a theorem in the section on the integers in Chapter 1, every subgroup of $\mathbf{Z}$ is of this form.

Order: Suppose G is a multiplicative group. If G has an infinite number of elements, we say that $o(G)$, which denotes the order of G, is infinite. If G has n elements, then $o(G) = n$. Suppose $a \in G$ and $H = \{a^i : i \in \mathbf{Z}\}$. H is an abelian subgroup of G called the subgroup generated by a. We define the order of the element a to be the order of H, i.e. the order of the subgroup generated by a. Let f: $\mathbf{Z} \to H$ be the surjective function defined by $f(m) = a^m$. Note that $f(k+l) = f(k) \cdot f(l)$ where the addition is in $\mathbf{Z}$ and the multiplication is in the group H. When the element a has finite order if f is not injective, we present the first real theorem in group theory as follows.

Theorem 2.2.2: Suppose a is an element of a multiplicative group G, and $H = \{a^i : i \in \mathbf{Z}\}$. If $\exists$ distinct integers i and j with $a^i = a^j$, then a has some finite order n. In this case, H has n distinct elements, $H = \{a^0, a^1, \ldots, a^{n-1}\}$, and $a^m = e$ if $n \mid m$. In particular, the order of a is the smallest positive integer n with $a^n = e$, and $f^{-1}(e) = n\mathbf{Z}$.

Proof: Suppose $j < i$ and $a^i = a^j$, then $a^{i-j} = e$ and thus $\exists$ a smallest positive integer n with $a^n = e$. This implies that the elements of $\{a^0, a^1, \ldots, a^{n-1}\}$ are distinct, and we must show they are all of H. If $m \in \mathbf{Z}$, the Euclidean algorithm states that $\exists$ integers q and r with $0 \leqslant r < n$ and $m = nq + r$. Thus $a^m = a^{nq} \cdot a^r$, and so $H = \{a^0, a^1, \ldots, a^{n-1}\}$, and $a^m = e$ if $n \mid m$. Later in this chapter we will see that f is a homomorphism from an additive group to a multiplicative group and that, in additive notation, H is isomorphic to $\mathbf{Z}$ or $\mathbf{Z}_n$.

Definition: A group G is cyclic if $\exists$ an element of G which generates G.

Theorem 2.2.3: If G is cyclic and H is a subgroup of G, then H is cyclic.

Proof: Suppose G is a cyclic group of order n, then $\exists$ $a \in G$ with $G = \{a^0, a^1, \ldots, a^{n-1}\}$. Suppose H is a subgroup of G with more than one element, let m be the smallest integer with $0 < m < n$ and $a^m \in H$, then $m \mid n$ and a^m generates H. The case where G is an infinite cyclic group is left as an exercise. Note that $\mathbf{Z}$ is an additive cyclic group and it was shown in the previous chapter that subgroups of $\mathbf{Z}$ are cyclic.

Coset: Suppose H is a subgroup of a group G. It will be shown below that H partitions G into right cosets. It also partitions G into left cosets, and in general these partitions are distinct.

Theorem 2.2.4: If H is a subgroup of a multiplicative group G, then $a \sim b$ defined by $a \sim b$ iff $a \cdot b^{-1} \in H$ is an equivalence relation. If $a \in G$, $cl(a) = \{b \in G : a \sim b\} = \{h \cdot a : h \in H\} = Ha$. Note that $a \cdot b^{-1} \in H$ iff $b \cdot a^{-1} \in H$.

If H is a subgroup of an additive group G, then $a \sim b$ defined by $a \sim b$ iff $(a - b) \in H$ is an equivalence relation. If $a \in G$, $cl(a) = \{b \in G : a \sim b\} = \{h + a : h \in H\} = H + a$. Note that $(a - b) \in H$ iff $(b - a) \in H$.

Definition: These equivalence classes are called right cosets, if the relation is defined by $a \sim b$ iff $b^{-1} \cdot a \in H$. If the equivalence classes are $cl(a) = aH$, where $cl(a) = \{b \in G : a \sim b\} = \{h \cdot a : h \in H\} = Ha$, they are called left cosets, where H is a left and right coset. If G is abelian, there is no distinction between right and left cosets. Note that $b^{-1} \cdot a \in H$ if $a^{-1} \cdot b \in H$.

Theorem 2.2.5: Suppose H is a subgroup of a multiplicative group G. If $a \in G$, we define the right coset containing a to be $Ha = \{h \cdot a : h \in H\}$, then we have the following properties.

1) $Ha = H$ if $a \in H$.

2) If $b \in Ha$, then $Hb = Ha$, i.e. if $h \in H$, then $H(h \cdot a) = (Hh)a = Ha$.

3) If $Hc \cap Ha \neq \varnothing$, then $Hc = Ha$.

4) The right cosets form a partition of G, i.e. each a in G belongs to one and only one right coset.

5) Elements a and b belong to the same right coset if $a \cdot b^{-1} \in H$ if $b \cdot a^{-1} \in H$.

Proof: The best way to prove this theorem is to develop facility with cosets. Also this theorem holds for G an additive group.

Theorem 2.2.6: Suppose H is a subgroup of a multiplicative group G.

1) Any two right cosets have the same number of elements. That is, if $a, b \in G, f: Ha \to Hb$ defined by $f(h \cdot a) = h \cdot b$ is a bijection. Also any two left cosets have the same number of elements. Since H is a right and left coset, any two cosets have the same number of elements.

2) G has the same number of right cosets as left cosets. The bijection is given by $F(Ha) = a^{-1}H$. The number of right (or left) cosets is called the index of H in G.

3) If G is finite, $o(H)$ (index of H) $= o(G)$ and so $o(H) \mid o(G)$. In other words, $o(G)/o(H)$ = the number of right cosets = the number of left cosets.

4) If G is finite, and $a \in G$, then $o(a) \mid o(G)$. (Proof: The order of a is the order of the subgroup generated by a, and by 3) this divides the order of G.)

5) If G has prime order, then G is cyclic, and any element (except e) is a generator. (Proof: Suppose $o(G) = p$ and $a \in G, a \neq e$, then $o(a) \mid p$ and thus $o(a) = p$).

6) If $o(G) = n$ and $a \in G$, then $a^n = e$. (Proof: $a^{o(a)} = e$ and $n = o(a)(o(G)/o(a))$.)

2.3 Normal Subgroups

Now we present a special kind of subgroup H which is called a normal subgroup. The definition and properties of normal subgroup are described below.

Theorem 2.3.1: If H is a subgroup of G, then the following are equivalent.

1) If $a \in G$, then $aHa^{-1} = H$.

2) If $a \in G$, then $aHa^{-1} \subset H$.

3) If $a \in G$, then $aH = Ha$.

4) Every right coset is a left coset, i. e. if $a \in G$, $\exists b \in G$ with $Ha = bH$.

Proof: 1) $\Rightarrow$ 2) is obvious. Suppose 2) is true and shows 3). We have $(aHa^{-1})a \subset Ha$, so $aH \subset Ha$. Also $a(a^{-1}Ha) \subset aH$, so $Ha \subset aH$. Thus $aH = Ha$. 3) $\Rightarrow$ 4) is obvious. Suppose 4) is true and shows 3). $Ha = bH$ contains a, so $bH = aH$

because a coset is an equivalence class. Finally, suppose 3) is true and shows 1). Multiply $aH = Ha$ on the right by a^{-1}.

Definition: If H satisfies any of the four conditions above, then H is said to be a normal subgroup of G.

Note: For any group G, G and e are normal subgroups. If G is an abelian group, then every subgroup of G is normal.

Quotient groups: Suppose N is a normal subgroup of G, and C and D are cosets. We wish to define a coset E which is the product of C and D. If $c \in C$ and $d \in D$, define E to be the coset containing $c \cdot d$, i. e. $E = N(c \cdot d)$. The coset E does not depend upon the choice of c or d. This is made precise in the next theorem.

Theorem 2. 3. 2: Suppose G is a multiplicative group, N is a normal subgroup, and G/N is the collection of all cosets. Then $(Na) \cdot (Nb) = N(a \cdot b)$ is a well defined multiplication (binary operation) on G/N, and with this multiplication, G/N is a group. Its identity is N and $(Na)^{-1} = (Na^{-1})$. Furthermore, if G is finite, $o(G/N) = o(G)/o(N)$.

Proof: Multiplication of elements in G/N is multiplication of subsets in G. $(Na) \cdot (Nb) = N(aN)b = N(Na)b = N(a \cdot b)$. Once multiplication is well defined, the group axioms are immediate.

Example 2. 3. 1: Suppose $G = \mathbf{Z}$ under $+$, $n > 1$, and $N = n\mathbf{Z}$. $\mathbf{Z}_n$, the group of integers mod n is defined by $\mathbf{Z}_n = \mathbf{Z}/n\mathbf{Z}$. If a is an integer, the coset $a + n\mathbf{Z}$ is denoted by $[a]$. Note that $[a] + [b] = [a + b]$, $-[a] = [-a]$, and $[a] = [a + nl]$ for any integer l. Any additive abelian group has a scalar multiplication over $\mathbf{Z}$, and in this case it is just $[a]m = [am]$. Note that $[a] = [r]$ where r is the remainder of a divided by n, and thus the distinct elements of $\mathbf{Z}_n$ are $[0]$, $[1], \ldots, [n-1]$. Also $\mathbf{Z}_n$ is cyclic because each of $[1]$ and $[-1] = [n-1]$ is a generator. We already know that if p is a prime, any non-zero element of $\mathbf{Z}_p$ is a generator, because $\mathbf{Z}_p$ has p elements.

Theorem 2. 3. 3: If $n > 1$ and a is any integer, then $[a]$ is a generator of $\mathbf{Z}_n$ if $(a, n) = 1$.

Proof: The element $[a]$ is a generator if the subgroup generated by $[a]$ contains $[1]$ if $\exists$ an integer k such that $[a]k = [1]$ if $\exists$ integer k and l such that $ak + nl = 1$.

2.4 Homomorphisms

Homomorphisms are functions between groups that commute with the group operations. It follows that they honor identities and inverses. In this section we list the basic properties of homomorphisms.

Definition: If G and $\bar{G}$ are multiplicative groups, a function $f: G \to \bar{G}$ is a homomorphism if, for all $a, b \in G$, $f(a \cdot b) = f(a) \cdot f(b)$. On the left side, the group operation is in G, while on the right side it is in $\bar{G}$. The kernel of f is defined by ker $(f) = f^{-1}(\bar{e}) = \{a \in G: f(a) = \bar{e}\}$. In other words, the kernel is the set of solutions to the equation $f(x) = \bar{e}$. (If $\bar{G}$ is an additive group, ker $(f) = f^{-1}(\underline{0})$.)

Example 2.4.1: The constant map $f: G \to \bar{G}$ defined by $f(a) = \bar{e}$ is a homomorphism. If H is a subgroup of G, the inclusion $i: H \to G$ is a homomorphism. The function $f: \mathbf{Z} \to \mathbf{Z}$ defined by $f(t) = 2t$ is a homomorphism of additive groups, while the function defined by $f(t) = t + 2$ is not a homomorphism. The function $h: \mathbf{Z} \to \mathbf{R} - 0$ defined by $h(t) = 2^t$ is a homomorphism from an additive group to a multiplicative group.

Theorem 2.4.1: Suppose G and $\bar{G}$ are groups and $f: G \to \bar{G}$ is a homomorphism.

1) $f(e) = \bar{e}$.

2) $f(a^{-1}) = f(a)^{-1}$.

3) f is injective $\Leftrightarrow$ ker $(f) = e$.

4) If H is a subgroup of G, $f(H)$ is a subgroup of $\bar{G}$. In particular, image (f) is a subgroup of $\bar{G}$.

5) If $\bar{H}$ is a subgroup of $\bar{G}$, $f^{-1}(\bar{H})$ is a subgroup of G. Furthermore, if $\bar{H}$ is normal in $\bar{G}$, then $f^{-1}(\bar{H})$ is normal in G.

6) The kernel of f is a normal subgroup of G.

7) If $\bar{g} \in \bar{G}$, $f^{-1}(\bar{g})$ is void or is a coset of ker (f), i. e. if $f(g) = \bar{g}$, then $f^{-1}(\bar{g}) = Ng$ where $N = $ ker (f). In other words, if the equation $f(x) = \bar{g}$ has a solution, then the set of all solutions is a coset of $N = $ ker (f).

8) The composition of homomorphisms is a homomorphism, i. e. if $h: \bar{G} \to \bar{\bar{G}}$ is a homomorphism, then $h \circ f: G \to \bar{\bar{G}}$ is a homomophism.

9) If $h \circ f: G \to \overline{G}$ is a bijection, then the function $f^{-1}: G \to \overline{G}$ is a homomorphism. In this case, f is called an isomorphism, and we write $G \approx \overline{G}$. In the case $G = \overline{G}$, f is also called an automorphism.

10) Isomorphisms preserve all algebraic properties. For example, if f is an isomorphism and $H \subset G$ is a subset, then H is a subgroup of G if $f(H)$ is a subgroup of $\overline{G}$, H is normal in G if $f(H)$ is normal in $\overline{G}$, G is cyclic if $\overline{G}$ is cyclic, etc.

11) Suppose H is a normal subgroup of G, then $\pi: G \to G/H$ defined by $\pi(a) = Ha$ is a surjective homomorphism with kernel H. Furthermore, if $f: G \to \overline{G}$ is a surjective homomorphism with kernel H, then $G/H \approx \overline{G}$.

12) Suppose H is a normal subgroup of G. If $H \subset \ker(f)$, then $\bar{f}: G/H \to \overline{G}$ defined by $\bar{f}(Ha) = f(a)$ is a well-defined homomorphism making the following diagram commute. The image of $\bar{f}$ is the image of f and the kernel of $\bar{f}$ is $\ker(f)/H$. Thus if $H = \ker(f)$, $\bar{f}$ is injective, and thus $G/H \approx \text{image}(f)$.

13) Given any group homomorphism f, domain $(f)/\ker(f) \approx$ image (f). This is the fundamental connection between quotient groups and homomorphisms.

14) Suppose K is a group, then K is an infinite cycle group if K is isomorphic to the integers under addition, i. e. $K \approx \mathbf{Z}$. K is a cyclic group of order n if $K \approx \mathbf{Z}_n$.

Proof of 14): Suppose $\overline{G} = K$ is generated by some element a. Then $f: \mathbf{Z} \to K$ defined by $f(m) = a^m$ is a homomorphism from an additive group to a multiplicative group. If $o(a)$ is infinite, f is an isomorphism. If $o(a) = n$, $\ker(f) = n\mathbf{Z}$ and $\bar{f}: \mathbf{Z}_n \to K$ is an isomorphism.

2.5 Permutations

Suppose X is a non-void set. A bijection $f: X \to X$ is called a permutation on X, and the collection of all these permutations is denoted by $S = S(X)$. In this sense, variables are written on the left, i. e. $f = (x)f$. Therefore the composition $f \circ g$ means "f followed by g." $S(X)$ forms a multiplicative group under composition.

The next theorem shows that the symmetric groups are incredibly rich and complex.

Theorem 2.5.1 (Cayley's theorem): Suppose G is a multiplicative group with n elements and S_n is the group of all permutations on the set G, then G is isomorphic to a subgroup of S_n.

Proof: Let $h: G \to S_n$ be the function which sends a to the bijection $h_a: G \to G$ defined by $(g)h_a = g \cdot a$. The simply proof procedure is as follows.

1) For each given a, h_a is a bijection from G to G.

2) h is a homomorphism, i. e. $h_{a \cdot b} = h_a \circ h_b$.

3) h is injective and thus G is isomorphic to image $(h) \subset S_n$.

The symmetric groups: Now let $n \geqslant 2$ and let S_n be the group of all permutations on $\{1, 2, \ldots, n\}$. The following definition shows that each element of S_n may be represented by a matrix.

Definition: Suppose $1 < k \leqslant n$, $\{a_1, a_2, \ldots, a_k\}$ is a collection of distinct integers with $1 \leqslant a_i \leqslant n$, and $(b_1, b_2, \ldots, b_k)$ is the same collection in some different order, then the matrix $\begin{pmatrix} a_1 a_2 \ldots a_k \\ b_1 b_2 \ldots b_k \end{pmatrix}$ represents $f \in S_n$ defined by $(a_i)f = b_i$ for $1 \leqslant i \leqslant k$, and $(a)f = a$ for all other a. The composition of two permutations is computed by applying the matrix on the left first and the matrix on the right second.

Definition: $\begin{pmatrix} a_1 a_2 \ldots a_{k-1} a_k \\ a_2 a_3 \ldots a_k a_1 \end{pmatrix}$ is called a k-cycle, and is denoted by $(a_1, a_2, \ldots, a_k)$. A 2-cycle is called a *transposition*. The cycles $(a_1, \ldots, a_k)$ and $(c_1, \ldots, c_l)$ are *disjoint* provided $a_i \neq c_j$ for all $1 \leqslant i \leqslant k$ and $1 \leqslant j \leqslant l$.

Here we list the seven basic properties of permutations by the next theorem.

Theorem 2.5.2:

1) Disjoint cycles commute.

2) Every permutation can be written uniquely (except for order) as the product of disjoint cycles.

3) Every permutation can be written (non-uniquely) as the product of transpositions. (Proof: $(a_1, \ldots, a_n) = (a_1, a_2)(a_1, a_3) \ldots (a_1, a_n)$.)

4) The parity of the number of these transpositions is unique. This means that if f is the product of p transpositions and also of q transpositions, then p is even iff q is even. In this case, f is said to be an even permutation. In the other case, f is an odd permutation.

5) A k-cycle is even (odd) iff k is odd (even). For example $(1, 2, 3) = (1, 2)(1, 3)$ is an even permutation.

6) Suppose $f, g \in S_n$. If one of f and g is even and the other is odd, then $g \circ f$ is odd. If f and g are both even or both odd, then $g \circ f$ is even.

7) The map $h: S_n \to \mathbf{Z}_2$ defined by h (even) $= [0]$ and h (odd) $= [1]$ is a homomorphism from a multiplicative group to an additive group. Its kernel (the subgroup of even permutations) is denoted by A_n and is called the *alternating* group. Thus A_n is a normal subgroup of index 2, and $S_n / A_n \approx \mathbf{Z}_2$.

Proof of 4): The proof presented here uses polynomials in n variables with real coefficients. Suppose $S = \{1, \ldots, n\}$. If σ is a permutation on S and $p = p(x_1, \ldots, x_n)$ is a polynomial in n variables, define $\sigma(p)$ to be the polynomial $p(x_{(1)\sigma}, \ldots, x_{(n)\sigma})$. Thus if $p = x_1 x_2^2 + x_1 x_3$, and σ is the transposition $(1,2)$, then $\sigma(p) = x_2 x_1^2 + x_2 x_3$. Note that if σ_1 and σ_2 are permutations, $\sigma_2(\sigma_1(p)) = (\sigma_1 \cdot \sigma_2)(p)$. Now let p be the product of all $(x_i - x_j)$ where $1 \leqslant i \leqslant j \leqslant n$. (For example, if $n = 3$, $p = (x_1 - x_2)(x_1 - x_3)(x_2 - x_3)$.) If σ is a permutation on S, then for each $1 \leqslant i, j \leqslant n$ with $i \neq j$, $\sigma(p)$ has $(x_i - x_j)$ or $(x_j - x_i)$ as a factor. Thus $\sigma(p) = \pm p$. A careful examination shows that if σ_i is a transposition, $\sigma_i(p) = -p$. Any permutation σ is the product of transpositions, $\sigma = \sigma_1 \cdot \sigma_2 \cdots \sigma_t$. Thus if $\sigma(p) = p$, t must be even, and if $\sigma(p) = -p$, t must be odd.

2.6 Product of Groups

The product of groups is usually presented for multiplicative groups. Because the case of infinite products is more difficult, we first consider the product of two groups for simplicity.

Theorem 2.6.1: Suppose G_1 and G_2 are additive groups. Define an addition on $G_1 \times G_2$ by $(a_1, a_2) + (b_1, b_2) = (a_1 + b_1, a_2 + b_2)$. This operation makes $G_1 \times G_2$ into a group. Its "zero" is $(\underline{0}_1, \underline{0}_2)$ and $-(a_1, a_2) = (-a_1, -a_2)$. The projections $\pi_1: G_1 \times G_2 \to G_1$ and $\pi_2: G_1 \times G_2 \to G_2$ are group homomorphisms. Suppose G is an additive group. We know there is a bijection from {function $f: G \to G_1 \times G_2$} to {ordered pairs of functions (f_1, f_2) where $f_1: G \to G_1$ and $f_2: G \to G_2$}. Under this bijection, f is a group homomorphism if each of f_1 and f_2 is a group homomorphism.

Proof: It is transparent that the product of groups is a group, so let's prove the last part. Suppose G, G_1 and G_2 are groups and $f=(f_1, f_2)$ is a function from G to $G_1 \times G_2$. Now $f(a+b)=(f_1(a+b), f_2(a+b))$ and $f(a)+f(b)=(f_1(a), f_2(a))+(f_1(b), f_2(b))=(f_1(a)+f_1(b), f_2(a)+f_2(b))$. An examination of these two equations shows that f is a group homomorphism if each of f_1 and f_2 is a group homomorphism.

Theorem 2.6.2: Suppose T is an index set, and for any $t \in T$, G_t is an additive group. Define an addition on $\prod_{t \in T} G_t = \prod G_t$ by $\{a_t\}+\{b_t\}=\{a_t+b_t\}$. This operation makes the product into a group. Its "zero" is $\{\underline{0}_t\}$ and $-\{a_t\}=\{-a_t\}$. Each projection $\pi_s: \prod G_t \rightarrow G_s$ is a group homomorphism. Suppose G is an additive group. Under the natural bijection from $\{$function $f: G \rightarrow \prod G_t\}$ to $\{$sequence of functions $\{f_t\}_{t \in T}$ where $f_t: G \rightarrow G_t\}$, f is a group homomorphism if each f_t is a group homomorphism. Finally, the scalar multiplication on $\prod G_t$ by integers is given coordinatewise, i.e. $\{a_t\} n=\{a_t n\}$.

Proof: The addition on $\prod G_t$ is coordinatewise.

Chapter 3

Rings

We introduce the concept of group in the last chapter, now another algebra notion "ring" is presented. Rings are additive abelian groups with a second operation called multiplication and it is also very important for the following study. We first introduce some properties of the additive abelian group from which we define the ring.

Suppose R is an additive abelian group, $R \neq \underline{0}$, and R has a second binary operation (i. e. map from $R \times R$ to R), which is denoted by multiplication. Consider the following properties.

1) If $a, b, c \in R$, $(a \cdot b) \cdot c = a \cdot (b \cdot c)$. (The associative property of multiplication.)

2) If $a, b, c \in R$, $a \cdot (b + c) = (a \cdot b) + (a \cdot c)$ and $(b + c) \cdot a = (b \cdot a) + (c \cdot a)$. (The distributive law, which connects addition and multiplication.)

3) R has a multiplicative identity, i. e. an element $\underline{1} = \underline{1}_R \in R$ such that if $a \in R$, $a \cdot \underline{1} = \underline{1} \cdot a = a$.

4) If $a, b \in R$, $a \cdot b = b \cdot a$. (The commutative property for multiplication.)

3.1 Commutative Rings

Definition: The definitions are as above. If 1), 2), and 3) are satisfied, R is said to be a ring. If in addition 4) is satisfied, R is said to be a commutative ring.

Example 3.1.1: The basic commutative rings in mathematics are the integers **Z**, the rational numbers **Q**, the real numbers **R**, and the complex numbers **C**. It will be shown later that $\mathbf{Z}_n$, the integers mod n, has a natural multiplication under which it is a commutative ring. Also if R is any commutative ring, we will define $R[x_1, x_2, \ldots, x_n]$, a polynomial ring in n variables. Now suppose R is any ring, $n \geqslant 1$ and R_n is the collection of all $n \times n$ matrices over R. Under these operations, R_n is a ring. This is a basic example of a non-commutative ring. If $n > 1$, R_n is never commutative, even if R is commutative.

Theorem 3.1.1: Suppose R is a ring and $a, b \in R$.

1) $a \cdot \underline{0} = \underline{0} \cdot a = \underline{0}$. Therefore $\underline{1} \neq \underline{0}$.

2) $(-a) \cdot b = a \cdot (-b) = -(a \cdot b)$.

Since R is an additive abelian group, it has a scalar multiplication over **Z**. This scalar multiplication can be written on the right or left, i. e. $na = an$, and the next theorem shows it relates nicely to the ring multiplication.

Theorem 3.1.2: Suppose $a, b \in R$ and $n, m \in \mathbf{Z}$.

1) $(na \cdot mb) = (nm)(a \cdot b)$. (This follows the distributive law and the previous theorem.)

2) Let $\underline{n} = n\,\underline{1}$. For example, $\underline{2} = \underline{1} + \underline{1}$. Then $na = \underline{n} \cdot a$, that is, scalar multiplication by n is the same as ring multiplication by $\underline{n}$. Of course, $\underline{n}$ may be $\underline{0}$ even though $n \neq 0$.

3.2 Units

Definition: An element a of a ring R is a unit provided $\exists$ an element $a^{-1} \in R$ with $a \cdot a^{-1} = a^{-1} \cdot a = \underline{1}$.

Theorem 3.2.1: $\underline{0}$ can never be a unit. $\underline{1}$ is always a unit. If a is a unit, a^{-1} is also a unit with $(a^{-1})^{-1} = a$. The product of units is a unit with $(a \cdot b)^{-1} = b^{-1} \cdot a^{-1}$. More generally, if $a_1, a_2, \ldots, a_n$ are units, then their product is a unit with $(a_1, a_2, \ldots, a_n)^{-1} = a_n{}^{-1} \cdot a_{n-1}{}^{-1} \ldots a_1{}^{-1}$. The set of all units of R forms a multiplicative group denoted by R^*. Finally if a is a unit, $(-a)$ is a unit and $(-a)^{-1} = -(a^{-1})$.

Note: If a is a unit, then it must have a two-sided inverse: the left inverse and the right inverse, just as shown in the following theorem.

Theorem 3.2.2: Suppose $a \in R$ and $\exists$ elements b and c with $b \cdot a = a \cdot c = \underline{1}$, then $b = c$ and so a is a unit with $a^{-1} = b = c$.

Proof: $b = b \cdot \underline{1} = b \cdot (a \cdot c) = (b \cdot a) \cdot c = \underline{1} \cdot c = c$.

Corollary: Inverses are unique.

Zero divisor: Suppose R is a commutative ring. A non-zero element $a \in R$ is called a zero divisor provided $\exists$ a non-zero element b with $a \cdot b = \underline{0}$. Note that if a is a unit, it cannot be a zero divisor.

Theorem 3.2.3: Suppose R is a commutative ring and $a \in (R - \underline{0})$ is not a zero divisor, then $(a \cdot b = a \cdot c) \Rightarrow b = c$. In other words, multiplication by a is an injective map from R to R. It is surjective if a is a unit.

Definition: A domain (or integral domain) is a commutative ring such that, if $a \neq 0$, a is not a zero divisor. A field is a commutative ring such that, if $a \neq 0$, a is a unit. In other words, R is a field if it is commutative and its non-zero elements form a group under multiplication.

Theorem 3.2.4: A field is a domain. A finite domain is a field.

Proof: A field is a domain because a unit cannot be a zero divisor. Suppose R is a finite domain and $a \neq 0$, then $f: R \to R$ defined by $f(b) = a \cdot b$ is injective and, by the pigeonhole principle, f is surjective. Thus a is a unit and so R is a field.

3.3 The Integers Mod N

Theorem 3.3.1: Suppose $n > 1$. Define a multiplication on $\mathbf{Z}_n$ by $[a] \cdot [b] = [ab]$. This is a well defined binary operation which makes $\mathbf{Z}_n$ into a commutative ring.

Proof: Since $[a + kn] \cdot [b + ln] = [ab + n(al + bk + kln)] = [ab]$, the multiplication is well defined. The ring axioms are easily verified.

Theorem 3.3.2: Suppose $n > 1$ and $a \in \mathbf{Z}$, then the following are equivalent.

1) $[a]$ is a generator of the additive group $\mathbf{Z}_n$.

2) $(a, n) = 1$.

3) $[a]$ is a unit of the ring $\mathbf{Z}_n$.

Proof: We already know that 1) and 2) are equivalent. Recall that if b is an integer, $[a]b = [a] \cdot [b] = [ab]$. Thus 1) and 3) are equivalent, because each says $\exists$ an integer b with $[a]b = [1]$.

Corollary: If $n > 1$, the following are equivalent.

1) $\mathbf{Z}_n$ is a domain.

2) $\mathbf{Z}_n$ is a field.

3) n is a prime.

Proof: We already know 1) and 2) are equivalent, because $\mathbf{Z}_n$ is finite. Suppose 3) is true, then by the previous theorem, each of $[1], [2], \ldots, [n-1]$ is a unit, and thus 2) is true. Now suppose 3) is false, then $n = ab$ where $1 < a < n, 1 < b < n, [a][b] = 0$, and thus $[a]$ is a zero divisor and 1) is false.

Subring: Suppose S is a subset of a ring R. The statement that S is a subring of R means that S is a subgroup of the group R, $\underline{1} \in S$ and $(a, b \in S \Rightarrow a \cdot b \in S)$, then clearly S is a ring and has the same multiplicative identity as R. Note that $\mathbf{Z}$ is a subring of $\mathbf{Q}$, $\mathbf{Q}$ is a subring of $\mathbf{R}$, and $\mathbf{R}$ is a subring of $\mathbf{C}$. Note that if S is a subring of R and $s \in S$, then s may be a unit in R but not in S. Note also that $\mathbf{Z}$ and $\mathbf{Z}_n$ have no proper subrings, and thus occupy a special place in ring theory, as well as in group theory.

3.4 Ideals and Quotient Rings

Definition: A subset I of a ring R is a $\begin{Bmatrix} \text{left} \\ \text{right} \\ \text{2-sided} \end{Bmatrix}$ ideal provided it is a subgroup of the additive group R and if $a \in R$ and $b \in I$, then $\begin{Bmatrix} a \cdot b \in I \\ b \cdot a \in I \\ a \cdot b \text{ and } b \cdot a \in I \end{Bmatrix}$.

The word "ideal" means "2-sided ideal." Of course, if R is commutative, every right or left ideal is an ideal.

Theorem 3.4.1: Suppose R is a ring.

1) R and $\underline{0}$ are ideals of R. These are called the improper ideals.

2) If $\{I_t\}_{t\in T}$ is a collection or right(left, 2-sided) ideals of R, then $\bigcap_{t\in T} I_t$ is a right(left, 2-sided) ideal of R.

3) Furthermore, if the collection is monotonic, then $\bigcup_{t\in T} I_t$ is a right(left, 2-sided) ideal of R.

4) If $a\in R, I=aR$ is a right ideal. Thus if R is commutative, aR is an ideal, called a principal ideal. Thus every subgroup of $\mathbf{Z}$ is a principal ideal, because it is of the form $n\mathbf{Z}$.

5) If R is a commutative ring and $I\subset R$ is an ideal, then the following are equivalent.

i) $I=R$.

ii) I contains some unit u.

iii) I contains $\underline{1}$.

Theorem 3.4.2: Suppose R is a ring and $I\subset R$ is an ideal, $I\neq R$. Since I is a normal subgroup of the additive group R, R/I is an additive abelian group. Multiplication of cosets defined by $(a+I)\cdot(b+I)=(ab+I)$ is well defined and makes R/I a ring.

Proof: $(a+I)\cdot(b+I)=a\cdot b+aI+Ib+II\subset a\cdot b+I$. Thus multiplication is well defined, and the ring axioms are easily verified. The multiplicative identity is $(\underline{1}+I)$.

Observation: If $R=\mathbf{Z}$ and $I=n\mathbf{Z}$, the ring structure on $\mathbf{Z}_n=\mathbf{Z}/n\mathbf{Z}$ is the same as the one previously defined.

3.5 Homomorphism

Definition: Suppose R and $\bar{R}$ are rings. A function $f: R\rightarrow\bar{R}$ is a ring homomorphism provided

1) f is a group homomorphism.

2) $f(\underline{1}_R)=\underline{1}_{\bar{R}}$.

3) If $a,b\in R$, then $f(a\cdot b)=f(a)\cdot f(b)$. (On the left, multiplication is in R, while on the right multiplication is in $\bar{R}$.)

The kernel of f is the kernel of f considered as a group homomorphism, namely $f^{-1}(\underline{0})$.

Here are the basic properties of ring homomorphism. And most of them have been presented in the theory of group.

Theorem 3. 5. 1: Suppose each of R and $\bar{R}$ is a ring.

1) The identity map $I_R:R\to R$ is a ring homomorphism.

2) The zero map from R to $\bar{R}$ is not a ring homomorphism (because it does not send $\underline{1}_R$ to $\underline{1}_{\bar{R}}$).

3) The composition of ring homomorphism is a ring homomorphism.

4) If $f:R\to\bar{R}$ is a bijection which is a ring homomorphism, then $f^{-1}:\bar{R}\to R$ is a ring homomorphism. Such an f is called a ring isomorphism. In the case $\bar{R}=R$, f is also called a ring automorphism.

5) The image of a ring homomorphism is a subring of the range.

6) The kernel of a ring homomorphism is an ideal of the domain. In fact, if $f: R\to\bar{R}$ is a homomorphism and $I\subset\bar{R}$ is an ideal, then $f^{-1}(I)$ is an ideal of R.

7) Suppose I is an ideal of R, $I\neq R$, and $\pi:R\to R/I$ is the natural projection, $\pi(a)=(a+I)$. Then π is a surjective ring homomorphism with kernel I. Furthermore, if $f:R\to\bar{R}$ is a surjective ring homomorphism with kernel I, then $R/I\approx\bar{R}$.

8) From now on the word "homomorphism" means "ring homomorphism." Suppose $f:R\to\bar{R}$ is a homomorphism and I is an ideal of R, $I\neq R$.

9) Given any ring homomorphism f, domain $(f)/\ker(f)\approx$ image (f).

3. 6 Polynomial Rings

In calculus, we consider real functions f which are polynomials, $f(x)=a_0+a_1x+...+a_nx^n$. The sum and product of polynomials are again polynomials, and it is easy to see that the collection of polynomial functions forms a commutative ring. We can do the same thing formally in a purely algebraic setting.

Definition: Suppose R is a commutative ring and x is a "variable" or "symbol." The polynomial ring $R[x]$ is the collection of all polynomials $f(x)=a_0+a_1x+...+a_nx^n$ where $a_i\in R$. Under the obvious addition and multiplication, $R[x]$ is a commutative ring. The degree of a non-zero polynomial f is the largest integer n such that $a_n\neq 0$, and is denoted by $n=\deg(f)$. If $a_n=\underline{1}$, then f is said to be monic.

Formally, think of a polynomial $a_0 + a_1x + ...$ as an infinite sequence $(a_0, a_1, ...)$ such that each $a_i \in R$ and only a finite number are non-zero. Then $(a_0, a_1, ...) + (b_0, b_1, ...) = (a_0 + b_0, a_1 + b_1, ...)$ and $(a_0, a_1, ...) \cdot (b_0, b_1, ...) = (a_0b_0, a_0b_1 + a_1b_0, a_0b_2 + a_1b_1 + a_2b_0, ...)$. Note that on the right, the ring multiplication $\boldsymbol{a} \cdot \boldsymbol{b}$ is written simply as $\boldsymbol{ab}$, as is often done for convenience.

3.6.1 The Division Algorithm

Theorem 3.6.1: If R is a domain, $R[x]$ is also a domain.

Proof: Suppose f and g are non-zero polynomials, then deg (f) + deg (g) = deg (fg) and thus fg is not $\underline{0}$. Another way to prove this theorem is to look at the bottom terms instead of the top terms. Let a_ix^i and b_jx^j be the first non-zero terms of f and g. Then $a_ib_jx^{i+j}$ is the first non-zero term of fg.

Theorem 3.6.2: (**The Division Algorithm**) Suppose R is a commutative ring, $f \in R[x]$ has degree $\geqslant 1$ and its top coefficient is a unit in R (If R is a field, the top coefficient of f will always be a unit), then for any $g \in R[x]$, $\exists!\ h, r \in R[x]$ such that $g = fh + r$ with $r = \underline{0}$ or deg $(r) <$ deg (f).

Proof: Here we only outline the proof of existence and leave uniqueness as an exercise. Suppose $f = a_0 + a_1x + ... + a_mx^m$ where $m \geqslant 1$ and a_m is a unit in R. For any g with deg $(g) < m$, set $h = \underline{0}$ and $r = g$. In the general case, the idea is to divide f into g until the remainder has degree less than m. The proof is by induction on the degree of g. Suppose $n \geqslant m$ and the result holds for any polynomial of degree less than n. Suppose g is a polynomial of degree n. Now $\exists$ a monomial bx^t with $t = n - m$ and deg $(g - fbx^t) < n$. By induction, $\exists$ h_1 and r with $fh_1 + r = (g - fbx^t)$ and deg $(r) < m$. The result follows the equation $f(h_1 + bx^t) + r = g$.

Note: In the above definition:

1) If $r = 0$, we say that f divides g.

2) If c is a root of g, i.e. $g(c) = 0$, then $f = x - c$ divides g.

3) More generally, $x - c$ divides g with remainder $g(c)$.

Theorem 3.6.3: Suppose R is a domain, $n > 0$ and $g(x) = a_0 + a_1x + ... + a_nx^n$ is a polynomial of degree n with at least one root in R, then g has at most n roots. Let $c_1, c_2, ..., c_k$ be the distinct roots of g in the ring R, then $\exists$ a unique sequence of positive integers $n_1, n_2, \ldots, n_k$ and a unique polynomial h with no root in R so that $g(x) = (x - c_1)^{n_1} \ldots (x - c_k)^{n_k} h(x)$. (If h has degree 0, i. e. if $h = a_n$, then "all the roots of g belong to R." If $g = a_n x^n$, then "all the roots of g are $\underline{0}$.")

Proof: Because the uniqueness is to prove, we prove the existence only. The theorem is clearly true for $n = 1$. Suppose $n > 1$ and the theorem is true for any polynomial of degree less than n. Now suppose g is a polynomial of degree n and c_1 is a root of g, then $\exists$ a polynomial h_1 with $g(x) = (x - c_1)h_1$. Since h_1 has degree less than n, the result follows by induction.

Note: If g is any non-constant polynomial in $\mathbf{C}[x]$, all the roots of g belong to $\mathbf{C}$, i. e. $\mathbf{C}$ is an algebraically closed field.

Definition: A domain T is a principal ideal domain if, given any ideal I, $\exists t \in T$ such that $I = tT$. Note that $\mathbf{Z}$ is a principal ideal domain and any field is principal ideal domain.

Theorem 3. 6. 4: Suppose F is a field, I is a proper ideal of $F[x]$, and n is the smallest positive integer such that I contains a polynomial of degree n, then I contains a unique polynomial of the form $f = a_0 + a_1 x + \ldots + a_{n-1} x^{n-1} + x^n$ and it has the property that $I = fF[x]$. Thus $F[x]$ is a principal ideal domain. Furthermore, each coset of I can be written uniquely in the form $(c_0 + c_1 x + \ldots + c_{n-1} x^{n-1} + I)$.

Proof: Note this is similar to showing that a subgroup of $\mathbf{Z}$ is generated by one element.

Theorem 3. 6. 5: Suppose R is a subring of a commutative ring $\mathbf{C}$ and $c \in \mathbf{C}$, then $\exists\,!$ homomorphism $h: R[x] \to \mathbf{C}$ with $h(x) = c$ and $h(r) = r$ for all $r \in R$. It is defined by $h(a_0 + a_1 x + \ldots + a_n x^n) = a_0 + a_1 c + \ldots + a_n c^n$, i. e. h sends $f(x)$ to $f(c)$. The image of h is the smallest subring of $\mathbf{C}$ containing R and c.

3. 6. 2 Associate

Definition: Suppose F is a field and $f \in F[x]$ has degree $\geqslant 1$. The statement that g is an associate of f means $\exists$ a unit $u \in F[x]$ such that $g = uf$. The statement that f is irreducible means that if h is a non-constant polynomial which divides f, then h is an associate of f.

Theorem 3. 6. 6: Suppose F is a field and $f \in F[x]$ has degree $\geqslant 1$, then f

factors as the product of irreducibles, and this factorization is unique up to order and associates. Also the following are equivalent.

1) $F[x]/(f)$ is a domain.

2) $F[x]/(f)$ is a field.

3) f is irreducible.

Definition: Now suppose x and y are "variables." If $a \in R$ and $n, m \geqslant 0$, then $ax^n y^m = ay^m x^n$ is called a monomial. Define an element of $R[x,y]$ to be any finite sum of monomials.

Theorem 3.6.7: $R[x,y]$ is a commutative ring and $(R[x])[y] \approx R[x,y] \approx (R[y])[x]$. In other words, any polynomial in x and y with coefficients in R may be written as a polynomial in y with coefficients in $R[x]$, or as a polynomial in x with coefficients in $R[y]$.

Side comment: It is true that if F is a field, each $f \in F[x,y]$ factors as the product of irreducibles. However $F[x,y]$ is not a principal ideal domain. For example, the ideal $I = xF[x,y] + yF[x,y] = \{f \in F[x,y] : f(\underline{0},\underline{0}) = \underline{0}\}$ is not principal.

If R is a commutative ring and $n \geqslant 2$, the concept of a polynomial ring in n variables works fine without a hitch. If $a \in R$ and $v_1, v_2, \ldots, v_n$ are non-negative integers, then $ax_1^{v_1} x_2^{v_2} \ldots x_n^{v_n}$ is called a monomial. Order does not matter here. Define an element of $R[x_1, x_2, \ldots, x_n]$ to be any finite sum of monomials. This gives a commutative ring and there is canonical isomorphism $R[x_1, x_2, \ldots, x_n] \approx (R[x_1, x_2, \ldots, x_{n-1}])[x_n]$. Using this and induction on n, it is easy to prove the following theorem.

Theorem 3.6.8: If R is a domain, $R[x_1, x_2, \ldots, x_n]$ is a domain and its units are just the units of R.

3.7 Product of Rings

The product of rings works just as the product of groups which we introduced in the above chapter. That is to say, they have the similar definition and properties. So here we present the product of rings simply.

Theorem 3.7.1: Suppose T is an index set and for each $t \in T$, R_t is a ring. On the additive abelian group $\prod_{t \in T} R_t = \prod R_t$,define multiplication by $\{r_t\} \cdot \{s_t\} = \{r_t \cdot s_t\}$,then $\prod R_t$ is a ring and each projection $\pi_s: \prod R_t \to R_s$ is a ring homomorphism. Suppose R is a ring. Under the natural bijection from {functions $f:R \to \prod R_t$} to {sequence of functions $\{\{f_t\}_{t \in T}\}$ where $f_t: R \to R_t$} ,f is a ring homomorphism if each f_t is a ring homomorphism.

Proof: We already know f is a group homomorphism if each f_t is a group homomorphism. Note that $\{\underline{1}_t\}$ is the multiplicative identity of $\prod R_t$,and $f(\underline{1}_R) = \{\underline{1}_t\}$ if $f_t(\underline{1}_R) = \underline{1}_t$ for each $t \in T$. Finally, since multiplication is defined coordinatewise, f is a ring homomorphism if each f is a ring homomorphism.

3.8 Characteristic

Theorem 3.8.1: If R is a ring, there is one and only one ring homomorphism $f: \mathbf{Z} \to R$. It is given by $f(m) = m\underline{1} = \underline{m}$. Thus the subgroup of R generated by $\underline{1}$ is a subring of R isomorphic to $\mathbf{Z}$ or isomorphic to $\mathbf{Z}_n$ for some positive integer n.

Definition: Suppose R is a ring and $f: \mathbf{Z} \to R$ is the natural ring homomorphism $f(m) = m\underline{1} = \underline{m}$. The non-negative integer n with ker $(f) = n\mathbf{Z}$ is called the *characteristic* of R. Thus f is injective if R has characteristic 0 and if $\underline{1}$ has infinite order. If f is not injective, the characteristic of R is the order of $\underline{1}$.

Theorem 3.8.2: Suppose R is a domain. If R has characteristic 0, then each non-zero $a \in R$ has infinite order. If R has finite characteristic n, then n is a prime and each non-zero $a \in R$ has order n.

Proof: Suppose R has characteristic 0, a is a non-zero element of R, and m is a positive integer, then $ma = \underline{m} \cdot a$ cannot be $\underline{0}$ because $\underline{m}, a \neq \underline{0}$ and R is a domain. Thus $o(a) = \infty$. Now suppose R has characteristic n, then R contains $\mathbf{Z}_n$ as a subring, and thus $\mathbf{Z}_n$ is a domain and n is a prime. If a is a non-zero element of R, $na = \underline{n} \cdot a = \underline{0} \cdot a = \underline{0}$ and thus $o(a) = n$.

3.9 Boolean Rings

Definition: A ring R is a Boolean ring if for each $a \in R, a^2 = a$, i. e. each element of R is an idempotent.

Theorem 3.9.1: Suppose R is a Boolean ring.

1) If $a \in R, 2a = a + a = \underline{0}$, and so $a = -a$.

2) R is commutative.

3) If R is a domain, $R \approx \mathbf{Z}_2$.

4) The image of a Boolean ring is a Boolean ring. That is, if I is an ideal of R with $I \neq R$, then every element of R/I is idempotent and thus R/I is a Boolean ring. It follows 3) that R/I is a domain if R/I is a field and if $R/I \approx \mathbf{Z}_2$.

Proof:

1) $(a + a) = (a + a)^2 = a^2 + 2a^2 + a^2 = 4a$. Thus $2a = \underline{0}$.

2) $(a + b) = (a + b)^2 = a^2 + (a \cdot b) + (b \cdot a) + b^2 = a + (a \cdot b) - (b \cdot a) + b$. Thus $a \cdot b = b \cdot a$.

3) Suppose $a \neq \underline{0}$, then $a \cdot (\underline{1} - a) = \underline{0}$ and so $a = \underline{1}$.

Suppose X is a non-void set. If a is a subset of X, let $a' = (X - a)$ be a complement of a in X. Now suppose R is a non-void collection of subsets of X. Consider the following properties for R.

1) $a \in R \Rightarrow a' \in R$.

2) $a, b \in R \Rightarrow (a \cap b) \in R$.

3) $a, b \in R \Rightarrow (a \cup b) \in R$.

4) $\varnothing \in R$ and $X \in R$.

Theorem 3.9.2: If 1) and 2) are satisfied, then 3) and 4) are satisfied. In this case, R is called a Boolean algebra of sets.

Proof: Suppose 1) and 2) are true, and $a, b \in R$, then $a \cup b = (a' \cap b')'$ belongs to R and so 3) is true. Since R is non-void, it contains some elements a. Then $\varnothing = a \cap a'$ and $X = a \cup a'$ belong to R, and so 4) is true.

Theorem 3.9.3: Suppose R is a Boolean algebra of sets. Define an addition on R by $a + b = (a \cup b) - (a \cap b)$. Under this addition, R is an abelian group with $\underline{0} = \varnothing$ and $a = -a$. Define a multiplication on R by $a \cdot b = a \cap b$. Under this multiplication R becomes a Boolean ring with $\underline{1} = X$.

Chapter 4

Matrices and Matrix Rings

We have learned the basic knowledge about matrices in former study. Here we first consider matrices in full generality, i. e. over an arbitrary ring R. In this chapter, we first introduce some classical concepts such as elementary matrices and determinant. But the highlight of this chapter lies in the matrix ring.

4.1 Elementary Operations and Elementary Matrices

There are 3 types of elementary row and column operations on a matrix $\boldsymbol{A}$. $\boldsymbol{A}$ need not be square.

Type 1: Multiply row i by some unit $a \in R$.

Multiply column i by some unit $a \in R$.

Type 2: Interchange row i and row j.

Interchange column i and column j.

Type 3: Add a times row j to row i where $i \neq j$ and a is any element of R.

Add a times column j to column i where $i \neq j$ and a is any element of R.

Elementary matrices: Elementary matrices are square and invertible. There are 3 types of elementary matrices which are shown in the following. They are obtained by performing row or column operations on the identity matrix.

Type 1: $A = \begin{pmatrix} 1 & & & & & \\ & 1 & & & 0 & \\ & & a & & & \\ & & & 1 & & \\ & 0 & & & 1 & \\ & & & & & 1 \end{pmatrix}$ where a is a unit in R, all the off-diagonal elements are zero.

Type 2: $A = \begin{pmatrix} 1 & & & & & \\ & 0 & & & 1 & \\ & & 1 & & & \\ & & & 1 & & \\ & 1 & & & 0 & \\ & & & & & 1 \end{pmatrix}$ there are two non-zero off-diagonal elements.

Type 3: $A = \begin{pmatrix} 1 & & & & & \\ & 1 & & & a_{i,j} & \\ & & 1 & & & \\ & & & 1 & & \\ & 0 & & & 1 & \\ & & & & & 1 \end{pmatrix}$ where $i \neq j$ and $a_{i,j}$ is any element of R, there is at most one non-zero off-diagonal element, and it may be above or below the diagonal.

Theorem 4.1.1: Suppose A is a matrix (which is not need to be square). To perform an elementary row (or column) operation on A, perform the operation on an identity matrix to obtain an elementary matrix B, and multiply on the left (or right). That is, BA = row operation on A and AB = column operation on A.

4.2 Systems of Equations

Suppose $A = (a_{i,j}) \in R_{m,n}$ and $C = \begin{pmatrix} c_1 \\ \vdots \\ c_m \end{pmatrix} \in R^m = R_{m,1}$. The system

$\begin{pmatrix} a_{1,1}x_1 + \ldots + a_{1,n}x_n = c_1 \\ \vdots \\ a_{m,1}x_1 + \ldots + a_{m,n}x_n = c_m \end{pmatrix}$ of m equations in n unknowns, can be written as one matrix equation in one unknown, namely as $(a_{i,j})\begin{pmatrix} x_1 \\ \vdots \\ x_n \end{pmatrix} = \begin{pmatrix} c_1 \\ \vdots \\ c_m \end{pmatrix}$ or $\boldsymbol{AX} = \boldsymbol{C}$.

Define $f: R^n \to R^m$ by $f(\boldsymbol{D}) = \boldsymbol{AD}$, then f is a group homomorphism and also $f(\boldsymbol{D}c) = f(\boldsymbol{D})c$ for any $c \in R$.

Theorem 4.2.1:

1) $\boldsymbol{AX} = \underline{\boldsymbol{0}}$ is called the homogeneous equation. Its solution set is $\ker(f)$.

2) $\boldsymbol{AX} = \boldsymbol{C}$ has a solution if $\boldsymbol{C} \in$ image (f). If $\boldsymbol{D} \in R^n$ is one solution, the solution set is the coset $\boldsymbol{D} + \ker(f)$ in R^n.

3) Suppose $\boldsymbol{B} \in R_m$ is invertible, then $\boldsymbol{AX} = \boldsymbol{C}$ and $(\boldsymbol{BA})\boldsymbol{X} = \boldsymbol{BC}$ have the same sets of solutions. Thus we may perform any row operation on both sides of the equation and not change the solution set.

4) If $\boldsymbol{A} \in R_m$ is invertible, then $\boldsymbol{AX} = \boldsymbol{C}$ has the unique solution $\boldsymbol{X} = \boldsymbol{A}^{-1}\boldsymbol{C}$.

4.3 Determinants

For each $n \geqslant 1$ and each commutative ring R, determinant is a function R_n to R. For example, if $n = 1$, $|(a)| = a$ and if $n = 2$, $\left|\begin{pmatrix} a & b \\ c & d \end{pmatrix}\right| = ad - bc$.

Definition: Let $\boldsymbol{A} = (a_{i,j}) \in R_n$. If σ is a permutation on $(1, 2, \ldots, n)$, let sign $(\sigma) = 1$ if σ is an even permutation, and $\text{sign}(\sigma) = -1$ if σ is an odd permutation. The determinant is defined by $|\boldsymbol{A}| = \sum \text{sign}(\sigma) a_{1,\sigma(1)} \cdot a_{2,\sigma(2)} \cdots a_{n,\sigma(n)}$. Check that for $n = 2$, this agrees with the definition above.

For each σ, $a_{1,\sigma(1)} \cdot a_{2,\sigma(2)} \cdots a_{n,\sigma(n)}$ contains exactly one factor from each row and one factor from each column. Since R is commutative, we may rearrange the factors so that the first comes from the first column, the second from the second column, etc. This means that there is a permutation τ on $(1, 2, \ldots, n)$ such that $a_{1,\sigma(1)} \cdot a_{2,\sigma(2)} \cdots a_{n,\sigma(n)} = a_{\tau(1),1} \cdots a_{\tau(n),n}$. We wish to show that $\tau = \sigma^{-1}$ and thus $\text{sign}(\sigma) = \text{sign}(\tau)$. To reduce the abstraction, suppose $\sigma(2) =$

5, then the first expression will contain the factor $a_{2,5}$. In the second expression, it will appear as $a_{\tau(5),5}$, and so $\tau(5)=2$. Anyway, τ is the inverse of σ and thus there are two ways to define determinant.

Corollary: Then the determinant of a matrix is equal to the determinant of its transpose. $|A| = |A^{\mathrm{T}}|$.

Theorem 4.3.1: $|A| = \sum \operatorname{sign}(\sigma) a_{1,\sigma(1)} \cdot a_{2,\sigma(2)} \cdots a_{n,\sigma(n)}$

$= \sum \operatorname{sign}(\tau) a_{\tau(1),1} \cdot a_{\tau(2),2} \cdots a_{\tau(n),n}$.

You may view an $n \times n$ matrix A as a sequence of n column vectors or as a sequence of n row vectors. Here we will use column vectors. This means we write the matrix A as $A = (A_1, A_2, \ldots A_n)$ where each $A_i \in R_{n,1} = R^n$.

Theorem 4.3.2: If two columns of A are equal, then $|A| = 0$.

Proof: For simplicity, assume the first two columns are equal, i. e. $A_1 = A_2$. Now $|A| = \sum \operatorname{sign}(\tau) a_{\tau(1),1} \cdot a_{\tau(2),2} \cdots a_{\tau(n),n}$ and this summation has $n!$ terms and $n!$ is an even number. Let γ be the transposition which interchanges one and two, then for any τ, $\sum \operatorname{sign}(\tau) a_{\tau(1),1} \cdot a_{\tau(2),2} \cdots a_{\tau(n),n} = a_{\tau\gamma(1),1} \cdot a_{\tau\gamma(2),2} \cdots a_{\tau\gamma(n),n}$. This pairs up the $n!$ terms of the summation, and since sign $(\tau) = -\operatorname{sign}(\tau_\gamma)$, these pairs are canceled in the summation. Therefore $|A| = 0$.

Theorem 4.3.3: Suppose $1 \leqslant r \leqslant n$, $C_r \in R_{n,1}$, and $a, c \in R$, then $|(A_1, \ldots, A_{r-1}, aA_r + cC_r, A_{r+1}, \ldots, A_n)| = a|(A_1, \ldots, A_n)| + c|(A_1, \ldots, A_{r-1}, C_r, A_{r+1}, \ldots, A_n)|$.

Proof: This is immediate from the definition of determinant and the distributive law of multiplication in the ring R.

Note: From the above definition and theorem, we get the summary that determinant is a function $d: R_n \to R$.

Theorem 4.3.4: Interchanging two columns of A multiplies the determinant by minus one.

Proof: For simplicity, show that $|(A_2, A_1, A_3, \ldots, A_n)| = -|A|$. We know $0 = |(A_1 + A_2, A_1 + A_2, A_3, \ldots, A_n)| = |(A_1, A_1, A_3, \ldots, A_n)| + |(A_1, A_2, A_3, \ldots, A_n)| + |(A_2, A_1, A_3, \ldots, A_n)| + |(A_2, A_2, A_3, \ldots, A_n)|$. Since the first and last of these four terms are zero, the result follows.

Theorem 4.3.5: If τ is a permutation of $(1, 2, \ldots, n)$, then $|A| = \operatorname{sign}(\tau)$

$|(A_{\tau(1)}, A_{\tau(2)}, \ldots, A_{\tau(n)})|$.

Proof: The permutation τ is the finite product of transpositions.

Theorem 4.3.6: Here are some summary about the determinant of a matrix.

1) Multiplying any row or column of matrix by a scalar $c \in R$, multiplies the determinant by c.

2) Interchanging two rows or two columns multiplies the determinant by -1.

3) Adding c times one row to another row, or adding c times one column to another column, does not change the determinant.

4) If a matrix has two rows equal or two columns equal, its determinant is zero.

5) More generally, if one row is c times another row, or one column is c times another column, then the determinant is zero.

Definition: Let $M_{i,j}$ be the determinant of the $(n-1) \times (n-1)$ matrix obtained by removing row i and column j from A. Let $C_{i,j} = (-1)^{i+j} M_{i,j}$. $M_{i,j}$ and $C_{i,j}$ are called the (i,j) minor and cofactor of A.

Theorem 4.3.7: For any $1 \leqslant i \leqslant n$, $|A| = a_{i,1}C_{i,1} + a_{i,2}C_{i,2} + \ldots + a_{i,n}C_{i,n}$. For any $1 \leqslant j \leqslant n$, $|A| = a_{1,j}C_{1,j} + a_{2,j}C_{2,j} + \ldots + a_{n,j}C_{n,j}$. Thus if any row or any column is zero, the determinant is zero.

Theorem 4.3.8:

1) If A is an upper or lower triangular matrix, $|A|$ is the product of the diagonal elements.

2) If A is an elementary matrix of type 2, $|A| = -1$.

3) If A is an elementary matrix of type 3, $|A| = 1$.

Proof:

1) We will prove the first statement for upper triangular matrices. If $A \in R_2$ is an upper triangular matrix, then its determinant is the product of the diagonal elements. Suppose $n > 2$ and the theorem is true for matrices in R_{n-1}. Suppose $A \in R_n$ is upper triangular. The result follows by expanding by the first column.

2) An elementary matrix of type 2 is obtained from the identity matrix by interchanging two rows or columns, and thus has determinant -1.

3) An elementary matrix of type 3 is a special type of upper or lower triangular matrix, so its determinant is 1.

Theorem 4.3.9: (Determinant by blocks) Suppose $A \in R_n$, $B \in R_{n,m}$, and

$D \in R_m$, then the determinant of $\begin{pmatrix} A & B \\ 0 & D \end{pmatrix}$ is $|A||D|$.

Proof: Expand by the first column and use induction on n.

Theorem 4.3.10: The determinant of the product is the product of the determinants, i. e. if $A, B \in R_n$, $|AB| = |A||B|$. Thus $|AB| = |BA|$ and if C is invertible, $|C^{-1}AC| = |ACC^{-1}| = |A|$.

Corollary: If A is a unit in R_n, then $|A|$ is a unit in R and $|A^{-1}| = |A|^{-1}$.

Proof: $1 = |I| = |AA^{-1}| = |A||A^{-1}|$.

Classical adjoint: Suppose R is a commutative ring and $A \in R_n$. The classical adjoint of A is $(C_{i,j})^{\mathrm{T}}$, i. e. the matrix's (j, i) term is the (i, j) cofactor. Before we consider the general case, let's examine 2×2 matrices.

If $A = \begin{pmatrix} a & b \\ c & d \end{pmatrix}$, then $(C_{i,j}) = \begin{pmatrix} d & -b \\ -c & a \end{pmatrix}$ and so $(C_{i,j})^{\mathrm{T}} = \begin{pmatrix} d & -b \\ -c & a \end{pmatrix}$.

Then $A(C_{i,j})^{\mathrm{T}} = (C_{i,j})^{\mathrm{T}} A = \begin{pmatrix} |A| & 0 \\ 0 & |A| \end{pmatrix} = |A|I$. Thus if $|A|$ is a unit in R, A is invertible and $A^{-1} = |A|^{-1}(C_{i,j})^{\mathrm{T}}$. In particular, if $|A| = 1$, $A^{-1} = \begin{pmatrix} d & -b \\ -c & a \end{pmatrix}$.

Theorem 4.3.11: If R is commutative and $A \in R_n$, then $A(C_{i,j})^{\mathrm{T}} = (C_{i,j})^{\mathrm{T}} A = |A|I$.

Proof: We must show that the diagonal elements of the product $A(C_{i,j})^{\mathrm{T}}$ are all $|A|$ and the other elements are $\underline{0}$. The (s, s) term is the dot product of row s of A with row s of $(C_{i,j})$ and is thus $|A|$ (computed by expansion by row s). For $s \neq t$, the (s, t) term is the dot product of rows s of A with row of $(C_{i,j})$. Since this is the determinant of a matrix with row s = row t, the (s, t) term is $\underline{0}$. The proof that $(C_{i,j})^{\mathrm{T}} A = |A|I$ is left as an exercise.

Theorem 4.3.12: Suppose R is a commutative ring and $A \in R_n$, then A is a unit in R_n if $|A|$ is a unit in R. (Thus if R is a field, A is invertible if $|A| \neq 0$.) If A is invertible, then $A^{-1} = |A|^{-1}(C_{i,j})^{\mathrm{T}}$. Thus if $|A| = 1$, $A^{-1} = (C_{i,j})^{\mathrm{T}}$, the classical adjoint of A.

4.4 Similarity

Definition: Suppose $A, B \in R_n$. B is said to be similar to A if $\exists$ an invertible $C \in R_n$ such that $B = C^{-1}AC$, i. e. B is similar to A if B is a conjugate of A.

Theorem 4.4.1:

1) B is similar to B.

2) B is similar to A if A is similar to B.

3) If D is similar to B and B is similar to A, then D is similar to A.

4) "Similarity" is an equivalence relation on R_n.

Here we leave the proof as an exercise.

Theorem 4.4.2: Suppose A and B are similar, then $|A| = |B|$ and thus A is invertible if B is invertible.

Proof: Suppose $B = C^{-1}AC$, then $|B| = |C^{-1}AC| = |ACC^{-1}| = |A|$.

Trace: Suppose $A = (a_{i,j}) \in R_n$, then the trace is defined by trace$(A) = a_{1,1} + a_{2,2} + \ldots + a_{n,n}$. That is, the trace of A is the sum of its diagonal terms.

Theorem 4.4.3: Suppose $A \in R_{m,n}$ and $B \in R_{n,m}$, then AB and BA are square matrices with trace(AB) = trace(BA).

Proof: This proof involves a change in the order of summation. By definition,

$$\text{trace}(AB) = \sum_{1 \leqslant i \leqslant m} a_{i,1}b_{1,i} + \ldots + a_{i,n}b_{n,i} = \sum_{\substack{1 \leqslant i \leqslant m \\ 1 \leqslant j \leqslant n}} a_{i,j}b_{j,i} = \sum_{1 \leqslant j \leqslant n} b_{j,1}a_{1,j} + \ldots + b_{j,m}a_{m,j} = \text{trace}(BA)$$

Theorem 4.4.4: If $A, B \in R_n$, trace$(A + B)$ = trace(A) + trace(B) and trace (AB) = trace(BA).

Theorem 4.4.5: If A and B are similar, then trace(A) = trace(B).

Proof: trace(B) = trace$(C^{-1}AC)$ = trace(ACC^{-1}) = trace(A).

Summary: Determinant and trace are functions from R_n to R. Determinant is a multiplicative homomorphism and trace is an additive homomorphism. Furthermore $|AB| = |BA|$ and trace(AB) = trace(AB). If A and B are similar, $|A| = |B|$ and trace(A) = trace(B).

Characteristic polynomials: If $A \in R_n$, the characteristic polynomial $CP_A(x) \in R[x]$ is defined by $CP_A(x) = |xI - A|$. Any $\lambda \in R$ which is a root of $CP_A(x)$ is called a characteristic root of A.

Theorem 4.4.6: $CP_A(x) = a_0 + a_1 x + \ldots + a_{n-1} x^{n-1} + x^n$ where trace $(\boldsymbol{A}) = -a_{n-1}$ and $|\boldsymbol{A}| = (-1)^n a_0$.

Proof: This follows a direct computation of the determinant.

Theorem 4.4.7: If $\boldsymbol{A}$ and $\boldsymbol{B}$ are similar, then they have the same characteristic polynomials.

Proof: Suppose $\boldsymbol{B} = \boldsymbol{C}^{-1}\boldsymbol{A}\boldsymbol{C}$. $CP_B(x) = |(x\boldsymbol{I} - \boldsymbol{C}^{-1}\boldsymbol{A}\boldsymbol{C})| = |\boldsymbol{C}^{-1}(x\boldsymbol{I} - \boldsymbol{A})\boldsymbol{C}| = |(x\boldsymbol{I} - \boldsymbol{A})| = CP_A(x)$.

Note: A square matrix over a field is nilpotent if all its characteristic roots are 0 and if it is similar to a strictly upper triangular matrix.

Part Ⅱ

Chapter 5

Vector Spaces

In this chapter, we introduce the notion vector which is the basis of the following chapters. All sections in this chapter are presented around this concept. Maybe you have studied the vector that is a directed line segment. This type of vector is only a special example of the more general vector presented in this chapter.

5.1 The Axioms for a Vector Space

The concept of a vector put forward here is purely algebraic. The definition given for a vector is that it is a member of a set that satisfies certain algebraic rules.

A vector space is a triple (V, F, f) consisting of

1) An additive Abelian group V.

2) A field F.

3) A function $f: F \times V \to V$ called scalar multiplication such that

$$\begin{aligned} f(\lambda, f(\mu, \boldsymbol{v})) &= f(\lambda\mu, \boldsymbol{v}) \\ f(\lambda + \mu, \boldsymbol{u}) &= f(\lambda, \boldsymbol{u}) + f(\mu, \boldsymbol{u}) \\ f(\lambda, \boldsymbol{u} + \boldsymbol{v}) &= f(\lambda, \boldsymbol{u}) + f(\lambda, \boldsymbol{v}) \end{aligned} \tag{5.1.1}$$

for all $\lambda, \mu \in F$ and $\boldsymbol{u}, \boldsymbol{v} \in V$. A vector is an element of a vector space. The notation (V, F, f) for a vector space will be shortened to simply V. The first of eq. (5.

1.1) is usually called the associative law for scalar multiplication, while the second and third equations are distributive laws, the second for scalar addition and the third for vector addition.

It is also customary to use a simplified notation for the scalar multiplication function f. We shall write $f(\lambda, \boldsymbol{v}) = \lambda \boldsymbol{v}$ and also regard $\lambda \boldsymbol{v}$ and $\boldsymbol{v}\lambda$ to be identical. In this simplified notation we shall now list in detail the axioms of a vector space. In this definition the vector $\boldsymbol{u} + \boldsymbol{v}$ in V is called the sum of $\boldsymbol{u}$ and $\boldsymbol{v}$ and the difference of $\boldsymbol{u}$ and $\boldsymbol{v}$ is written $\boldsymbol{u} - \boldsymbol{v}$ and is defined by

$$\boldsymbol{u} - \boldsymbol{v} = \boldsymbol{u} + (-\boldsymbol{v}) \tag{5.1.2}$$

Definition: Let V be a set and F a field. V is a vector space if it satisfies the following rules:

1) There exists a binary operation in V called addition and denoted by + such that

a) $(\boldsymbol{u} + \boldsymbol{v}) + \boldsymbol{w} = \boldsymbol{u} + (\boldsymbol{v} + \boldsymbol{w})$ for all $\boldsymbol{u}, \boldsymbol{v}, \boldsymbol{w} \in V$.

b) $\boldsymbol{u} + \boldsymbol{v} = \boldsymbol{v} + \boldsymbol{u}$ for all $\boldsymbol{u}, \boldsymbol{v} \in V$.

c) There exists an element $\boldsymbol{0} \in V$ such that $\boldsymbol{u} + \boldsymbol{0} = \boldsymbol{u}$ for all $\boldsymbol{u} \in V$.

d) For every $\boldsymbol{u} \in V$ there exists an element $-\boldsymbol{u} \in V$, such that $\boldsymbol{u} + (-\boldsymbol{u}) = \boldsymbol{0}$.

2) There exists an operation called scalar multiplication in which every scalar $\lambda \in F$ can be combined with every element $\boldsymbol{u} \in V$ to give an element $\lambda \boldsymbol{u} \in V$ such that

a) $\lambda(\mu \boldsymbol{u}) = (\lambda\mu)\boldsymbol{u}$

b) $(\lambda + \mu)\boldsymbol{u} = \lambda\mu + \lambda\boldsymbol{u}$

c) $\lambda(\boldsymbol{u} + \boldsymbol{v}) = \lambda\boldsymbol{u} + \lambda\boldsymbol{v}$

d) $1\boldsymbol{u} = \boldsymbol{u}$

For all $\lambda, \mu \in F$ and all $\boldsymbol{u}, \boldsymbol{v} \in V$.

If the field F employed in a vector space is actually the field of real numbers $\mathbf{R}$, the space is called a real vector space. A complex vector space is defined similarly.

In fact, the real vector space is a trivial special case. But the complex case is more useful in the material on spectral decompositions. So a vector space should be understood to be a complex vector space unless we provide some qualifying statement to the contrary.

There are many varied sets of objects that qualify as vector spaces. The following is a list of examples of vector spaces:

1) The vector space ζ^N is the set of all N-tuples of the form $\boldsymbol{u} = (\lambda_1, \lambda_2, \ldots, \lambda_N)$, where N is a positive integer and $\lambda_1, \lambda_2, \ldots, \lambda_N \in \zeta$. Since an N-tuples is ordered set, if $\boldsymbol{v} = (\mu_1, \mu_2, \ldots, \mu_N)$ is a second N-tuple, then $\boldsymbol{u}$ and $\boldsymbol{v}$ are equal iff

$$\mu_k = \lambda_k \text{ for all } k = 1, 2, \ldots, N$$

The zero N-tuple is $\mathbf{0} = (0, 0, \ldots, 0)$ and the negative of the N-tuple $\boldsymbol{u}$ is $-\boldsymbol{u} = (-\lambda_1, -\lambda_2, \ldots, -\lambda_N)$. Addition and scalar multiplication of N-tuple are defined by the formulas

$$\boldsymbol{u} + \boldsymbol{v} = (\mu_1 + \lambda_1, \mu_2 + \lambda_2, \ldots, \mu_N + \lambda_N)$$

and

$$\mu\boldsymbol{u} = (\mu\lambda_1, \mu\lambda_2, \ldots, \mu\lambda_N)$$

respectively. The notation ζ^N is used for this vector space because it can be considered to be an N-th Cartesian product of ζ.

2) The set V of all $N \times M$ complex matrices is a vector space with respect to the usual operation of matrix addition and multiplication by a scalar.

3) Let H be a vector space whose vectors are actually functions defined on a set A with values in ζ. Thus, if $\boldsymbol{k} \in H, x \in A$, then $\boldsymbol{k}(x) \in \zeta$ and $\boldsymbol{k}: A \to \zeta$. If $\boldsymbol{k}$ is another vector of H, then the equality of vectors is defined by

$$\boldsymbol{h} = \boldsymbol{k} \Leftrightarrow \boldsymbol{h}(x) = \boldsymbol{k}(x) \text{ for all } x \in A$$

The zero vector is the zero function whose value is zero for all x. Addition and scalar multiplication are defined by

$$(\boldsymbol{h} + \boldsymbol{k})(x) = \boldsymbol{h}(x) + \boldsymbol{k}(x)$$

and

$$(\lambda\boldsymbol{h})(x) = \lambda(\boldsymbol{h}(x))$$

respectively.

4) Let P denote the set of all polynomials u of degree N of the form

$$u = \lambda_0 + \lambda_1 x + \ldots + \lambda_N x^N$$

where $\lambda_0, \lambda_1, \ldots, \lambda_N \in \boldsymbol{C}$. The set P forms a vector space over the complex numbers $\boldsymbol{C}$ if addition and scalar multiplication of polynomials are defined in the usual way.

5) The set of complex numbers $\boldsymbol{C}$, the usual definitions of addition and multiplication by a real number, forms a real vector space.

6) The zero element $\mathbf{0}$ of any vector space forms a vector space by itself.

Theorem 5.1.1: $\lambda \boldsymbol{u} = \boldsymbol{0} \Leftrightarrow \lambda = 0$ or $\boldsymbol{u} = \boldsymbol{0}$

Proof: The proof of this theorem is actually equivalent to the proof of the following three assertions:

$$(a) 0\boldsymbol{u} = \boldsymbol{0}, (b) \lambda\boldsymbol{0} = \boldsymbol{0}, (c) \lambda\boldsymbol{u} = \boldsymbol{0} \Rightarrow \lambda = 0 \text{ or } \boldsymbol{u} = \boldsymbol{0}.$$

(a): take $\mu = 0$ in Axiom[2)b)] for a vector space; then $\lambda\boldsymbol{u} = \lambda\boldsymbol{u} + 0\boldsymbol{u}$. Therefore $\lambda\boldsymbol{u} - \lambda\boldsymbol{u} = \lambda\boldsymbol{u} - \lambda\boldsymbol{u} + 0\boldsymbol{u}$ and by Axioms[1)c)]and[1)d)] $\boldsymbol{0} = \boldsymbol{0} + 0\boldsymbol{u} = 0\boldsymbol{u}$ which proves(a).

(b): set $\boldsymbol{v} = \boldsymbol{0}$ in Axiom[2)c)] for a vector space; then $\lambda\boldsymbol{u} = \lambda\boldsymbol{u} + \lambda\boldsymbol{0}$. Therefore $\lambda\boldsymbol{u} - \lambda\boldsymbol{u} = \lambda\boldsymbol{u} - \lambda\boldsymbol{u} + \lambda\boldsymbol{0}$ and by Axiom[1)d)] $\boldsymbol{0} = \boldsymbol{0} + \lambda\boldsymbol{0} = \lambda\boldsymbol{0}$.

(c): we assume $\lambda\boldsymbol{u} = 0$. If $\lambda = 0$, we know from(a)that the equation $\lambda\boldsymbol{u} = \boldsymbol{0}$ is satisfied. If $\lambda \neq 0$, then we show that $\boldsymbol{u}$ must be zero as follows: $\boldsymbol{u} = 1\boldsymbol{u} = \lambda\left(\frac{1}{\lambda}\right)\boldsymbol{u} = \frac{1}{\lambda}(\lambda\boldsymbol{u}) = \frac{1}{\lambda}(\boldsymbol{0}) = \boldsymbol{0}$.

Theorem 5.1.2: $(-\lambda)\boldsymbol{u} = -\lambda\boldsymbol{u}$ for all $\lambda \in \mathbf{C}, \boldsymbol{u} \in V$.

Proof: Let $\mu = 0$ and replace λ by $-\lambda$ in Axiom[2)b)] for a vector space and this result follows directly.

Theorem 5.1.3: $-\lambda\boldsymbol{u} = \lambda(-\boldsymbol{u})$ for all $\lambda \in \mathbf{C}, \boldsymbol{u} \in V$.

5.2 Linear Independence, Dimension and Basis

In this section the concept of linear independence is introduced by first defining what is meant by linear dependence of a set of vectors and then defining a set of vectors that is not linearly dependent to be linearly independent. And also the notion of dimension and basis are introduced.

Definition: A finite set of $N(N \geqslant 1)$ vectors $\{\boldsymbol{v}_1, \boldsymbol{v}_2, \ldots, \boldsymbol{v}_N\}$ in a vector space V is said to be linearly dependent if there exists a set of scalars $\{\lambda^1, \lambda^2, \ldots, \lambda^N\}$, not all zero, such that

$$\sum_{j=1}^{N} \lambda^j \boldsymbol{v}_j = \boldsymbol{0} \tag{5.2.1}$$

The essential content of this definition is that at least one of the vectors $\{\boldsymbol{v}_1, \boldsymbol{v}_2, \ldots, \boldsymbol{v}_N\}$ can be expressed as a linear combination of the other vectors. This means that if $\lambda^1 \neq 0$, then $\boldsymbol{v}_1 = \sum_{i=2}^{N} \mu^j \boldsymbol{v}_j$, where $\mu^j = -\lambda^j/\lambda^1$ for $j = 2, 3, \ldots,$

N. As a numerical example, consider the two vectors

$$\boldsymbol{v}_1 = (1,2,3), \boldsymbol{v}_2 = (2,6,9)$$

from $\mathbf{R}^3$. These vectors are linearly dependent since

$$\boldsymbol{v}_2 = 3\boldsymbol{v}_1$$

Theorem 5.2.1: If the set of vectors $\{\boldsymbol{v}_1, \boldsymbol{v}_2, \ldots, \boldsymbol{v}_N\}$ is linearly dependent, then every other finite set of vectors containing $\{\boldsymbol{v}_1, \boldsymbol{v}_2, \ldots, \boldsymbol{v}_N\}$ is linearly dependent.

Theorem 5.2.2: Every set of vectors containing the zero vector is linearly dependent.

A set of $N(N \geqslant 1)$ vectors that is not linearly dependent is said to be linearly independent. Equivalently, a set of $N(N \geqslant 1)$ vectors $\{\boldsymbol{v}_1, \boldsymbol{v}_2, \ldots, \boldsymbol{v}_N\}$ is linearly independent if eq. (2.1) implies $\lambda^1 = \lambda^2 = \ldots = \lambda^N = 0$. For example, there are two vectors $\boldsymbol{v} = (1,2)$ and $\boldsymbol{v}_2 = (2,1)$ from $\mathbf{R}^2$. These two vectors are linearly independent because

$$\lambda \boldsymbol{v}_1 = \lambda \boldsymbol{v}_2 = \lambda(1,2) + \mu(2,1) = \mathbf{0} \Leftrightarrow \lambda + 2\mu = 0, 2\lambda + \mu = 0$$

and

$$\lambda + 2\mu = 0, 2\lambda + \mu = 0 \Leftrightarrow \lambda = 0, \mu = 0$$

Theorem 5.2.3: Every nonempty subset of a linearly independent set is linearly independent.

A linearly independent set in a vector space is said to be maximal if it is not a proper subset of any other linearly independent set. A vector space that contains a (finite) maximal, linearly independent set is then said to be finite dimensional. Of course, if V is not finite dimensional, then it is called infinite dimensional. In this text we shall be concerned only with finite dimensional vector spaces.

Theorem 5.2.4: Any two maximal, linearly independent sets of a finite-dimensional vector space must contain exactly the same number of vectors.

Proof: Let $\{\boldsymbol{v}_1, \boldsymbol{v}_2, \ldots, \boldsymbol{v}_N\}$ and $\{\boldsymbol{u}_1, \boldsymbol{u}_2, \ldots, \boldsymbol{u}_M\}$ be two maximal, linearly independent sets of V. Then we must show that $N = M$. Suppose that $N \neq M$, say $N < M$. By the fact that $\{\boldsymbol{v}_1, \boldsymbol{v}_2, \ldots, \boldsymbol{v}_N\}$ is maximal, the sets $\{\boldsymbol{v}_1, \ldots, \boldsymbol{v}_N, \boldsymbol{u}_1\}, \ldots, \{\boldsymbol{v}_1, \ldots, \boldsymbol{v}_N, \boldsymbol{u}_M\}$ are all linearly dependent. Hence there exist relations

$$\begin{gathered}\lambda_{11}\boldsymbol{v}_1 + \ldots + \lambda_{1N}\boldsymbol{v}_N + \mu_1\boldsymbol{u}_1 = 0 \\ \vdots \\ \lambda_{M1}\boldsymbol{v}_1 + \ldots + \lambda_{MN}\boldsymbol{v}_N + \mu_M\boldsymbol{u}_M = 0\end{gathered} \tag{5.2.2}$$

where the coefficients of each equation are not all equal to zero. In fact, the coefficients $\mu_1, \ldots, \mu_M$ of the vectors $\boldsymbol{u}_1, \ldots, \boldsymbol{u}_M$ are all nonzero, for if $\mu_i = 0$ for any i, then

$$\lambda_{i1}\boldsymbol{v}_1 + \ldots + \lambda_{iN}\boldsymbol{v}_N = 0$$

for some nonzero $\lambda_{i1}, \ldots, \lambda_{iN}$ contradicting the assumption that $\{\boldsymbol{v}_1, \ldots, \boldsymbol{v}_N\}$ is a linearly independent set. Hence we can solve eq. (5.2.2) for the vectors $\{\boldsymbol{u}_1, \ldots, \boldsymbol{u}_M\}$ in terms of the vectors $\{\boldsymbol{v}_1, \ldots, \boldsymbol{v}_N\}$, obtaining

$$\begin{aligned} \boldsymbol{u}_1 &= \mu_{11}\boldsymbol{v}_1 + \ldots + \mu_{1N}\boldsymbol{v}_N \\ &\vdots \\ \boldsymbol{u}_M &= \mu_{M1}\boldsymbol{v}_1 + \ldots + \mu_{MN}\boldsymbol{v}_N \end{aligned} \tag{5.2.3}$$

where $\mu_{ij} = -\lambda_{ij}/\mu_i$ for $i = 1, \ldots, M; j = 1, \ldots, N$.

Now we claim that the first N equations in the above system can be inverted in such a way that the vectors $\{\boldsymbol{v}_1, \ldots, \boldsymbol{v}_N\}$ are given by linear combinations of the vectors $\{\boldsymbol{u}_1, \ldots, \boldsymbol{u}_N\}$. Indeed, inversion is possible if the coefficient matrix $[\mu_{ij}]$ for $i, j = 1, \ldots, N$ is nonsingular. But this is clearly the case, for if that matrix were singular, there would be nontrivial solutions $\{\alpha_1, \ldots, \alpha_N\}$ to the linear system

$$\sum_{i=1}^{N} \alpha_i \mu_{ij} = 0 \qquad j = 1, \ldots, N \tag{5.2.4}$$

Then from eq. (5.2.3) and eq. (5.2.4) we would have

$$\sum_{i=1}^{N} \alpha_i \boldsymbol{u}_i = \sum_{j=1}^{N} \left(\sum_{i=1}^{N} \alpha_i \mu_{ij} \right) \boldsymbol{v}_j = 0$$

contradicting the assumption that set $\{\boldsymbol{u}_1, \ldots, \boldsymbol{u}_N\}$, being a subset of the linearly independent set $\{\boldsymbol{u}_1, \ldots, \boldsymbol{u}_M\}$, is linearly independent.

Now if the inversion of the first N equations of the system eq. (5.2.3) gives

$$\begin{aligned} \boldsymbol{v}_1 &= \xi_{11}\boldsymbol{u}_1 + \ldots + \xi_{1N}\boldsymbol{u}_N \\ &\vdots \\ \boldsymbol{v}_N &= \xi_{N1}\boldsymbol{u}_1 + \ldots + \xi_{NN}\boldsymbol{u}_N \end{aligned} \tag{5.2.5}$$

where $[\xi_{ij}]$ is the inverse of $[\mu_{ij}]$ for $j = 1, \ldots, N$, we can substitute eq. (5.2.5) into the remaining $M - N$ equations in the system eq. (5.2.3), obtaining

$$\begin{aligned} \boldsymbol{u}_{N=1} &= \sum_{j=1} \left(\sum_{i=1}^{N} \mu_{N+1,j}\xi_{ji} \right) \boldsymbol{u}_i \\ &\vdots \\ \boldsymbol{u}_M &= \sum_{j=1}^{N} \left(\sum_{i=1}^{N} \mu_{Mj}\xi_{ji} \right) \boldsymbol{u}_i \end{aligned}$$

But these equations contradict the assumption that the set $\{\boldsymbol{u}_1,\ldots,\boldsymbol{u}_M\}$ is linearly independent. Hence $M>N$ is impossible and the proof is complete.

Theorem 5.2.5: Let $\{\boldsymbol{u}_1,\ldots,\boldsymbol{u}_N\}$ be a maximal linearly independent set in V, and suppose that $\{\boldsymbol{v}_1,\ldots,\boldsymbol{v}_N\}$ is given by eq. (5.2.5), then $\{\boldsymbol{v}_1,\ldots,\boldsymbol{v}_N\}$ is also a maximal, linearly independent set iff the coefficient matrix $[\xi_{ij}]$ in eq. (5.2.5) is nonsingular. In particular, if $\boldsymbol{v}_i=\boldsymbol{u}_i$ for $i=1,\ldots,k-1,k+1,\ldots,N$ but $\boldsymbol{v}_k\neq\boldsymbol{u}_k$, then $\{\boldsymbol{v}_1,\ldots,\boldsymbol{v}_{k-1},\boldsymbol{v}_k,\boldsymbol{v}_{k+1},\ldots,\boldsymbol{v}_N\}$ is a maximal, linearly independent set if and only if the coefficient ξ_{kk} in the expansion of $\boldsymbol{v}_k$ in terms of $\{\boldsymbol{u}_1,\ldots,\boldsymbol{u}_N\}$ in eq. (5.2.5) is nonzero.

A list of examples of bases for vector spaces follows:

1) The set of N vectors

$$
\begin{gathered}
(1,0,0,\ldots,0)\\
(0,1,0,\ldots,0)\\
(0,0,1,\ldots,0)\\
\vdots\\
(0,0,0,\ldots,1)
\end{gathered}
$$

is linearly independent and constitutes a basis for $\mathbf{C}^N$, called the standard basis.

2) If $U^{2\times2}$ denotes the vector space of all 2×2 matrices with elements from the complex numbers $\mathbf{C}$, then the four matrices

$$
\begin{pmatrix}1&0\\0&0\end{pmatrix},\begin{pmatrix}0&1\\0&0\end{pmatrix},\begin{pmatrix}0&0\\1&0\end{pmatrix},\begin{pmatrix}0&0\\0&1\end{pmatrix}
$$

form a basis for $U^{2\times2}$ called the standard basis.

3) The elements 1 and $i=\sqrt{-1}$ form a basis for the vector space of complex numbers C over the field of real numbers.

Theorem 5.2.6: If $\{\boldsymbol{e}_1,\boldsymbol{e}_2,\ldots,\boldsymbol{e}_N\}$ is a basis for V, then every vector in V has the representation

$$
\boldsymbol{v}=\sum_{j=1}^{N}\xi^j\boldsymbol{e}_j \tag{5.2.6}
$$

where $\{\xi^1,\xi^2,\ldots,\xi^N\}$ are elements of C which depend upon the vector $\boldsymbol{v}\in V$.

Theorem 5.2.7: The N scalars $\{\xi^1,\xi^2,\ldots,\xi^N\}$ in eq. (5.2.6) are unique.

Proof: To proof the theorem we assume a lack of uniqueness. Thus we say that $\boldsymbol{v}$ has two representations

$$\boldsymbol{v} = \sum_{j=1}^{N} \xi^j \boldsymbol{e}_j , \boldsymbol{v} = \sum_{j=1}^{N} \mu^j \boldsymbol{e}_j$$

Subtracting the two representations, we have

$$\sum_{j=1}^{N} (\xi^j - \mu^j) \boldsymbol{e}_j = 0$$

and the linear independence of the basis requires that

$$\xi^j = \mu^j, j = 1, 2, \ldots, N$$

The coefficients $\{\xi^1, \xi^2, \ldots, \xi^N\}$ in the representation eq. (5.2.6) are the components of $\boldsymbol{v}$ with respect to the basis $\{\boldsymbol{e}_1, \boldsymbol{e}_2, \ldots, \boldsymbol{e}_N\}$.

Theorem 5.2.8: If $\{\boldsymbol{e}_1, \boldsymbol{e}_2, \ldots, \boldsymbol{e}_N\}$ is a basis for V and if $\boldsymbol{B} = \{\boldsymbol{b}_1, \boldsymbol{b}_2, \ldots, \boldsymbol{b}_K\}$ is a linearly independent set of $K(N \geqslant K)$ in V, then it is possible to exchange a certain K of the original base vectors with $\boldsymbol{b}_1, \boldsymbol{b}_2, \ldots, \boldsymbol{b}_K$ so that the new set is a basis for V.

Proof: We select b_1 from the set B and order the basis vectors such that the component ξ^1 of b_1 is nonzero in the representation

$$b_1 = \sum_{j=1}^{N} \xi^j \boldsymbol{e}_j$$

By Theorem 5.2.8, the vectors $\{\boldsymbol{b}_1, \boldsymbol{e}_2, \ldots, \boldsymbol{e}_N\}$ form a basis for V. A second vector $\boldsymbol{b}_2$ is selected from V and we again order the basis vectors so that this time the component λ^2 is not zero in the formula

$$\boldsymbol{b}_2 = \lambda^1 \boldsymbol{b}_1 + \sum_{j=2}^{N} \lambda^j \boldsymbol{e}_j$$

Again, by Theorem 5.2.8 the vectors $\{\boldsymbol{b}_1, \boldsymbol{b}_2, \boldsymbol{e}_3, \ldots, \boldsymbol{e}_N\}$ form a basis for V. The proof is completed by simply repeating the above construction $K-2$ times.

We now know that when a basis for V is given, every vector in V has a representation in the form of eq. (5.2.4). Inverting this condition somewhat, we now want to consider a set of vectors $B = \{\boldsymbol{b}_1, \boldsymbol{b}_2, \ldots, \boldsymbol{b}_M\}$ of V with the property that every vector $\boldsymbol{v} \in V$ can be written as

$$\boldsymbol{v} = \sum_{j=1}^{M} \lambda^j \boldsymbol{b}_j$$

Such a set is called a generating set of V. In some sense a generating set is a counterpart of a linearly independent set. The following theorem is the counter part of Theorem 5.2.8.

Theorem 5.2.9: If $B = \{\boldsymbol{b}_1, \ldots, \boldsymbol{b}_M\}$ is a generating set of V, then every other finite set of vectors containing B is also a generating set of V.

In view of this theorem, we see that the counterpart of a maximal, linearly independent set is a minimal generating set, which is defined by the condition that a generating set $\{\boldsymbol{b}_1, \ldots, \boldsymbol{b}_M\}$ is minimal if it contains no proper subset that is still a generating set. The following theorem shows the relation between a maximal, linearly independent set and a minimal generating set.

Theorem 5.2.10: Let $B = \{\boldsymbol{b}, \ldots, \boldsymbol{b}_M\}$ be a finite subset of a finite dimensional vector space V. Then the following conditions are equivalent:

1) B is a maximal linearly independent set.

2) B is a linearly independent generating set.

3) B is a minimal generating set.

Proof: We shall show that 1)⇒2)⇒3)⇒1).

1)⇒2). This implication is a direct consequence of the representation eq. (5.2.4).

2)⇒3). This implication is obvious. For if B is a linearly independent generating set but not a minimal generating set, then we can remove at least one vector, say $\boldsymbol{b}_M$, and the remaining set is still a generating set. But this is impossible because of $\boldsymbol{b}_M$. It can then be expressed as a linear combination of $\{\boldsymbol{b}_1, \ldots, \boldsymbol{b}_{M-1}\}$, contradicting the linear independence of B.

3)⇒1). If B is a minimal generating set, then B must be linearly independent because otherwise one of the vectors of B, say $\boldsymbol{b}_M$, can be written as a linear combination of the vectors $\{\boldsymbol{b}_1, \ldots, \boldsymbol{b}_{M-1}\}$. It follows then that $\{\boldsymbol{b}_1, \ldots \boldsymbol{b}_{M-1}\}$ is still a generating set, contradicting the assumption that $\{\boldsymbol{b}_1, \ldots, \boldsymbol{b}_M\}$ is minimal. Now a linearly independent generating set must be maximal, otherwise there exists a vector $\boldsymbol{b} \in V$ such that $\{\boldsymbol{b}_1, \ldots, \boldsymbol{b}_M, \boldsymbol{b}\}$ is linearly independent. Then $\boldsymbol{b}$ cannot be expressed as a linear combination of $\{\boldsymbol{b}_1, \ldots, \boldsymbol{b}_M\}$, thus contradicting the assumption that B is a generating set.

5.3 Intersection, Sum and Direct Sum of Subspaces

In this section operations such as "summing" and "intersecting" vector spaces are discussed. A nonempty subset U of a vector space V is a subspace if:

1) $\boldsymbol{u},\boldsymbol{w} \in U \Rightarrow \boldsymbol{u}+\boldsymbol{w} \in U$ for all $\boldsymbol{u},\boldsymbol{w} \in U$.

2) $\boldsymbol{u} \in U \Rightarrow \lambda\boldsymbol{u} \in U$ for all $\lambda \in \mathbf{C}$.

Conditions 1) and 2) in this definition can be replaced by the equivalent condition:

(a') $\boldsymbol{u},\boldsymbol{w} \in U \Rightarrow \lambda\boldsymbol{u}+\lambda\boldsymbol{w} \in U$ for all $\lambda \in \mathbf{C}$.

Examples of subspaces of vector spaces are given in the following list:

1) The subset of the vector space C^N of all N-tuples of the form $(0,\lambda_2,\ldots,\lambda_N)$ is a subspace of C^N.

2) Any vector space V is a subspace of itself.

3) The set consisting of the zero vector $\{\mathbf{0}\}$ is a subspace of V.

4) The set of real numbers $\mathbf{R}$ can be viewed as a subspace of the real space of complex numbers $\mathbf{C}$.

The vector spaces $\{\mathbf{0}\}$ and V itself are considered to be trivial subspaces of the vector space V. If U is not a trivial subspace, it is said to be a proper subspace of V.

The following theorem presents several properties of subspaces.

Theorem 5.3.1:

1) If U is a subspace of V, then $\mathbf{0} \in U$.

2) If U is a subspace of V, then $\dim U \leqslant \dim V$.

3) If U is a subspace of V, then $\dim U = \dim V$ if and only if $U = V$.

Proof:

1) The proof of this theorem follows easily condition 2) in the definition of a subspace above by setting $\lambda = 0$.

2) By Theorem 5.2.8 we know that any basis of U can be enlarged to a basis of V; it follows that $\dim U \leqslant \dim V$.

3) If $U = V$, then clearly $\dim U = \dim V$. Conversely, if $\dim U = \dim V$, then a basis for U is also a basis for V. Thus, any vector $\boldsymbol{v} \in V$ is also in U, and this implies $U = V$.

Operations that combine vector spaces to form other vector spaces are simple extensions of elementary operations defined on sets. If U and W are subspaces of a vector space V, the sum of U and W, written $U + W$, is the set

$$U + W = \{\boldsymbol{u}+\boldsymbol{w} \mid \boldsymbol{u} \in U, \boldsymbol{w} \in W\}$$

Similarly, if U and W are subspaces of a vector space V, the intersection of U and W, denoted by $U \cap W$, is the set

$$U \cap W = \{\boldsymbol{u} \mid \boldsymbol{u} \in U \text{ and } \boldsymbol{u} \in W\}$$

The union of two subspaces U and W of V, denoted by $U \cup W$, is the set

$$U \cup W = \{\boldsymbol{u} \mid \boldsymbol{u} \in U \text{ or } \boldsymbol{u} \in W\}$$

Some properties of the above operations are stated in the following theorems.

Theorem 5.3.2:

1) If U and W are subspaces of V, then $U + W$ is a subspace of V.

2) If U and W are subspaces of V, then U and W are also subspaces of $U + W$.

Theorem 5.3.3:

1) If U and W are subspaces of V, then the intersection $U \cap W$ is a subspace of V.

2) If U and W are subspaces of V, then the union $U \cup W$ is not generally subspace of V.

Proof:

1) Let $\boldsymbol{u}, \boldsymbol{w} \in U \cap W$, then $\boldsymbol{u}, \boldsymbol{w} \in W$ and $\boldsymbol{u}, \boldsymbol{w} \in U$. Since U and W are subspaces, $\boldsymbol{u} + \boldsymbol{w} \in W$ and $\boldsymbol{u} + \boldsymbol{w} \in U$ which means that $\boldsymbol{u} + \boldsymbol{w} \in U \cap W$ also. Similarly, if $\boldsymbol{u} \in U \cap W$, then for all $\lambda \in \mathbf{R}, \lambda \boldsymbol{u} \in U \cap W$.

2) Let $\boldsymbol{u} \in U, u \notin W$, and let $\boldsymbol{w} \in W$ and $\boldsymbol{w} \notin U$, then the union $\boldsymbol{u} + \boldsymbol{w} \notin W$ and $\boldsymbol{u} + \boldsymbol{w} \notin U$, which means that $\boldsymbol{u} + \boldsymbol{w} \notin U \cup W$.

Theorem 5.3.4: Let $\boldsymbol{v}$ be a vector in $U + W$, where U and W are subspaces of V. The decomposition of $\boldsymbol{v} \in U + W$ into the form $\boldsymbol{v} = \boldsymbol{u} + \boldsymbol{w}$, where $\boldsymbol{u} \in U$ and $\boldsymbol{w} \in W$, is unique if and only if $U \cap W = \{\mathbf{0}\}$.

Proof: Suppose there are two ways of decomposing $\boldsymbol{v}$; for example, let there be a $\boldsymbol{u}$ and $\boldsymbol{u}'$ in U and a $\boldsymbol{w}$ and $\boldsymbol{w}'$ in W such that

$$\boldsymbol{v} = \boldsymbol{u} + \boldsymbol{w} \text{ and } \boldsymbol{v} = \boldsymbol{u}' + \boldsymbol{w}'$$

The decomposition of $\boldsymbol{v}$ is unique if it can be shown that the vector $\boldsymbol{b}$,

$$\boldsymbol{b} = \boldsymbol{u} - \boldsymbol{u}' = \boldsymbol{w}' - \boldsymbol{w}$$

vanishes. The vector $\boldsymbol{b}$ is contained in $U \cap W$ since $\boldsymbol{u}$ and $\boldsymbol{u}'$ are known to be in U, and $\boldsymbol{w}$ and $\boldsymbol{w}'$ are known to be in W. Therefore $U \cap W = \{\mathbf{0}\}$ implies uniqueness. Conversely, if we have uniqueness, then $U \cap W = \{\mathbf{0}\}$, otherwise any nonzero vector $\boldsymbol{y} \in U \cap W$ has at least two decompositions $\boldsymbol{y} = \boldsymbol{y} + \mathbf{0} = \mathbf{0} + \boldsymbol{y}$.

Definition: The sum of two subspaces U and W is called the direct sum of U and W and denoted by $U \oplus W$ if $U \cap W = \{\mathbf{0}\}$.

This definition is motivated by the result proved in Theorem 5.3.4. If $U \oplus$

$W = V$, then U is called the direct complement of W in V. The operation of direct summing can be extended to a finite number of subspaces $V_1, V_2, \ldots, V_Q$ of V. The direct sum $V_1 \oplus V_2 \oplus \ldots \oplus V_Q$ is required to satisfy the conditions

$$V_R \cap \sum_{K=1}^{k=R-1} V_K + V_R \cap \sum_{K=R+1}^{Q} V_K = \{\mathbf{0}\} \text{ for } R = 1, 2, \ldots, Q$$

Theorem 5.3.5: If U and W are subspaces of V, then

$$\dim U \oplus \dim W = \dim U + \dim W$$

Proof: Let $\{\boldsymbol{u}_1, \boldsymbol{u}_2, \ldots, \boldsymbol{u}_R\}$ be a basis for U and $\{\boldsymbol{w}_1, \boldsymbol{w}_2, \ldots, \boldsymbol{w}_Q\}$ be a basis of W, then the set of vectors $\{\boldsymbol{u}_1, \boldsymbol{u}_2, \ldots, \boldsymbol{u}_R, \boldsymbol{w}_1, \boldsymbol{w}_2, \ldots, \boldsymbol{w}_Q\}$ is linearly independent since $U \cap W = \{\mathbf{0}\}$. This set of vectors generates $U \oplus W$ since for any $U \oplus W$ we can write

$$\boldsymbol{v} = \boldsymbol{u} + \boldsymbol{w} = \sum_{j=1}^{R} \lambda^j \boldsymbol{u}_j + \sum_{j=1}^{Q} \mu^j \boldsymbol{w}_j$$

where $\boldsymbol{u} \in U$ and $\boldsymbol{w} \in W$. Therefore by Theorem 5.3.5, $\dim U \oplus \dim W = R + Q$.

The result of this theorem can easily be generalized to

$$\dim(V_1 \oplus V_2 \oplus \ldots V_Q) = \sum_{j=1}^{Q} \dim V_i$$

5.4 Factor Space

In this section an equivalence relation is introduced and the vector space is partitioned into equivalence sets. The class of all equivalence sets is itself a vector space called the factor space.

If U is a subset of a vector space V, then two vectors $\boldsymbol{w}$ and $\boldsymbol{v}$ in V are said to be equivalent with respect to U, written $\boldsymbol{v} - \boldsymbol{w}$. If $\boldsymbol{w} - \boldsymbol{v}$ is a vector contained in U, it is easy to see that this relation is an equivalence relation and it induces a partition of V into equivalence sets of vectors. If $\boldsymbol{v} \in V$, then the equivalence set of $\boldsymbol{v}$, denoted by $\bar{\boldsymbol{v}}$, is the set of all vectors of the form $\boldsymbol{v} + \boldsymbol{u}$, where $\boldsymbol{u}$ is any vector of U,

$$\bar{\boldsymbol{v}} = \{\boldsymbol{v} + \boldsymbol{u} \mid \boldsymbol{u} \in U\}$$

To illustrate the equivalence relation and its decomposition of a vector space into equivalence sets, we will consider the real vector space $\mathbf{R}^2$, which we can represent by the Euclidean plane. Let $\boldsymbol{u}$ be a fixed vector in $\mathbf{R}^2$, and define the subspace U of $\mathbf{R}^2$ by $\{\lambda \boldsymbol{u} \mid \lambda \in \mathbf{R}\}$. This subspace consists of all vectors of the form $\lambda \boldsymbol{u}, \lambda \in \mathbf{R}$, which are all parallel to the same straight line. From the definition of equivalence, the vector $\boldsymbol{v}$ is seen to be equivalent to the vector $\boldsymbol{w}$ if the vector $\boldsymbol{v} - \boldsymbol{w}$ is parallel to the line representing U. Therefore all vectors that differ from the vector $\boldsymbol{v}$ by a vector that is parallel to the line representing U are equivalent. The set of all vectors equivalent to the vector $\boldsymbol{v}$ is therefore the set of all vectors that terminate on the dashed line parallel to the line representing U. The equivalence set of $\boldsymbol{v}$, i. e. $\bar{\boldsymbol{v}}$, is the set of all such vectors.

Theorem 5.4.1: If $\bar{\boldsymbol{v}} \neq \bar{\boldsymbol{w}}$, then $\bar{\boldsymbol{v}} \cap \bar{\boldsymbol{w}} = \varnothing$.

Proof: Assume that $\bar{\boldsymbol{v}} \neq \bar{\boldsymbol{w}}$ but that there exists a vector $\boldsymbol{x}$ in $\bar{\boldsymbol{v}} \cap \bar{\boldsymbol{w}}$. Then $\boldsymbol{x} \in \bar{\boldsymbol{v}}, \boldsymbol{x} - \boldsymbol{v}$, and $\boldsymbol{x} \in \bar{\boldsymbol{w}}, \boldsymbol{x} - \boldsymbol{w}$. By the transitive property of the equivalence relation we have $\boldsymbol{x} - \boldsymbol{w}$, which implies $\bar{\boldsymbol{v}} = \bar{\boldsymbol{w}}$ and which is a contradiction. Therefore $\bar{\boldsymbol{v}} \cap \bar{\boldsymbol{w}}$ contains no vector unless $\bar{\boldsymbol{v}} = \bar{\boldsymbol{w}}$, in which case $\bar{\boldsymbol{v}} \cap \bar{\boldsymbol{w}} = \bar{\boldsymbol{v}}$.

We shall now develop the structure of the factor space. The factor class of U, denoted by V/U is the class of all equivalence sets in V formed by using a subspace U of V. The factor class is sometimes called a quotient class. Addition and scalar multiplication of equivalence sets are denoted by

$$\bar{\boldsymbol{v}} + \bar{\boldsymbol{w}} = \overline{\boldsymbol{v} + \boldsymbol{w}}$$

and

$$\lambda \bar{\boldsymbol{v}} = \overline{\lambda \boldsymbol{v}}$$

respectively. It is easy to verify that the addition and multiplication operations defined above depend only on the equivalence sets but not on any particular vectors used in representing the sets.

Theorem 5.4.2: The factor set V/U forms a vector space, called a factor space, with respect to the operations of addition and scalar multiplication of equivalence classes defined above.

The theorem is easy to prove, here we leave it as an exercise.

The factor space is also called a quotient space. The subspace U in V/U plays the role of the zero vector of the factor space. In the trivial case when $U = V$, there is only one equivalence set and it plays the role of the zero vector. On the other extreme when $U = \{\mathbf{0}\}$, then each equivalence set is a single vector and $V/\{\mathbf{0}\} = V$.

5.5 Inner Product Spaces

We define the concept of length through the concept of an inner product. An inner product on a complex vector space V is a function $f: V \times V \to \mathbf{C}$ with the following properties:

1) $f(\boldsymbol{u},\boldsymbol{v}) = \overline{f(\boldsymbol{v},\boldsymbol{u})}$.

2) $\lambda f(\boldsymbol{u},\boldsymbol{v}) = f(\lambda \boldsymbol{u},\boldsymbol{v})$.

3) $f(\boldsymbol{u}+\boldsymbol{w},\boldsymbol{v}) = f(\boldsymbol{u},\boldsymbol{v}) + f(\boldsymbol{w},\boldsymbol{v})$.

4) $f(\boldsymbol{u},\boldsymbol{u}) \geqslant 0$ and $f(\boldsymbol{u},\boldsymbol{u}) = 0 \Leftrightarrow \boldsymbol{u} = \mathbf{0}$.

for all $\boldsymbol{u},\boldsymbol{v},\boldsymbol{w} \in V$ and $\lambda \in \mathbf{C}$. In 1) the bar denotes the complex conjugate. 2) and 3) require that f be linear in its first argument, i. e. $f(\lambda \boldsymbol{u} + \mu \boldsymbol{v}, \boldsymbol{w}) = \lambda f(\boldsymbol{u},\boldsymbol{w}) + \mu f(\boldsymbol{v},\boldsymbol{w})$ for all $\boldsymbol{u},\boldsymbol{v},\boldsymbol{w} \in V$ and all $\lambda, \mu \in \mathbf{C}$. 1) and the linearity implied by 2) and 3) insure that f is conjugate linear in its second argument, i. e. $f(\boldsymbol{u},\lambda \boldsymbol{v} + \mu \boldsymbol{w}) = \bar{\lambda} f(\boldsymbol{u},\boldsymbol{v}) + \bar{\mu} f(\boldsymbol{u},\boldsymbol{w})$ for all $\boldsymbol{u},\boldsymbol{v},\boldsymbol{w} \in V$ and all $\lambda, \mu \in \mathbf{C}$. Since 1) ensures that $f(\boldsymbol{u},\boldsymbol{u})$ is real, 4) is meaningful, and it requires that f should be positive definite. There are many notations for the inner product. We shall employ the notation of the "dot product" and write

$$f(\boldsymbol{u},\boldsymbol{v}) = \boldsymbol{u} \cdot \boldsymbol{v}$$

An inner product space is simply a vector space with an inner product.

Definition: A complex inner product space, or simply an inner product space, is a set V and a field $\mathbf{C}$ such that:

1) There exists a binary operation in V called addition and denoted by + such that

(1) $(\boldsymbol{u}+\boldsymbol{v}) + \boldsymbol{w} = \boldsymbol{u} + (\boldsymbol{v}+\boldsymbol{w})$ for all $\boldsymbol{u},\boldsymbol{v},\boldsymbol{w} \in V$.

(2) $\boldsymbol{u}+\boldsymbol{v} = \boldsymbol{v}+\boldsymbol{u}$ for all $\boldsymbol{u},\boldsymbol{v} \in V$.

(3) There exists an element $\mathbf{0} \in V$ such that $\boldsymbol{u} + \mathbf{0} = \boldsymbol{u}$ for all $\boldsymbol{u} \in V$.

(4) For every $\boldsymbol{u} \in V$ there exists an element $-\boldsymbol{u} \in V$ such that $\boldsymbol{u} + (-\boldsymbol{u}) = \mathbf{0}$.

2) There exists an operation called scalar multiplication in which every scalar $\lambda \in \mathbf{C}$ can be combined with every element $\boldsymbol{u} \in V$ to give an element $\lambda \boldsymbol{u} \in V$ such that:

(1) $\lambda(\mu \boldsymbol{u}) = (\lambda \mu)\boldsymbol{u}$.

(2) $(\lambda+\mu)\boldsymbol{u}=\lambda\boldsymbol{u}+\mu\boldsymbol{u}$.

(3) $\lambda(\boldsymbol{u}+\boldsymbol{v})=\lambda\boldsymbol{u}+\lambda\boldsymbol{v}$.

(4) $1\boldsymbol{u}=\boldsymbol{u}$; for all $\lambda\in\mathbf{C}$ and all $\boldsymbol{u},\boldsymbol{v}\in V$.

3) There exists an operation called inner product by which any ordered pair of vectors $\boldsymbol{u}$ and $\boldsymbol{v}$ in V determines an element of $\mathbf{C}$ denoted by $\boldsymbol{u}\cdot\boldsymbol{v}$ such that

(1) $\boldsymbol{u}\cdot\boldsymbol{v}=\overline{\boldsymbol{v}\cdot\boldsymbol{u}}$.

(2) $\lambda\boldsymbol{u}\cdot\boldsymbol{v}=(\lambda\boldsymbol{u})\cdot\boldsymbol{v}$.

(3) $(\boldsymbol{u}+\boldsymbol{w})\cdot\boldsymbol{v}=\boldsymbol{u}\cdot\boldsymbol{v}+\boldsymbol{w}\cdot\boldsymbol{v}$.

(4) $\boldsymbol{u}\cdot\boldsymbol{u}\geqslant\mathbf{0}$ and $\boldsymbol{u}\cdot\boldsymbol{u}=\mathbf{0}\Leftrightarrow\boldsymbol{u}=\mathbf{0}$; for all $\boldsymbol{u},\boldsymbol{v},\boldsymbol{w}\in V$ and $\lambda\in\mathbf{C}$.

A real inner product space is defined similarly. The vector space C^N becomes an inner product space if, for any two vectors $\boldsymbol{u},\boldsymbol{v}\in C^N$, where $\boldsymbol{u}=(\lambda_1,\lambda_2,\ldots,\lambda_N)$ and $\boldsymbol{v}=(\mu_1,\mu_2,\ldots,\mu_N)$, we define the inner product of $\boldsymbol{u}$ and $\boldsymbol{v}$ by

$$\boldsymbol{u}\cdot\boldsymbol{v}=\sum_{j=1}^{N}\lambda_j\overline{\mu_j}$$

The length of a vector is an operation, denoted by $\|\ \|$, which assigns to each nonzero vector $\boldsymbol{v}\in V$ a positive real number by the following rule:

$$\|\boldsymbol{v}\|=\sqrt{\boldsymbol{v}\cdot\boldsymbol{v}}\tag{5.5.1}$$

Of course the length of the zero vector is zero.

Theorem 5.5.1: The Schwarz inequality,

$$|\boldsymbol{u}\cdot\boldsymbol{v}|\leqslant\|\boldsymbol{u}\|\|\boldsymbol{v}\|\tag{5.5.2}$$

is valid for any two vectors $\boldsymbol{u},\boldsymbol{v}$ in an inner product space.

Proof: The Schwarz inequality is easily seen to be trivially true when either $\boldsymbol{u}$ or $\boldsymbol{v}$ is the $\mathbf{0}$ vector, so we shall assume that neither $\boldsymbol{u}$ nor $\boldsymbol{v}$ is zero. Construct the vector $(\boldsymbol{u}\cdot\boldsymbol{u})\boldsymbol{v}-(\boldsymbol{v}\cdot\boldsymbol{u})\boldsymbol{u}$ and above properties, which require that every vector have a nonnegative length, hence

$$(\|\boldsymbol{u}\|^2\|\boldsymbol{v}\|^2-(\boldsymbol{u}\cdot\boldsymbol{v})\overline{(\boldsymbol{u}\cdot\boldsymbol{v})})\|\boldsymbol{u}\|^2\geqslant 0$$

Since $\boldsymbol{u}$ must not be zero, it follows that

$$\|\boldsymbol{u}\|^2\|\boldsymbol{v}\|^2\geqslant(\boldsymbol{u}\cdot\boldsymbol{v})\overline{(\boldsymbol{u}\cdot\boldsymbol{v})}=|\boldsymbol{u}\cdot\boldsymbol{v}|^2$$

and the positive square root of this equation is Schwarz's inequality.

Theorem 5.5.2: The triangle inequality

$$\|\boldsymbol{u}+\boldsymbol{v}\|\leqslant\|\boldsymbol{u}\|+\|\boldsymbol{v}\|\tag{5.5.3}$$

is valid for any two vectors $\boldsymbol{u},\boldsymbol{v}$ in an inner product space.

Proof: The squared length of $\boldsymbol{u}+\boldsymbol{v}$ can be written in the form

$$\|\boldsymbol{u}+\boldsymbol{v}\|^2=(\boldsymbol{u}+\boldsymbol{v})\cdot(\boldsymbol{u}+\boldsymbol{v})=\|\boldsymbol{u}\|^2+\|\boldsymbol{v}\|^2+\boldsymbol{u}\cdot\boldsymbol{v}+\boldsymbol{v}\cdot\boldsymbol{u}$$

$$= \| \boldsymbol{u} \|^2 + \| \boldsymbol{v} \|^2 + 2\mathrm{Re}(\boldsymbol{u} \cdot \boldsymbol{v})$$

where Re signifies the real part. By use of the Schwarz inequality this can be rewritten as

$$\| \boldsymbol{u} + \boldsymbol{v} \|^2 \leqslant \| \boldsymbol{u} \|^2 + \| \boldsymbol{v} \|^2 + 2 \| \boldsymbol{u} \| \| \boldsymbol{v} \| = (\| \boldsymbol{u} \| + \| \boldsymbol{v} \|)^2$$

Taking the positive square root of this equation, we obtain the triangular inequality.

For a real inner product space the concept of angle is defined as follows. The angle between two vectors $\boldsymbol{u}$ and $\boldsymbol{v}$, denoted by θ, is defined by

$$\cos\theta = \frac{\boldsymbol{u} \cdot \boldsymbol{v}}{\| \boldsymbol{u} \| \| \boldsymbol{v} \|} \tag{5.5.4}$$

This definition of a real-valued angle is meaningful because the Schwarz inequality in this case shows that the quantity on the right-hand side of eq. (5.5.4) must have a value lying between 1 and -1, i. e.

$$-1 \leqslant \frac{\boldsymbol{u} \cdot \boldsymbol{v}}{\| \boldsymbol{u} \| \| \boldsymbol{v} \|} \leqslant +1$$

Returning to complex inner product spaces in general, we say that two vectors $\boldsymbol{u}$ and $\boldsymbol{v}$ are orthogonal if $\boldsymbol{u} \cdot \boldsymbol{v}$ or $\boldsymbol{v} \cdot \boldsymbol{u}$ is zero. Clearly, this definition is consistent with the real case, since orthogonality then means $|\theta| = \pi/2$.

The inner product space is a very substantial algebraic structure and parts of the structure can be given slightly different interpretations. In particular it can be shown that the length $\| \boldsymbol{v} \|$ of a vector $\boldsymbol{v} \in V$ is a particular example of a mathematical concept known as a norm, and thus an inner product space is a normal space. A norm on V is a real-valued function defined on V whose value is denoted by $\| \boldsymbol{v} \|$ and which satisfies the following axioms:

(1) $\| \boldsymbol{v} \| \geqslant 0$ and $\| \boldsymbol{v} \| = 0 \Leftrightarrow \boldsymbol{v} = \boldsymbol{0}$,

(2) $\| \lambda \boldsymbol{u} \| = |\lambda| \, \| \boldsymbol{u} \|$,

(3) $\| \boldsymbol{u} \| + \| \boldsymbol{v} \| \geqslant \| \boldsymbol{u} + \boldsymbol{v} \|$,

for all $\boldsymbol{u}, \boldsymbol{v} \in V$ and all $\lambda \in \mathbf{C}$. In defining the norm of $\boldsymbol{v}$, we have employed the same notation as that for the length of $\boldsymbol{v}$ because we will show that the length defined by an inner product is a norm, but the converse need not be true.

Theorem 5.5.3: The operation of determining the length $\| \boldsymbol{v} \|$ of $\boldsymbol{v} \in V$ is a norm on V.

Proof: The properties of an inner product imply Axioms 1 and 2 of a norm, and the triangle inequality is proved by Theorem 5.5.2.

We will now show that the inner product space is also a metric space. A metric space is a nonempty set U equipped with a positive real-valued function $U \times U \to \mathbf{R}$, called the distance function that satisfies the following axioms

(1) $d(\boldsymbol{u},\boldsymbol{v}) \geqslant 0$ and $d(\boldsymbol{u},\boldsymbol{v}) = 0 \Leftrightarrow \boldsymbol{u} = \boldsymbol{v}$,

(2) $d(\boldsymbol{u},\boldsymbol{v}) = d(\boldsymbol{v},\boldsymbol{u})$,

(3) $d(\boldsymbol{u},\boldsymbol{w}) \leqslant d(\boldsymbol{u},\boldsymbol{v}) + d(\boldsymbol{v},\boldsymbol{w})$,

for all $\boldsymbol{u},\boldsymbol{v},\boldsymbol{w} \in U$. In other words, a metric space is a set U of objects and a positive-definite, symmetric distance function d satisfying the triangle inequality. The boldface notation for the elements of U is simply to save the introduction of yet another notation.

Theorem 5.5.4: An inner product space V is a metric space with the distance function given by

$$d(\boldsymbol{u},\boldsymbol{v}) = \| \boldsymbol{u} - \boldsymbol{v} \| \tag{5.5.5}$$

Proof: Let $\boldsymbol{w} = \boldsymbol{u} - \boldsymbol{v}$; then from the requirement that $\| \boldsymbol{w} \| \geqslant 0$ and $\| \boldsymbol{w} \| = 0 \Leftrightarrow \boldsymbol{w} = \mathbf{0}$ it follows that

$$\| \boldsymbol{u} - \boldsymbol{v} \| \geqslant 0 \text{ and } \| \boldsymbol{u} - \boldsymbol{v} \| = 0 \Leftrightarrow \boldsymbol{u} = \boldsymbol{v}$$

Similarly, from the requirement that $\| \boldsymbol{w} \| = \| -\boldsymbol{w} \|$, it follows that

$$\| \boldsymbol{u} - \boldsymbol{v} \| = \| \boldsymbol{v} - \boldsymbol{u} \|$$

Finally, let $\boldsymbol{u}$ be replaced by $\boldsymbol{u} - \boldsymbol{v}$ and $\boldsymbol{v}$ by $\boldsymbol{v} - \boldsymbol{w}$ in the triangle inequality eq. (5.5.3); then

$$\| \boldsymbol{u} - \boldsymbol{w} \| \leqslant \| \boldsymbol{u} - \boldsymbol{v} \| + \| \boldsymbol{v} - \boldsymbol{w} \|$$

which is the third and last requirement for a distance function in a metric space.

The inner product as well as the expressions eq. (5.5.1) for the length of a vector can be expressed in terms of any basis and the components of the vectors relative to that basis. Let $\{\boldsymbol{e}_1, \boldsymbol{e}_2, \ldots, \boldsymbol{e}_N\}$ be a basis of V and denote the inner product of any two base vectors by e_{jk},

$$e_{jk} = \boldsymbol{e}_j \cdot \boldsymbol{e}_k = \overline{e_{kj}} \tag{5.5.6}$$

Thus, if the vectors $\boldsymbol{u}$ and $\boldsymbol{v}$ have the representations

$$\boldsymbol{u} = \sum_{j=1}^{N} \lambda^j \boldsymbol{e}_j, \boldsymbol{v} = \sum_{k=1}^{N} \mu^k \boldsymbol{e}_k \tag{5.5.7}$$

Relative to the basis $\{\boldsymbol{e}_1, \boldsymbol{e}_2, \ldots, \boldsymbol{e}_N\}$, then the inner product of $\boldsymbol{u}$ and $\boldsymbol{v}$ is given by

$$\boldsymbol{u} \cdot \boldsymbol{v} = \sum_{j=1}^{N} \lambda^j \boldsymbol{e}_j \cdot \sum_{k=1}^{N} \mu^k \boldsymbol{e}_k = \sum_{j=1}^{N} \sum_{k=1}^{N} \lambda^j \overline{\mu}^k e_{jk} \tag{5.5.8}$$

Eq. (5.5.8) is the component expression for the inner product.

From the definition eq. (5.5.1) for the length of a vector $\mathbf{v}$, $\mathbf{v}$ and from eq. (5.5.6) and eq. (5.5.7), we can write

$$\|\mathbf{v}\| = \left(\sum_{j=1}^{N}\sum_{k=1}^{N} e_{jk}\,\mu^{j}\overline{\mu}^{k}\right)^{1/2} \tag{5.5.9}$$

This equation gives the component expression for the length of a vector. For a real inner product space, it easily follows that

$$\cos\theta = \frac{\sum_{j=1}^{n}\sum_{k=1}^{N} e_{jk}\,\mu^{j}\lambda^{k}}{\left(\sum_{p=1}^{N}\sum_{r=1}^{N} e_{pr}\,\mu^{p}\mu^{r}\right)^{1/2}\left(\sum_{l=1}^{N}\sum_{s=1}^{N} e_{sl}\lambda^{s}\lambda^{l}\right)^{1/2}} \tag{5.5.10}$$

In formulas eq. (5.5.8) – eq. (5.5.10) we notice that we have changed the particular indices that indicate summation so that no more than two indices occur in any summand. The reason for changing these dummy indices of summation can be made apparent by failing to change them. For example, in order to write eq. (5.5.9) in its present form, we can write the component expressions for $\mathbf{v}$ in two equivalent ways,

$$\mathbf{v} = \sum_{j=1}^{N}\mu^{j}\,\mathbf{e}_{j}, \quad \mathbf{v} = \sum_{k=1}^{N}\mu^{k}\,\mathbf{e}_{k}$$

and then take the dot product. If we had used the same indices to represent summation in both cases, then that index would occur four times in eq. (5.5.9) and all the cross terms in the inner product would be left out.

5.6 Orthonormal Bases and Orthogonal Complements

Definition:

1) Vectors with a magnitude of 1 are called unit vectors or normalized vectors.

2) A set of vectors in an inner product space V is said to be an orthogonal set if all the vectors in the set are mutually orthogonal, and it is said to be an orthonormal set if the set is orthogonal and if all the vectors are unit vectors.

3) In equation form, an orthonormal set $\{\mathbf{i}_1, \ldots, \mathbf{i}_M\}$ satisfies the conditions

$$\mathbf{i}_j \cdot \mathbf{i}_k = \delta_{jk} \equiv \begin{cases} 1 & \text{if } j = k \\ 0 & \text{if } j \neq k \end{cases} \tag{5.6.1}$$

The symbol δ_{jk} introduced in the equation above is called the Kronecker delta.

Theorem 5.6.1: An orthonormal set is linearly independent.

Proof: Assume that the orthonormal set $\{\boldsymbol{i}_1, ..., \boldsymbol{i}_M\}$ is linearly dependent, that is to say there exists a set of scalars $\{\lambda^1, \lambda^2, ... \lambda^M\}$, not all zero, such that

$$\sum_{j=1}^{M} \lambda^j \boldsymbol{i}_j = \boldsymbol{0}$$

The inner product of this sum with the unit $\boldsymbol{i}_k$ vector gives the expression

$$\sum_{j=1}^{M} \lambda^j \delta_{jk} = \lambda^1 \delta_{1k} + ... + \lambda^M \delta_{Mk} = \lambda^k = 0$$

for $k = 1, 2, ..., M$. A contradiction has therefore been achieved and the theorem is proved.

Corollary: An orthonormal set in an inner product space V can have no more than $N = \dim V$ elements. An orthonormal set is said to be complete in an inner product space V if it is not a proper subset of another orthonormal set in the same space.

Theorem 5.6.2: A complete orthonormal set is a basis for V; such a basis is called an orthonormal basis.

Any set of linearly independent vectors can be used to construct an orthonormal set, and likewise, any basis can be used to construct an orthonormal basis. The process by which this is done is called the Gram-Schmidt orthogonalization process and it is developed in the proof of the following theorem.

Theorem 5.6.3: Given a basis $\{\boldsymbol{e}_1, \boldsymbol{e}_2, ..., \mathbf{e}_N\}$ of an inner product space V, then there exists an orthonormal basis $\{\boldsymbol{i}_1, ..., \boldsymbol{i}_N\}$ such that $\{\boldsymbol{e}_1, ..., \boldsymbol{e}_k\}$ and $\{\boldsymbol{i}_1, ..., \boldsymbol{i}_k\}$ generate the same subspace U_k of V, for each $k = 1, ..., N$.

Proof: The construction proceeds in two steps: first, a set of orthogonal vectors is constructed, then this set is normalized. Let $\{\boldsymbol{d}_1, \boldsymbol{d}_2, ..., \boldsymbol{d}_N\}$ denote a set of orthogonal but not unit, vectors. This set is constructed from $\{\boldsymbol{e}_1, \boldsymbol{e}_2, ..., \boldsymbol{e}_N\}$ as follows: Let $\boldsymbol{d}_1 = \boldsymbol{e}_1$ and put

$$\boldsymbol{d}_2 = \boldsymbol{e}_2 + \xi \boldsymbol{d}_1$$

The scalar ξ will be selected so that $\boldsymbol{d}_2$ is orthogonal to $\boldsymbol{d}_1$; orthogonality of $\boldsymbol{d}_1$ and $\boldsymbol{d}_2$ requires that their inner product be zero; hence

$$\boldsymbol{d}_2 \cdot \boldsymbol{d}_1 = 0 = \boldsymbol{e}_2 \cdot \boldsymbol{d}_1 + \xi \boldsymbol{d}_1 \cdot \boldsymbol{d}_1$$

implies

$$\xi = -\frac{\boldsymbol{e}_2 \cdot \boldsymbol{d}_1}{\boldsymbol{d}_1 \cdot \boldsymbol{d}_1}$$

where $\boldsymbol{d}_1 \cdot \boldsymbol{d}_1 \neq 0$ since $\boldsymbol{d}_1 \neq \boldsymbol{0}$. The vector $\boldsymbol{d}_2$ is not zero, because $\boldsymbol{e}_1$ and $\boldsymbol{e}_2$ are linearly independent. The vector $\boldsymbol{d}_3$ is defined by

$$\boldsymbol{d}_3 = \boldsymbol{e}_3 + \xi^2 \boldsymbol{d}_2 + \xi^1 \boldsymbol{d}_1$$

The scalars ξ^2 and ξ^1 are determined by the requirement that $\boldsymbol{d}_3$ be orthogonal to both $\boldsymbol{d}_1$ and $\boldsymbol{d}_2$; thus

$$\boldsymbol{d}_3 \cdot \boldsymbol{d}_1 = \boldsymbol{e}_3 \cdot \boldsymbol{d}_1 + \xi^1 \boldsymbol{d}_1 \cdot \boldsymbol{d}_1 = 0$$

$$\boldsymbol{d}_3 \cdot \boldsymbol{d}_2 = \boldsymbol{d}_3 \cdot \boldsymbol{d}_1 + \xi^2 \boldsymbol{d}_2 \cdot \boldsymbol{d}_2 = 0$$

and, as a result,

$$\xi^1 = -\frac{\boldsymbol{e}_1 \cdot \boldsymbol{d}_1}{\boldsymbol{d}_1 \cdot \boldsymbol{d}_1}, \xi^2 = -\frac{\boldsymbol{e}_3 \cdot \boldsymbol{d}_2}{\boldsymbol{d}_2 \cdot \boldsymbol{d}_2}$$

The linear independence of $\boldsymbol{e}_1, \boldsymbol{e}_2$, and $\boldsymbol{e}_3$ requires that $\boldsymbol{d}_3$ be nonzero. It is easy to see that this scheme can be repeated until a set of N orthogonal vectors $\{\boldsymbol{d}_1, \boldsymbol{d}_2, \ldots, \boldsymbol{d}_N\}$ has been obtained. The orthonormal set is then obtained by defining

$$\boldsymbol{i}_k = \boldsymbol{d}_k / \| \boldsymbol{d}_k \|, k = 1, 2, \ldots, N$$

It is easy to see that $\{\boldsymbol{e}_1, \ldots, \boldsymbol{e}_k\}$, $\{\boldsymbol{d}_1, \ldots, \boldsymbol{d}_k\}$, and hence, $\{\boldsymbol{i}_1, \ldots, \boldsymbol{i}_k\}$ generate the same subspace for each k.

Definition: If U is a subspace of an inner product space, then the orthogonal complement of U is a subset of V, denoted by $U^{\perp}$, such that

$$U^{\perp} = \{\boldsymbol{v} \mid \boldsymbol{v} \cdot \boldsymbol{u} = 0 \text{ for all } \boldsymbol{u} \in U\} \qquad (5.6.2)$$

The properties of orthogonal complements are developed in the following theorem.

Theorem 5.6.4: If U is a subspace of V, then (a) $U^{\perp}$ is a subspace of V; (b) $V = U \oplus U^{\perp}$.

Proof:

(a) If $\boldsymbol{u}_1$ and $\boldsymbol{u}_2$ are in $U^{\perp}$, then

$$\boldsymbol{u}_1 \cdot \boldsymbol{v} = \boldsymbol{u}_2 \cdot \boldsymbol{v} = 0$$

for all $\boldsymbol{v} \in V$. Therefore for any $\lambda_1, \lambda_2 \in \mathbf{C}$,

$$(\lambda_1 \boldsymbol{u}_1 + \lambda_2 \boldsymbol{u}_2) \cdot \boldsymbol{v} = \lambda_1 \boldsymbol{u}_1 \cdot \boldsymbol{v} + \lambda_2 \boldsymbol{u}_2 \cdot \boldsymbol{v} = 0$$

Thus $\lambda_1 \boldsymbol{u}_1 + \lambda_2 \boldsymbol{u}_2 \in U^{\perp}$.

(b) Consider the vector space $U + U^{\perp}$. Let $\boldsymbol{v} \in U \cap U^{\perp}$, then by eq. (5.6.2), $\boldsymbol{v} \cdot \boldsymbol{v} = 0$, which implies $\boldsymbol{v} = \boldsymbol{0}$. Thus $U + U^{\perp} = U \oplus U^{\perp}$. To establish that $V = U \oplus U^{\perp}$, let $\{\boldsymbol{i}_1, \ldots, \boldsymbol{i}_R\}$ be an orthonormal basis for V, then consider the decomposition

$$v = \{v - \sum_{q=1}^{R} (v \cdot i_q) i_q\} + \sum_{q=1}^{R} (v \cdot i_q) i_q$$

The term in the brackets is orthogonal to each i_q and it thus belongs to $U^{\perp}$. Also, the second term is in U. Therefore $V = U + U^{\perp} = U \oplus U^{\perp}$, and the proof is complete.

As a result of the last theorem, any vector $v \in V$ can be written uniquely in the form

$$v = u + w \tag{5.6.3}$$

where $u \in U$ and $w \in U^{\perp}$. If v is a vector with the decomposition indicated in eq. (5.6.3), then

$$\| v \|^2 = \| u + w \|^2 = \| u \|^2 + \| w \|^2 + u \cdot w + w \cdot u = \| u \|^2 + \| w \|^2$$

It follows this equation

$$\| v \|^2 \geqslant \| u \|^2, \quad \| v \|^2 \geqslant \| w \|^2$$

The equation is the well-known Pythagorean theorem, and the inequalities are special cases of an inequality known as Bessel's inequality.

5.7 Reciprocal Basis and Change of Basis

In this section we define a special basis, called the reciprocal basis, associated with each basis of the inner product space V. Components of vectors relative to both the basis and the reciprocal basis are defined and formulas for the change of basis are developed.

The set of N vectors $\{e^1, e^2, \ldots, e^N\}$ is said to be the reciprocal basis relative to the basis $\{e_1, e_2, \ldots, e_N\}$ of an inner product space V if

$$e^k \cdot e_s = \delta_s^k \quad k, s = 1, 2, \ldots, N \tag{5.7.1}$$

where the symbol δ_s^k is the Kronecker delta defined by

$$\delta_s^k = \begin{cases} 1, & k = s \\ 0, & k \neq s \end{cases} \tag{5.7.2}$$

Thus each vector of the reciprocal basis is orthogonal to $N - 1$ vectors of the basis and when its inner product is taken with the N-th vector of the basis, the inner product has the value one. The following two theorems show that the reciprocal basis just defined exists uniquely and is actually a basis for V.

Theorem 5.7.1: The reciprocal basis relative to a given basis exists and is unique.

Proof: Existence. We prove only the existence of the vector $\boldsymbol{e}^1$; the existence of the remaining vectors can be proved similarly. Let U be the subspace generated by the vectors $\boldsymbol{e}_2, \ldots, \boldsymbol{e}_N$, and suppose that $U^{\perp}$ is the orthogonal complement of U, then $\dim U = N - 1$, and from Theorem 5.3.9, $\dim U^{\perp} = 1$. Hence we can choose a nonzero vector $\boldsymbol{w} \in U^{\perp}$. Since $\boldsymbol{e}_1 \notin U$, $\boldsymbol{e}_1$ and $\boldsymbol{w}$ are not orthogonal,

$$\boldsymbol{e}_1 \cdot \boldsymbol{w} \neq 0$$

we can simply define

$$\boldsymbol{e}^1 = \frac{1}{\boldsymbol{e}_1 \cdot \boldsymbol{w}} \boldsymbol{w}$$

Then $\boldsymbol{e}^1$ obeys eq. (5.7.1) for $k = 1$.

Uniqueness. Assume that there are two reciprocal bases, $\{\boldsymbol{e}^1, \boldsymbol{e}^2, \ldots, \boldsymbol{e}^N\}$ and $\{\boldsymbol{d}^1, \boldsymbol{d}^2, \ldots, \boldsymbol{d}^N\}$ relative to the basis $\{\boldsymbol{e}_1, \boldsymbol{e}_2, \ldots, \boldsymbol{e}_N\}$ of V, then from eq. (5.7.1)

$$\boldsymbol{e}^k \cdot \boldsymbol{e}_s = \delta_s^k \text{ and } \boldsymbol{d}^k \cdot \boldsymbol{e}_s = \delta_s^k \text{ for } k, s = 1, 2, \ldots, N$$

Subtracting these two equations, we obtain

$$(\boldsymbol{e}^k - \boldsymbol{d}^k) \cdot \boldsymbol{e}_s = 0 \qquad k, s = 1, 2, \ldots, N \tag{5.7.3}$$

Thus the vector $\boldsymbol{e}^k - \boldsymbol{d}^k$ must be orthogonal to $\{\boldsymbol{e}_1, \boldsymbol{e}_2, \ldots, \boldsymbol{e}_N\}$. Since the basis generates V, eq. (5.7.3) is equivalent to

$$(\boldsymbol{e}^k - \boldsymbol{d}^k) \cdot \boldsymbol{v} = 0 \qquad \text{for all } \boldsymbol{v} \in V \tag{5.7.4}$$

In particular, we can choose $\boldsymbol{v}$ in eq. (5.7.4) to be equal to $\boldsymbol{e}^k - \boldsymbol{d}^k$, and it follows then the definition of an inner product space that

$$\boldsymbol{e}^k = \boldsymbol{d}^k \qquad k = 1, 2, \ldots, N \tag{5.7.5}$$

Therefore the reciprocal basis relative to $\{\boldsymbol{e}_1, \boldsymbol{e}_2, \ldots, \boldsymbol{e}_N\}$ is unique.

Theorem 5.7.2: The reciprocal basis $\{\boldsymbol{e}^1, \boldsymbol{e}^2, \ldots, \boldsymbol{e}^N\}$ with respect to the basis $\{\boldsymbol{e}_1, \boldsymbol{e}_2, \ldots, \boldsymbol{e}_N\}$ of the inner product space V is itself a basis for V.

Proof: Consider the linear relation

$$\sum_{q=1}^{N} \lambda_q \boldsymbol{e}^q = \boldsymbol{0}$$

If we compute the inner product of this equation with $\boldsymbol{e}_k$, $k = 1, 2, \ldots N$, then

$$\sum_{q=1}^{N} \lambda_q \boldsymbol{e}^q \cdot \boldsymbol{e}_k = \sum_{q=1}^{N} \lambda_q \delta_k^q = \lambda_k = 0$$

Thus the reciprocal basis is linearly independent. But since the reciprocal basis contains the same number of vectors as that of a basis, it is itself a basis for V.

Since $\boldsymbol{e}^k, k=1,2,\ldots,N$, is in V, we can always write

$$\boldsymbol{e}^k = \sum_{q=1}^{N} e^{kq}\,\boldsymbol{e}_q \tag{5.7.6}$$

where, by eq. (5.7.1) and eq. (5.5.6),

$$\boldsymbol{e}^k \cdot \boldsymbol{e}_s = \delta_s^k = \sum_{q=1}^{N} e^{kq} e_{qs} \tag{5.7.7}$$

and

$$\boldsymbol{e}^k \cdot \boldsymbol{e}^j = \sum_{q=1}^{N} e^{kq}\delta_q^j = e^{kj} = \overline{e^{jk}} \tag{5.7.8}$$

From a matrix viewpoint, eq. (5.7.7) shows that the N^2 quantities e_{kq} ($k,q=1,2,\ldots,N$) are the elements of the inverse of the matrix whose elements are e_{qs}. In particular, the matrices $[e^{kq}]$ and $[e_{qs}]$ are nonsingular. This remark is a proof of Theorem 5.7.2. It is possible to establish from eq. (5.7.6) and eq. (5.7.7) that

$$\boldsymbol{e}_s = \sum_{k=1}^{N} e_{sk}\,\boldsymbol{e}^k \tag{5.7.9}$$

To illustrate the construction of a reciprocal basis by an algebraic method, consider the real vector space $\mathbf{R}^2$, which we shall represent by the Euclidean plane. Let a basis be given for $\mathbf{R}^2$ which consists of two vectors $45°$degrees apart, the first one, $\boldsymbol{e}_1$, two units long and the second one, $\boldsymbol{e}_2$, one unit long.

To construct the reciprocal basis, we first note that from the given information and eq. (5.5.6) we can write

$$e_{11}=4, e_{12}=e_{21}=\sqrt{2}, e_{22}=1 \tag{5.7.10}$$

Writing eq. (5.7.7) out explicitly for the case $N=2$, we have

$$\begin{aligned} e^{11}e_{11}+e^{12}e_{21}&=1, & e^{21}e_{11}+e^{22}e_{21}&=0\\ e^{11}e_{12}+e^{12}e_{22}&=0, & e^{21}e_{12}+e^{22}e_{22}&=1 \end{aligned} \tag{5.7.11}$$

Substituting eq. (5.7.10) into eq. (5.7.11), we find that

$$e^{11} = =\frac{1}{2}, e^{12}=e^{21} = -1/\sqrt{2}, e^{22}=2 \tag{5.7.12}$$

When the results eq. (5.7.2) are put into the special case of eq. (5.7.6) for $N=2$, we obtain the explicit expressions for the reciprocal basis $\{\boldsymbol{e}^1, \boldsymbol{e}^2\}$

$$\boldsymbol{e}^1 = \frac{1}{2}\boldsymbol{e}_1 - \frac{1}{\sqrt{2}}\boldsymbol{e}_2, \boldsymbol{e}^2 = -\frac{1}{\sqrt{2}}\boldsymbol{e}_1 + 2\boldsymbol{e}_2 \tag{5.7.13}$$

Henceforth we shall use the same kernel letter to denote the components of a vector relative to a basis as we use for the vector itself. Thus, a vector $\boldsymbol{v}$ has components $v^k, k=1,2,...,N$ relative to the basis $\{\boldsymbol{e}_1,\boldsymbol{e}_2,...,\boldsymbol{e}_N\}$ and components $v^k, k=1,2,...,N$, relative to its reciprocal basis,

$$\boldsymbol{v} = \sum_{k=1}^{N} v^k \boldsymbol{e}_k, \boldsymbol{v} = \sum_{k=1}^{N} v_k \boldsymbol{e}^k \tag{5.7.14}$$

Definition: Definition as above the components $v^1, v^2, ..., v^N$ with respect to the basis $\{\boldsymbol{e}_1,\boldsymbol{e}_2,...,\boldsymbol{e}_N\}$ are often called the contravariant components of $\boldsymbol{v}$, while the components $v_1, v_2, ..., v_N$ with respect to the reciprocal basis $\{\boldsymbol{e}^1, \boldsymbol{e}^2,...,\boldsymbol{e}^N\}$ are called covariant components.

Theorem 5.7.3: If $\{\boldsymbol{e}^1,\boldsymbol{e}^2,...,\boldsymbol{e}^N\}$ is the reciprocal basis of $\{\boldsymbol{e}_1,\boldsymbol{e}_2,..., \boldsymbol{e}_N\}$, then $\{\boldsymbol{e}_1,\boldsymbol{e}_2,...,\boldsymbol{e}_N\}$ is also the reciprocal basis of $\{\boldsymbol{e}^1,\boldsymbol{e}^2,...,\boldsymbol{e}^N\}$.

For this reason we simply say that the bases $\{\boldsymbol{e}_1,\boldsymbol{e}_2,...,\boldsymbol{e}_N\}$ and $\{\boldsymbol{e}^1, \boldsymbol{e}^2,...,\boldsymbol{e}^N\}$ arc (mutually) reciprocal. The contravariant and covariant components of $\boldsymbol{v}$ are related to one another by the formulas

$$\begin{aligned} v^k &= \boldsymbol{v} \cdot \boldsymbol{e}^k = \sum_{q=1}^{N} e^{qk} v_q \\ v_k &= \boldsymbol{v} \cdot \boldsymbol{e}_k = \sum_{q=1}^{N} e_{qk} v^q \end{aligned} \tag{5.7.15}$$

where eq. (5.5.6) and eq. (5.7.8) have been employed. More generally, if $\boldsymbol{u}$ has contravariant components $\boldsymbol{u}^i$ and covariant components $\boldsymbol{u}_i$ relative to $\{\boldsymbol{e}_1,\boldsymbol{e}_2,..., \boldsymbol{e}_N\}$ and $\{\boldsymbol{e}^1,\boldsymbol{e}^2,...,\boldsymbol{e}^N\}$, respectively, then the inner product of $\boldsymbol{u}$ and $\boldsymbol{v}$ can be computed by the formulas

$$\boldsymbol{u} \cdot \boldsymbol{v} = \sum_{i=1}^{N} \boldsymbol{u}_i \overline{\boldsymbol{v}^i} = \sum_{i=1}^{N} \boldsymbol{u}^i \overline{\boldsymbol{v}_i} = \sum_{i=1}^{N} \sum_{j=1}^{N} e^{ij} \boldsymbol{u}_i \overline{\boldsymbol{v}_j} = \sum_{i=1}^{N} \sum_{j=1}^{N} e_{ij} \boldsymbol{u}^i \overline{\boldsymbol{v}^j} \tag{5.7.16}$$

which generalize the formulas eq. (5.6.4) and eq. (5.5.8).

As an example of the covariant and contravariant components of a vector, consider a vector $\boldsymbol{v}$ in $\mathbf{R}^2$ which has the representation

$$\boldsymbol{v} = \frac{3}{2}\boldsymbol{e}_1 + 2\boldsymbol{e}_2$$

The contravariant components of $\boldsymbol{v}$ are then $(3/2, 2)$; to compute the covariant components, the eq. (5.7.16) is written out for the case $N=2$,

$$v_1 = e_{11}v^1 + e_{21}v^2, v_2 = e_{12}v^1 + e_{22}v^2$$

Then from eq. (5.7.10) and the values of the contravariant components the covariant components are given by

$$v_1 = 6 + 2\sqrt{2}, v_2 = (3/\sqrt{2}) + 2$$

hence

$$\boldsymbol{v} = (6 + 2\sqrt{2})\boldsymbol{e}^1 + (3/\sqrt{2} + 2)\boldsymbol{e}^2$$

A substantial part of the usefulness of orthonormal bases is that each orthonormal basis is self-reciprocal; hence contravariant and covariant components of vectors coincide. The self-reciprocity of orthonormal bases follows by comparing the condition eq. (5.6.1) for an orthonormal basis with the condition eq. (5.7.1) for a reciprocal basis. In orthonormal systems indices are written only as subscripts because there is no need to distinguish between covariant and contravariant components.

In formulas which transferring from one basis to another basis, one basis is the set of linearly independent vectors $\{\boldsymbol{e}_1, \ldots, \boldsymbol{e}_N\}$, while the second basis is the set $\{\hat{\boldsymbol{e}}, \ldots, \hat{\boldsymbol{e}}_N\}$. From the fact that both bases generate V, we can write

$$\hat{\boldsymbol{e}}_k = \sum_{s=1}^{N} T_k^s \boldsymbol{e}_s, \quad k = 1, 2, \ldots, N \tag{5.7.17}$$

and

$$\boldsymbol{e}_q = \sum_{k=1}^{N} \hat{T}_q^k \hat{\boldsymbol{e}}_k, \quad q = 1, 2, \ldots, N \tag{5.7.18}$$

where $\hat{T}_q^k$ and T_k^s are both sets of N^2 scalars that are related to one another. From Theorem 5.2 the $N \times N$ matrices $[T_k^s]$ and $[\hat{T}_q^k]$ must be nonsingular.

Substituting eq. (5.7.17) into eq. (5.7.18) and replacing $\boldsymbol{e}_q$ by $\sum_{s-1}^{N} \delta_q^s \boldsymbol{e}_s$, we obtain

$$\boldsymbol{e}_q = \sum_{k=1}^{N} \sum_{s=1}^{N} \hat{T}_q^k T_k^s \boldsymbol{e}_s = \sum_{s=1}^{N} \delta_q^s \boldsymbol{e}_s$$

which can be rewritten as

$$\sum_{s=1}^{N} \left(\sum_{k=1}^{N} \hat{T}_q^k T_k^s - \delta_q^s \right) \boldsymbol{e}_s = \boldsymbol{0} \tag{5.7.19}$$

The linear independence of the basis $\{\boldsymbol{e}_s\}$ requires that

$$\sum_{k=1}^{N} \hat{T}_q^k T_k^s = \delta_q^s \tag{5.7.20}$$

A similar argument yields

$$\sum_{k=1}^{N} \hat{T}_k^q T_s^k = \delta_s^q \tag{5.7.21}$$

In matrix language we see that eq. (5.7.20) or eq. (5.7.21) requires that the matrix of element T_q^k be the inverse of the matrix of element $\hat{T}_k^q$. It is easy to verify that the reciprocal bases are related by

$$e^k = \sum_{q=1}^{N} \overline{T_q^k}\, \hat{e}^q, \quad \hat{e}^q = \sum_{k=1}^{N} \overline{\hat{T}_k^q}\, e^k \tag{5.7.22}$$

The covariant and contravariant components of $v \in V$ relative to the two pairs of reciprocal bases are given by

$$v = \sum_{k=1}^{N} v_k\, e^k = \sum_{k=1}^{N} v^k e_k = \sum_{q=1}^{N} \hat{v}^q\, \hat{e}_q = \sum_{q=1}^{N} \hat{v}_q\, \hat{e}^q \tag{5.7.23}$$

To obtain a relationship between, say, covariant components of v, one can substitute eq. (5.7.22) into eq. (5.7.23), thus

$$\sum_{k=1}^{N} \sum_{q=1}^{N} v_k\, \overline{T_q^k} \hat{e}^q = \sum_{q=1}^{N} \hat{v}_q \hat{e}^q$$

This equation can be rewritten in the form

$$\sum_{q=1}^{N} \left(\sum_{k=1}^{N} T_q^k\, v_k - \hat{v}_q \right) \hat{e}^q = \mathbf{0}$$

and it follows the linear independence of the basis

$$\hat{v}_q = \sum_{k=1}^{N} \overline{T_q^k}\, v_k \tag{5.7.24}$$

In a similar manner the following formulas can be obtained:

$$\hat{v}_q = \sum_{k=1}^{N} \overline{T_k^q}\, v^k \tag{5.7.25}$$

$$v^k = \sum_{q=1}^{N} T_q^k\, \hat{v}^q \tag{5.7.26}$$

and

$$v_k = \sum_{q=1}^{N} \overline{\hat{T}_k^q}\, \hat{v}_q \tag{5.7.27}$$

Chapter 6

Linear Transformations

We have learned the concepts of vector and vector space in last chapter, here we describe a new notion transformation. Linear transformation is a special class of functions defined on a vector space. In this chapter we present some knowledge about linear transformation, such as the sums and adjoint and so on.

6.1 Definition of Linear Transformation

If V and U are vector spaces, a linear transformation is a function $\boldsymbol{A}: V \to U$ such that

1) $\boldsymbol{A}(\boldsymbol{u}+\boldsymbol{v}) = \boldsymbol{A}(\boldsymbol{u}) + \boldsymbol{A}(\boldsymbol{v})$,

2) $\boldsymbol{A}(\lambda \boldsymbol{u}) = \lambda \boldsymbol{A}(\boldsymbol{u})$,

for all $\boldsymbol{u}, \boldsymbol{v} \in V$ and $\lambda \in \mathbf{C}$. Condition 1) asserts that $\boldsymbol{A}$ is a homomorphism on V with respect to the operation of addition. Condition 2) shows that $\boldsymbol{A}$, in addition to being a homomorphism, is also homogeneous with respect to the operation of scalar multiplication. Observe that the + symbol on the left side of 1) denotes addition in V, while on the right side it denotes addition in U. It would be extremely cumbersome to adopt different symbols for these quantities. Further, it is customary to omit the parentheses and write simply $\boldsymbol{Au}$ for $\boldsymbol{A}(\boldsymbol{u})$ when $\boldsymbol{A}$ is a linear transformation.

Theorem 6.1.1: If $\boldsymbol{A}: V \to U$ is a function from a vector space V to a vector space U, then $\boldsymbol{A}$ is a linear transformation if and only if

$$\boldsymbol{A}(\lambda \boldsymbol{u} + \mu \boldsymbol{v}) = \lambda \boldsymbol{A}(\boldsymbol{u}) + \mu \boldsymbol{A}(\boldsymbol{v})$$

for all $\boldsymbol{u}, \boldsymbol{v} \in V$ and $\lambda, \mu \in \mathbf{C}$.

Proof: The proof of this theorem is an elementary application of the definition.

It is also possible to show that

$$\boldsymbol{A}(\lambda^1 \boldsymbol{v}_1 + \ldots + \lambda^R \boldsymbol{v}_R) = \lambda^1 \boldsymbol{A}\boldsymbol{v}_1 + \ldots + \lambda^R \boldsymbol{A}\boldsymbol{v}_R \tag{6.1.1}$$

for all $\boldsymbol{v}_1, \ldots, \boldsymbol{v}_R \in V$ and $\lambda^1, \ldots, \lambda^R \in \mathbf{C}$.

We see that for a linear transformation $\boldsymbol{A}: V \to U$

$$\boldsymbol{A}\mathbf{0} = \mathbf{0} \quad \text{and} \quad \boldsymbol{A}(-\boldsymbol{v}) = -\boldsymbol{A}\boldsymbol{v} \tag{6.1.2}$$

Note that in eq. (6.1.2) we have used the same symbol for the zero vector in V and U.

Theorem 6.1.2: If $\{\boldsymbol{v}_1, \boldsymbol{v}_2, \ldots, \boldsymbol{v}_R\}$ is a linearly dependent set in V and if $\boldsymbol{A}: V \to U$ is a linear transformation, then $\{\boldsymbol{A}\boldsymbol{v}_1, \boldsymbol{A}\boldsymbol{v}_2, \ldots, \boldsymbol{A}\boldsymbol{v}_R\}$ is a linearly dependent set in U.

Proof: Since the vectors $\boldsymbol{v}_1, \ldots, \boldsymbol{v}_R$ are linearly dependent, we can write

$$\sum_{j=1}^{R} \lambda^j \boldsymbol{v}_j = \mathbf{0}$$

where at least one coefficient is not zero. Therefore

$$\boldsymbol{A}\left(\sum_{j=1}^{R} \lambda^j \boldsymbol{v}_j\right) = \sum_{j=1}^{R} \lambda^j \boldsymbol{A}\boldsymbol{v}_j = \mathbf{0}$$

where eq. (6.1.1) and eq. (6.1.2) have been used. The last equation proves the theorem.

If the vectors $\boldsymbol{v}_1, \ldots, \boldsymbol{v}_R$ are linearly independent, then their image set $\{\boldsymbol{A}\boldsymbol{v}_1, \boldsymbol{A}\boldsymbol{v}_2, \ldots, \boldsymbol{A}\boldsymbol{v}_R\}$ may or may not be linearly independent. For example, $\boldsymbol{A}$ might map all vectors into $\mathbf{0}$. The kernel of a linear transformation $\boldsymbol{A}: V \to U$ is the set

$$K(\boldsymbol{A}) = \{\boldsymbol{v} \mid \boldsymbol{A}\boldsymbol{v} = \mathbf{0}\}$$

In other words, $K(\boldsymbol{A})$ is the preimage of the set $\{\mathbf{0}\}$ in U. Since $\{\mathbf{0}\}$ is a subgroup of the additive group U, $K(\boldsymbol{A})$ is a subgroup of the additive group V.

Theorem 6.1.3: The set $K(\boldsymbol{A})$ is a subspace of V.

Proof: Since $K(\boldsymbol{A})$ is a subgroup, we only need to prove that if $\boldsymbol{v} \in K(\boldsymbol{A})$, then $\lambda \boldsymbol{v} \in K(\boldsymbol{A})$ for all $\lambda \in \mathbf{C}$. This is clear since

$$\boldsymbol{A}(\lambda \boldsymbol{v}) = \lambda \boldsymbol{A}\boldsymbol{v} = \mathbf{0}$$

for all $\boldsymbol{v}$ in $K(\boldsymbol{A})$.

Definition: The kernel of a linear transformation is sometimes called the null space. The nullity of a linear transformation is the dimension of the kernel, i. e. $\dim K(\boldsymbol{A})$ since $K(\boldsymbol{A})$ is a subspace of V, we have

$$\dim K(\boldsymbol{A}) \leqslant \dim V \tag{6.1.3}$$

Theorem 6.1.4: A linear transformation $\boldsymbol{A}: V \to U$ is one-to-one if and only if $K(\boldsymbol{A}) = \{\mathbf{0}\}$.

Proof: This theorem is just a special case of Theorem 6.1.1. For ease of reference, we shall repeat the proof. If $\boldsymbol{Au} = \boldsymbol{Av}$, then by the linearity of $\boldsymbol{A}$, $\boldsymbol{A}(\boldsymbol{u}-\boldsymbol{v}) = \mathbf{0}$. Thus if $K(\boldsymbol{A}) = \{\mathbf{0}\}$, then $\boldsymbol{Au} = \boldsymbol{Av}$ implies $\boldsymbol{u} = \boldsymbol{v}$, so that $\boldsymbol{A}$ is one-to-one. Now assume $\boldsymbol{A}$ is one-to-one. Since $K(\boldsymbol{A})$ is a subspace, it must contain the zero in V and therefore $\boldsymbol{A}\mathbf{0} = \mathbf{0}$. If $K(\boldsymbol{A})$ contained any other element $\boldsymbol{v}$, we would have $\boldsymbol{Av} = \mathbf{0}$, which contradicts the fact that $\boldsymbol{A}$ is one-to-one.

Linear transformations that are one-to-one are called regular linear transformations. The following theorem gives another condition for such linear transformations.

Theorem 6.1.5: A linear transformation $\boldsymbol{A}: V \to U$ is regular iff it maps linearly independent sets in V to linearly independent sets in U.

Proof:

Necessity. Let $\{\boldsymbol{v}_1, \boldsymbol{v}_2, \ldots, \boldsymbol{v}_R\}$ be a linearly independent set in V and $\boldsymbol{A}: V \to U$ be a regular linear transformation. Consider the sum

$$\sum_{j=1}^{R} \lambda^j \boldsymbol{A}\boldsymbol{v}_j = \mathbf{0}$$

This equation is equivalent to

$$\boldsymbol{A}\left(\sum_{j=1}^{R} \lambda^j \boldsymbol{v}_j\right) = \mathbf{0}$$

Since $\boldsymbol{A}$ is regular, we must have

$$\sum_{j=1}^{R} \lambda^j \boldsymbol{v}_j = \mathbf{0}$$

Since the vectors $\boldsymbol{v}_1, \boldsymbol{v}_2, \ldots, \boldsymbol{v}_R$ are linearly independent, this equation shows $\lambda^1 = \lambda^2 = \ldots = \lambda^R = 0$, which implies that the set $\{\boldsymbol{A}\boldsymbol{v}_1, \boldsymbol{A}\boldsymbol{v}_2, \ldots, \boldsymbol{A}\boldsymbol{v}_R\}$ is linearly independent.

Sufficiency. The assumption that $\boldsymbol{A}$ preserves linear independence implies, in particular, that $\boldsymbol{Av} \neq \mathbf{0}$ for every nonzero vector $\boldsymbol{v} \in V$ since such a vector forms a linearly independent set. Therefore $K(\boldsymbol{A})$ consists of the zero vector only, and thus $\boldsymbol{A}$ is regular.

For a linear transformation $\boldsymbol{A}: V \to U$ we denote the range of $\boldsymbol{A}$ by

$$R(\boldsymbol{A}) = \{\boldsymbol{Av} \mid \boldsymbol{v} \in V\}$$

We know that $R(\boldsymbol{A})$ is a subgroup of U. We leave it to the reader to prove the stronger results stated below.

Theorem 6.1.6: The range $R(\boldsymbol{A})$ is a subspace of U.

We have from Theorems 6.1.6 and 5.3.2 that

$$\dim R(\boldsymbol{A}) \leqslant \dim U \tag{6.1.4}$$

The rank of a linear transformation is defined as the dimension of $R(\boldsymbol{A})$, i. e. $\dim R(\boldsymbol{A})$. A stronger statement than eq. (6.1.4) can be made regarding the rank of linear transformation $\boldsymbol{A}$.

Theorem 6.1.7: $\dim R(\boldsymbol{A}) \leqslant \min(\dim V, \dim U)$.

Proof: Clearly, it suffices to prove that $\dim R(\boldsymbol{A}) \leqslant \dim V$, since this inequality and eq. (6.1.4) imply the assertion of the theorem. Let $\{\boldsymbol{e}_1, \boldsymbol{e}_2, \ldots, \boldsymbol{e}_N\}$ be a basis for V, where $N = \dim V$. Then $\boldsymbol{v} \in V$ can be written as

$$\boldsymbol{v} = \sum_{j=1}^{N} v^j \boldsymbol{e}_j$$

Therefore any vector $\boldsymbol{Av} \in R(\boldsymbol{A})$ can be written as

$$\boldsymbol{Av} = \sum_{j=1}^{N} v^j \boldsymbol{Ae}_j$$

Hence the vectors $\{\boldsymbol{Ae}_1, \boldsymbol{Ae}_2, \ldots, \boldsymbol{Ae}_N\}$ generate $R(\boldsymbol{A})$. By Theorem 5.2.10, we can conclude

$$\dim R(\boldsymbol{A}) \leqslant \dim V \tag{6.1.5}$$

A result that improves on eq. (6.1.5) is the following important theorem.

Theorem 6.1.8: If $\boldsymbol{A}: V \to U$ is a linear transformation, then

$$\dim V = \dim R(\boldsymbol{A}) + \dim K(\boldsymbol{A}) \tag{6.1.6}$$

Proof:

Let $P = \dim K(\boldsymbol{A})$, $R = \dim R(\boldsymbol{A})$, and $N = \dim V$. We must prove that $N = P + R$. Select N vectors in V such that $\{\boldsymbol{e}_1, \boldsymbol{e}_2, \ldots, \boldsymbol{e}_P\}$ is a basis for $K(\boldsymbol{A})$ and $\{\boldsymbol{e}_1, \boldsymbol{e}_2, \ldots, \boldsymbol{e}_P, \boldsymbol{e}_{P+1}, \ldots, \boldsymbol{e}_N\}$ a basis for V. As in the proof of Theorem 6.1.7, the vectors $\{\boldsymbol{Ae}_1, \boldsymbol{Ae}_2, \ldots, \boldsymbol{Ae}_P, \boldsymbol{Ae}_{P+1}, \ldots, \boldsymbol{Ae}_N\}$ generate $R(\boldsymbol{A})$. But, by the properties of the kernel $\boldsymbol{Ae}_1 = \boldsymbol{Ae}_2 = \ldots = \boldsymbol{Ae}_P = \boldsymbol{0}$. Thus the vectors $\{\boldsymbol{Ae}_{P+1}, \ldots, \boldsymbol{Ae}_N\}$ generate $R(\boldsymbol{A})$. If we can establish that these vectors are linearly independent, we can conclude from Theorem 5.2.10 that the vectors $\{\boldsymbol{Ae}_{P+1}, \ldots, \boldsymbol{Ae}_N\}$

form a basis for $R(\boldsymbol{A})$ and that $\dim R(\boldsymbol{A}) = R = N - P$. Consider the sum

$$\sum_{j=1}^{R} \lambda^j \boldsymbol{A}\,\boldsymbol{e}_{P+j} = \boldsymbol{0}$$

Therefore

$$\boldsymbol{A}\left(\sum_{j=1}^{R} \lambda^j \boldsymbol{e}_{P+j}\right) = \boldsymbol{0}$$

which implies that the vector $\sum_{j=1}^{R} \lambda^j \boldsymbol{e}_{P+j} \in K(\boldsymbol{A})$. This fact requires that $\lambda^1 = \lambda^2 = \ldots = \lambda^R = 0$, or otherwise the vector $\sum_{j=1}^{R} \lambda^j \boldsymbol{e}_{P+j}$ could be expanded in the basis $\{\boldsymbol{e}_1, \boldsymbol{e}_2, \ldots, \boldsymbol{e}_P\}$, contradicting the linear independence of $\{\boldsymbol{e}_1, \boldsymbol{e}_2, \ldots, \boldsymbol{e}_N\}$. Thus the set $\{\boldsymbol{A}\boldsymbol{e}_{P+1}, \ldots, \boldsymbol{A}\boldsymbol{e}_N\}$ is linearly independent and the proof of the theorem is complete.

As usual, a linear transformation $\boldsymbol{A}: V \to U$ is said to be surjective if $R(\boldsymbol{A}) = U$, i. e. for every vector $\boldsymbol{u} \in U$ there exists $\boldsymbol{v} \in V$ such that $\boldsymbol{A}\boldsymbol{v} = \boldsymbol{u}$.

Theorem 6.1.9: $\dim R(\boldsymbol{A}) = \dim U$ iff $\boldsymbol{A}$ is surjective.

In the special case when $\dim V = \dim U$, it is possible to state the following important theorem.

Theorem 6.1.10: If $\boldsymbol{A}: V \to U$ is a linear transformation and if $\dim V = \dim U$, then $\boldsymbol{A}$ is a linear transformation onto U iff $\boldsymbol{A}$ is regular.

Proof: Assume that $\boldsymbol{A}: V \to U$ is onto U, then eq. (6.1.6) and Theorem 6.1.9 show that

$$\dim V = \dim U = \dim U + \dim K(\boldsymbol{A})$$

Therefore $\dim K(\boldsymbol{A}) = 0$ and thus $K(\boldsymbol{A}) = \{\boldsymbol{0}\}$ and $\boldsymbol{A}$ is one-to-one. Next assume that $\boldsymbol{A}$ is one-to-one. By Theorem 6.1.4, $K(\boldsymbol{A}) = \{\boldsymbol{0}\}$ and thus $\dim K(\boldsymbol{A}) = 0$. Then eq. (6.1.6) shows that

$$\dim V = \dim R(\boldsymbol{A}) = \dim U$$

By using Theorem 5.3.3, we can conclude that

$$R(\boldsymbol{A}) = \boldsymbol{u}$$

and thus $\boldsymbol{A}$ is surjective.

6.2 Sums and Products of Liner Transformations

In this section we shall assign meaning to the operations of addition and scalar multiplication for linear transformations. If $\boldsymbol{A}$ and $\boldsymbol{B}$ are linear transformations $V \to U$, then their sum $\boldsymbol{A}+\boldsymbol{B}$ is a linear transformation defined by

$$(\boldsymbol{A}+\boldsymbol{B})\boldsymbol{v}=\boldsymbol{A}\boldsymbol{v}+\boldsymbol{B}\boldsymbol{v} \tag{6.2.1}$$

for all $\boldsymbol{v} \in V$. In a similar fashion, if $\lambda \in \mathbf{C}$, then $\lambda \boldsymbol{A}$ is a linear transformation $V \to U$ defined by

$$(\lambda \boldsymbol{A})\boldsymbol{v}=\lambda(\boldsymbol{A}\boldsymbol{v}) \tag{6.2.2}$$

for all $\boldsymbol{v} \in V$. If we write $L(V,U)$ for the set of linear transformations from $V \to U$, then eq. (6.2.1) and eq. (6.2.2) make $L(V,U)$ a vector space. The zero element in $L(V,U)$ is the linear transformation $\mathbf{0}$ defined by

$$\mathbf{0}\boldsymbol{v}=\mathbf{0} \tag{6.2.3}$$

for all $\boldsymbol{v} \in V$. The negative of $\boldsymbol{A} \in L(V,U)$ is a linear transformation $-\boldsymbol{A} \in L(V,U)$ defined by

$$-\boldsymbol{A}=-1\boldsymbol{A} \tag{6.2.4}$$

It follows eq. (6.2.4) that $-\boldsymbol{A}$ is the additive inverse of $\boldsymbol{A} \in L(V,U)$. This assertion follows

$$\boldsymbol{A}+(-\boldsymbol{A})=\boldsymbol{A}+(-1)\boldsymbol{A}=1\boldsymbol{A}+(-1)\boldsymbol{A}=(1-1)\boldsymbol{A}=0\boldsymbol{A}=\mathbf{0} \tag{6.2.5}$$

where eq. (6.2.1) and eq. (6.2.2) have been used. Consistent with our previous notation, we shall write $\boldsymbol{A}-\boldsymbol{B}$ for the sum $\boldsymbol{A}+(-\boldsymbol{B})$ formed from the linear transformations $\boldsymbol{A}$ and $\boldsymbol{B}$. The formal proof that $L(V,U)$ is a vector space is left as an exercise to the reader.

Theorem 6.2.1: $\dim L(V,U)=\dim V \dim U$.

Proof: Let $\{\boldsymbol{e}_1,\ldots,\boldsymbol{e}_N\}$ be a basis for V and $\{\boldsymbol{b}_1,\ldots,\boldsymbol{b}_M\}$ be a basis for U. Define NM linear transformations $\boldsymbol{A}_\alpha^k: V \to U$ by

$$\begin{aligned} &\boldsymbol{A}_\alpha^k \boldsymbol{e}_k=\boldsymbol{b}_\alpha, \quad k=1,\ldots,N;\alpha=1,\ldots,M \\ &\boldsymbol{A}_\alpha^k \boldsymbol{e}_p=\mathbf{0}, \quad k \neq p \end{aligned} \tag{6.2.6}$$

If $\boldsymbol{A}$ is an arbitrary member of $L(V,U)$, then $\boldsymbol{A}\boldsymbol{e}_k \in U$, and thus

$$\boldsymbol{A}\boldsymbol{e}_k=\sum_{\alpha=1}^{M} A_k^\alpha \boldsymbol{b}_\alpha, \quad k=1,\ldots,N$$

Based upon the properties of $\boldsymbol{A}_\alpha^k$, we can write the above equation as

$$\boldsymbol{A}\boldsymbol{e}_k=\sum_{\alpha=1}^{M} A_k^\alpha \boldsymbol{A}_\alpha^k \boldsymbol{e}_k=\sum_{\alpha=1}^{M}\sum_{s=1}^{N} A_s^\alpha \boldsymbol{A}_\alpha^s \boldsymbol{e}_k$$

Therefore, since the vectors $\boldsymbol{e}_1,\ldots,\boldsymbol{e}_N$ generate V, we find

$$\left(\boldsymbol{A} - \sum_{\alpha=1}^{M}\sum_{s=1}^{N} A_s^{\alpha}\boldsymbol{A}_{\alpha}^{s}\right)\boldsymbol{v} = \boldsymbol{0}$$

for all vectors $\boldsymbol{v} \in V$. Thus, from eq. (6.2.3),

$$\boldsymbol{A} = \sum_{\alpha=1}^{M}\sum_{s=1}^{N} A_s^{\alpha}\boldsymbol{A}_{\alpha}^{s} \tag{6.2.7}$$

This equation means that the MN linear transformations $\boldsymbol{A}_{\alpha}^{s}$ ($s = 1, \ldots, N; \alpha = 1, \ldots, M$) generate $L(V, U)$. If we can prove that these linear transformations are linearly independent, then the proof of the theorem is complete. To this end, set

$$\sum_{\alpha=1}^{M}\sum_{s=1}^{N} A_s^{\alpha}\boldsymbol{A}_{\alpha}^{s} = \boldsymbol{0}$$

Then, from eq. (6.2.6),

$$\sum_{\alpha=1}^{M}\sum_{s=1}^{N} A_s^{\alpha}\boldsymbol{A}_{\alpha}^{s}(\boldsymbol{e}_p) = \sum_{\alpha=1}^{M} A_p^{\alpha}\boldsymbol{b}_{\alpha} = \boldsymbol{0}$$

Thus $A_p^{\alpha} = 0$, ($p = 1, \ldots, N; \alpha = 1, \ldots, M$) because the vectors $\boldsymbol{b}_1, \ldots, \boldsymbol{b}_M$ are linearly independent in U. Hence $\{\boldsymbol{A}_{\alpha}^{s}\}$ is a basis of $L(V, U)$. As a result, we have

$$\dim L(V, U) = MN = \dim U \dim V \tag{6.2.8}$$

If $\boldsymbol{A}: V \to U$ and $\boldsymbol{B}: U \to W$ are linear transformations, their product is a linear transformation $V \to W$, written as $\boldsymbol{BA}$, defined by

$$\boldsymbol{BAv} = \boldsymbol{B}(\boldsymbol{Av}) \tag{6.2.9}$$

for all $\boldsymbol{v} \in V$.

The following theorem summarizes the properties of the product operation.

Theorem 6.2.2:

$$\begin{aligned} \boldsymbol{C}(\boldsymbol{BA}) &= (\boldsymbol{CB})\boldsymbol{A} \\ (\lambda\boldsymbol{A} + \mu\boldsymbol{B})\boldsymbol{C} &= \lambda\boldsymbol{AC} + \mu\boldsymbol{BC} \\ \boldsymbol{C}(\lambda\boldsymbol{A} + \mu\boldsymbol{B}) &= \lambda\boldsymbol{CA} + \mu\boldsymbol{CB} \end{aligned} \tag{6.2.10}$$

for all $\lambda, \mu \in \mathbf{C}$ and where it is understood that $\boldsymbol{A}, \boldsymbol{B}$, and $\boldsymbol{C}$ are defined on the proper vector spaces so as to make the indicated products defined.

The proof of Theorem 6.2.2 is left as an exercise to the reader.

6.3 Special Types of Linear Transformations

In this section we shall examine the properties of several special types of linear transformations. The first of these is one called an isomorphism. A vector space isomorphism is a regular surjective linear transformation $\boldsymbol{A}: V \to U$. It immediately follows Theorem 6.1.8 that if $\boldsymbol{A}: V \to U$ is an isomorphism, then

$$\dim V = \dim U \tag{6.3.1}$$

An isomorphism $\boldsymbol{A}: U \to U$ establishes a one-to-one correspondence between the elements of V and U. Thus there exists a unique inverse function $\boldsymbol{B}: U \to V$ with the property that if

$$\boldsymbol{u} = \boldsymbol{A}\boldsymbol{v} \tag{6.3.2}$$

then

$$\boldsymbol{v} = \boldsymbol{B}(\boldsymbol{u}) \tag{6.3.3}$$

for all $\boldsymbol{u} \in U$ and $\boldsymbol{v} \in V$. We shall now show that $\boldsymbol{B}$ is a linear transformation. Consider the vectors $\boldsymbol{u}_1$ and $\boldsymbol{u}_2 \in U$ and the corresponding vectors [as a result of eq. (6.3.2) and eq. (6.3.3)] $\boldsymbol{v}_1$ and $\boldsymbol{v}_2 \in V$, then by eq. (6.3.2), eq. (6.3.3), and the properties of the linear transformation $\boldsymbol{A}$,

$$\begin{aligned}
\boldsymbol{B}(\lambda \boldsymbol{u}_1 + \mu \boldsymbol{u}_2) &= \boldsymbol{B}(\lambda \boldsymbol{A}\boldsymbol{v}_1 + \mu \boldsymbol{A}\boldsymbol{v}_2) \\
&= \boldsymbol{B}(\boldsymbol{A}(\lambda \boldsymbol{v}_1 + \mu \boldsymbol{v}_2)) \\
&= \lambda \boldsymbol{v}_1 + \mu \boldsymbol{v}_2 \\
&= \lambda \boldsymbol{B}(\boldsymbol{u}_1) + \mu \boldsymbol{B}(\boldsymbol{u}_2)
\end{aligned}$$

Thus $\boldsymbol{B}$ is a linear transformation. This linear transformation shall be written as $\boldsymbol{A}^{-1}$. Clearly the linear transformation $\boldsymbol{A}^{-1}$ is also an isomorphism whose inverse is $\boldsymbol{A}$; i. e.

$$(\boldsymbol{A}^{-1})^{-1} = \boldsymbol{A} \tag{6.3.4}$$

Theorem 6.3.1: If $\boldsymbol{A}: V \to U$ and $\boldsymbol{B}: U \to W$ are isomorphism, then $\boldsymbol{B}\boldsymbol{A}: V \to W$ is an isomorphism whose inverse is computed by

$$(\boldsymbol{B}\boldsymbol{A})^{-1} = \boldsymbol{A}^{-1}\boldsymbol{B}^{-1} \tag{6.3.5}$$

Proof: The fact that $\boldsymbol{B}\boldsymbol{A}$ is an isomorphism follows directly the corresponding properties of $\boldsymbol{A}$ and $\boldsymbol{B}$. The fact that the inverse of $\boldsymbol{B}\boldsymbol{A}$ is computed by eq. (6.3.5) follows directly because if

$$\boldsymbol{u} = \boldsymbol{A}\boldsymbol{v} \quad \text{and} \quad \boldsymbol{w} = \boldsymbol{B}\boldsymbol{u}$$

Then

$$\boldsymbol{v} = \boldsymbol{A}^{-1}\boldsymbol{u} \quad \text{and} \quad \boldsymbol{u} = \boldsymbol{B}^{-1}\boldsymbol{w}$$

Thus

$$\boldsymbol{v} = (\boldsymbol{B}\boldsymbol{A})^{-1}\boldsymbol{w} = \boldsymbol{A}^{-1}\boldsymbol{B}^{-1}\boldsymbol{w}$$

Therefore $((\boldsymbol{B}\boldsymbol{A})^{-1} - \boldsymbol{A}^{-1}\boldsymbol{B}^{-1})\boldsymbol{w} = \boldsymbol{0}$ for all $\boldsymbol{w} \in W$, which implies eq. (6.3.5).

The identity linear transformation $\boldsymbol{I}: V \to V$ is defined by

$$\boldsymbol{I}\boldsymbol{v} = \boldsymbol{v} \tag{6.3.6}$$

for all $\boldsymbol{v}$ in V. Often it is desirable to distinguish the identity linear transformations on different vector spaces. In these cases we shall denote the identity linear transformation by $\boldsymbol{I}_V$. It follows eq. (6.3.1) and eq. (6.3.3) that if $\boldsymbol{A}$ is an isomorphism, then

$$\boldsymbol{A}\boldsymbol{A}^{-1} = \boldsymbol{I}_U \quad \text{and} \quad \boldsymbol{A}^{-1}\boldsymbol{A} = \boldsymbol{I}_V \tag{6.3.7}$$

Conversely, if $\boldsymbol{A}$ is a linear transformation from V to U, and if there exists a linear transformation $\boldsymbol{B}: U \to V$ such that $\boldsymbol{A}\boldsymbol{B} = \boldsymbol{I}_U$ and $\boldsymbol{B}\boldsymbol{A} = \boldsymbol{I}_V$, then $\boldsymbol{A}$ is an isomorphism and $\boldsymbol{B} = \boldsymbol{A}^{-1}$.

The proof of this assertion is left as an exercise to the reader.

A vector space V and a vector space U are said to be isomorphic if there exists at least one isomorphism from V to U.

Theorem 6.3.2: Two finite-dimensional vector spaces V and U are isomorphic iff they have the same dimension.

Proof: Clearly, if V and U are isomorphic, by virtue of the properties of isomorphism, $\dim V = \dim U$. If U and V have the same dimension, we can construct a regular onto linear transformation $\boldsymbol{A}: V \to U$ as follows. If $\{\boldsymbol{e}_1, \ldots, \boldsymbol{e}_N\}$ are a basis for V and $\{\boldsymbol{b}_1, \ldots, \boldsymbol{b}_N\}$ is a basis for U, define $\boldsymbol{A}$ by

$$\boldsymbol{A}\boldsymbol{e}_k = \boldsymbol{b}_k, \quad k = 1, \ldots, N \tag{6.3.8}$$

Or, equivalently, if

$$\boldsymbol{v} = \sum_{k=1}^{n} v^k \boldsymbol{e}^k$$

Then define $\boldsymbol{A}$ by

$$\boldsymbol{A}\boldsymbol{v} = \sum_{k=1}^{N} v^k \boldsymbol{b}_k \tag{6.3.9}$$

$\boldsymbol{A}$ is regular because if $\boldsymbol{A}\boldsymbol{v} = \boldsymbol{0}$, then $\boldsymbol{v} = \boldsymbol{0}$. Theorem 6.1.10 tells us $\boldsymbol{A}$ is surjective and thus is an isomorphism.

As a corollary to Theorem 6.3.2, we see that V and the vector space C^N, where $N = \dim V$, are isomorphic.

We introduced the notation $L(V, U)$ for the vector space of linear before transformations from V to U. The set $L(V, V)$ corresponds to the vector space of linear transformations $V \to V$. An element of $L(V, V)$ is called an endomorphism of V. This nomenclature parallels the previous usage of the word endomorphism introduced in Chapter 2. If an endomorphism is regular(and thus onto), it is called an automorphism. The identity linear transformation defined by eq. (6.3.6) is an example of an automorphism. If $\boldsymbol{A} \in L(V,V)$, then it is easily seen that

$$\boldsymbol{AI} = \boldsymbol{IA} = \boldsymbol{A} \tag{6.3.10}$$

Also, if $\boldsymbol{A}$ and $\boldsymbol{B}$ are in $L(V,V)$, it is meaningful to compute the products $\boldsymbol{AB}$ and $\boldsymbol{BA}$; however,

$$\boldsymbol{AB} \neq \boldsymbol{BA} \tag{6.3.11}$$

in general. For example, let V be a two-dimensional vector space with basis $\{\boldsymbol{e}_1, \boldsymbol{e}_2\}$ and define $\boldsymbol{A}$ and $\boldsymbol{B}$ by the rules

$$\boldsymbol{Ae}_k = \sum_{j=1}^{2} \boldsymbol{A}_k^j \boldsymbol{e}_j \quad \text{and} \quad \boldsymbol{Be}_k = \sum_{j=1}^{2} \boldsymbol{B}_k^j \boldsymbol{e}_j$$

Where $\boldsymbol{A}_k^j$ and $\boldsymbol{B}_k^j$, $k,j = 1,2$, are prescribed, then

$$\boldsymbol{BAe}_k = \sum_{j,l=1}^{2} \boldsymbol{A}_k^j \boldsymbol{B}_j^l \boldsymbol{e}_l \quad \text{and} \quad \boldsymbol{ABe}_k = \sum_{j,l=1}^{2} \boldsymbol{B}_k^j \boldsymbol{A}_j^l \boldsymbol{e}_l$$

An examination of these formulas shows that it is only for special values of $\boldsymbol{A}_k^j$ and $\boldsymbol{B}_k^j$ that $\boldsymbol{AB} = \boldsymbol{BA}$.

The set $L(V,V)$ has defined on it three operations. They are (a) addition of elements of $L(V,V)$, (b) multiplication of an element of $L(V,V)$ by a scalar, and (c) the product of a pair of elements of $L(V,V)$. The operations (a) and (b) make $L(V,V)$ into a vector space, while it is easily shown that the operations (a) and (c) make $L(V,V)$ into a ring. The structure of $L(V,V)$ is an example of an associative algebra.

The subset of $L(V,V)$ that consists of all automorphisms of V is denoted by $YL(V)$. It is immediately apparent that $YL(V)$ is not a subspace of $L(V,V)$, because the sum of two of its elements need not be an automorphism. However, this set is easily shown to be a group with respect to the product operation. This group is called the general linear group. Its identity element is $\boldsymbol{I}$ and if $\boldsymbol{A} \in YL(V)$, its inverse is $\boldsymbol{A}^{-1} \in YL(V)$.

A projection is an endomorphism $\boldsymbol{P} \in L(V,V)$ which satisfies the condition

$$\boldsymbol{P}^2 = \boldsymbol{P} \tag{6.3.12}$$

The following theorem gives an important property of a projection.

Theorem 6.3.3: If $\boldsymbol{P}: V \to V$ is a projection, then

$$V = R(\boldsymbol{P}) \oplus K(\boldsymbol{P}) \tag{6.3.13}$$

Proof: Let $\boldsymbol{v}$ be an arbitrary vector in V. Let

$$w = v - Pv \tag{6.3.14}$$

then, by eq. (6.3.12), $Pw = Pv - P(Pv) = Pv - Pv = 0$, thus, $w \in K(P)$. Since $Pv \in R(P)$, eq. (6.3.14) implies that

$$V = R(P) + K(P)$$

To show that $R(P) \cap K(P) = \{0\}$, *let* $u \in R(P) \cap K(P)$, then, since $u \in R(P)$ for some $v \in V, u = Pv$. But, since u is also in $K(P)$,

$$0 = Pu = P(Pv) = Pv = u$$

which completes the proof.

The name projection arises from the geometric interpretation of eq. (6.3.13). Given any $v \in V$, then there are unique vectors $u \in R(P)$ and $w \in K(P)$ such that

$$v = u + w \tag{6.3.15}$$

where

$$Pu = u \quad \text{and} \quad Pw = 0 \tag{6.3.16}$$

Geometrically, P takes v and projects onto the subspace $R(P)$ along the subspace $K(P)$.

Given a projection P, the linear transformation $I - P$ is also a projection. It is easily shown that

$$V = R(I - P) \oplus K(I - P)$$

and

$$R(I - P) = K(P), \quad K(I - P) = R(P)$$

It follows eq. (6.3.16) that the restriction of P to $R(P)$ is the identity linear transformation on the subspace $R(P)$. Likewise, the restriction of $I - P$ to $K(P)$ is the identity linear transformation on $K(P)$. Theorem 6.3.3 is a special case of the following theorem.

Theorem 6.3.4: If $P_k, k = 1, \ldots, R$ are projection operators with the properties that

$$\begin{aligned} P_k^2 &= P_k, k = 1, \ldots, R \\ P_k P_q &= 0, k \neq q \end{aligned} \tag{6.3.17}$$

and

$$I = \sum_{k=1}^{R} P_k \tag{6.3.18}$$

then

$$V = R(P_1) \oplus R(P_2) \oplus \ldots \oplus R(P_R) \tag{6.3.19}$$

The proof of this theorem is left as an exercise for the reader. As a converse of Theorem 6.3.4, if V has the decomposition

$$V = V_1 \oplus \ldots \oplus V_R \tag{6.3.20}$$

then the endomorphisms $\boldsymbol{P}_k : V_k \rightarrow V_k$ defined by

$$\boldsymbol{P}_k \boldsymbol{v} = \boldsymbol{v}_k, \quad k = 1, \ldots, R \tag{6.3.21}$$

where

$$\boldsymbol{v} = \boldsymbol{v}_1 + \boldsymbol{v}_2 + \ldots + \boldsymbol{v}_R \tag{6.3.22}$$

are projections and satisfy eq. (6.3.18). Moreover, $V_k = R(\boldsymbol{P}_k)$, $k = 1, \ldots, R$.

6.4 The Adjoint of a Linear Transformation

In this section a particular inner product is needed to study the adjoint of a linear transformation as well as other ideas associated with the adjoint.

Given a linear transformation $\boldsymbol{A}: V \rightarrow U$, a function $\boldsymbol{A}^*: U \rightarrow V$ is called the adjoint of $\boldsymbol{A}$ if

$$\boldsymbol{u} \cdot (\boldsymbol{A}\boldsymbol{v}) = (\boldsymbol{A}^* \boldsymbol{u}) \cdot \boldsymbol{v} \tag{6.4.1}$$

for all $\boldsymbol{v} \in V$ and $\boldsymbol{u} \in U$. Observe that in eq. (6.4.1) the inner product on the left side is the one in U, while the one for the right side is the one in V. Next we will examine the properties of the adjoint. It is probably worthy of note here that for linear transformations defined on real inner product spaces what we have called the adjoint is often called the transpose. Since our later applications are for real vector spaces, the name transpose is actually more important.

Theorem 6.4.1: For every linear transformation $\boldsymbol{A}: V \rightarrow U$, there exists a unique adjoint $\boldsymbol{A}^*: U \rightarrow V$ satisfying the condition eq. (6.4.1).

Proof:

1) Existence: Choose a basis $\{\boldsymbol{e}_1, \ldots, \boldsymbol{e}_N\}$ for V and a basis $\{\boldsymbol{b}_1, \ldots, \boldsymbol{b}_M\}$ for U. Then $\boldsymbol{A}$ can be characterized by the $M \times N$ matrix $[A_k^\alpha]$ in such a way that

$$\boldsymbol{A}\boldsymbol{e}_k = \sum_{\alpha=1}^{M} A_k^\alpha \boldsymbol{b}_\alpha, \quad k = 1, \ldots, N \tag{6.4.2}$$

This system suffices to define $\boldsymbol{A}$ since for any $\boldsymbol{v} \in V$ with the representation

$$\boldsymbol{v} = \sum_{k=1}^{N} V^k \boldsymbol{e}_k$$

the corresponding representation of $\boldsymbol{A}\boldsymbol{v}$ is determined by $[A_k^\alpha]$ and $[v^k]$ by

$$Av = \sum_{\alpha=1}^{M} \left(\sum_{k=1}^{N} A_k^\alpha v^k \right) b_\alpha$$

Now let $\{e^1, \ldots, e^N\}$ and $\{b^1, \ldots, b^M\}$ be the reciprocal bases of $\{e_1, \ldots, e_N\}$ and $\{b_1, \ldots, b_M\}$, respectively. We shall define a linear transformation A^* by a system similar to eq. (6.4.2) except that we shall use the reciprocal bases. Thus we put

$$A^* b^\alpha = \sum_{k=1}^{N} A^{*\alpha}_{\ k} e^k \tag{6.4.3}$$

where the matrix $[A^{*\alpha}_{\ k}]$ is defined by

$$A^{*\alpha}_{\ k} \equiv \overline{A_k^\alpha} \tag{6.4.4}$$

for all $\alpha = 1, \ldots, M$ and $k = 1, \ldots, N$. For any vector $u \in U$ the representation of $A^* u$ relative to $\{e^1, \ldots, e^N\}$ is then given by

$$A^* u = \sum_{k=1}^{N} \left(\sum_{\alpha=1}^{M} A^{*\alpha}_{\ k} u_\alpha \right) e^k$$

Where the representation of u itself relative to $\{b^1, \ldots, b^M\}$ is

$$u = \sum_{\alpha=1}^{M} u_\alpha b^\alpha$$

Having defined the linear transformation A^*, we now verify that A and A^* satisfy the relation eq. (6.4.1). Since A and A^* are both linear transformations, it suffices to check eq. (6.4.1) for u equal to an arbitrary element of the basis $\{b^1, \ldots, b^M\}$, say $u = b^\alpha$, and for v equal to an arbitrary element of the basis $\{e_1, \ldots, e_N\}$, say $v = e_k$. For this choice of u and v we obtain from eq. (6.4.2)

$$b^\alpha \cdot Ae_k = b^\alpha \cdot \sum_{\beta=1}^{M} A_k^\beta b_\beta = \sum_{\beta=1}^{M} \overline{A_k^\beta} \delta_\beta^\alpha = \overline{A_k^\alpha} \tag{6.4.5}$$

and likewise from eq. (6.4.3)

$$A^* b^\alpha \cdot e_k = \left(\sum_{l=1}^{N} A^{*\alpha}_{\ l} e^l \right) \cdot e_k = \sum_{l=1}^{N} A^{*\alpha}_{\ l} \delta_k^l = A_k^{*\alpha} \tag{6.4.6}$$

Comparing eq. (6.4.5) and eq. (6.4.6) with eq. (6.4.4), we see that the linear transformation A^* defined by eq. (6.4.3) satisfies the condition

$$b^\alpha \cdot Ae_k = A^* b^\alpha \cdot e_k$$

for all $\alpha = 1, \ldots, M$ and $k = 1, \ldots, N$, and hence also the condition eq. (6.4.1) for all $u \in U$ and $v \in V$.

2) Uniqueness: Assume that there are two functions $A_1^* : U \to V$ and $A_2^* : U \to V$ which satisfy eq. (6.4.1), then

$$(A_1^* u) \cdot v = (A_2^* u) \cdot v = u \cdot (Av)$$

Thus

$$(A_1^* u - A_2^* u) \cdot v = 0$$

Since the last formula must hold for all $v \in V$, the inner product properties show that

$$A_1^* u = A_2^* u$$

This formula must hold for every $u \in U$ and thus

$$A_1^* = A_2^*$$

As a corollary to the preceding theorem we see that the adjoint A^* of a linear transformation A is a linear transformation. Further, the matrix $[A_k^{*\alpha}]$ that characterizes A^* by eq. (6.4.3) is related to the matrix $[A_\alpha^k]$ that characterizes A satisfying $b^\alpha \cdot Ae_k = A^* b^\alpha e_k$. Notice the choice of bases in eq. (6.4.2) and eq. (6.4.3), however.

The following theorem summarizes other properties of the adjoint.

Theorem 6.4.2:

(a) $(A+B)^* = A^* + B^*$ $\quad(6.4.7)_1$

(b) $(AB)^* = B^* A^*$ $\quad(6.4.7)_2$

(c) $(\lambda A)^* = \bar{\lambda} A^*$ $\quad(6.4.7)_3$

(d) $0^* = 0$ $\quad(6.4.7)_4$

(e) $I^* = I$ $\quad(6.4.7)_5$

(f) $(A^*)^* = A$ $\quad(6.4.7)_6$

(g) *If* A is nonsingular, so is A^* and in addition,

$$A^{*-1} = A^{-1*} \tag{6.4.8}$$

In (a), A and B are in $L(V,U)$; in (b), $B \in L(V,U)$ and $A \in L(U,W)$; in (c), $\lambda \in \mathbf{C}$; in (d), 0 is the zero element; and in (e), I is the identity element in $L(V,V)$.

The proof of the above theorem is straightforward and is left as an exercise for the reader.

Theorem 6.4.3: If $A: V \to U$ is a linear transformation, then V and U have the orthogonal decompositions:

$$V = R(A^*) \oplus K(A) \tag{6.4.9}$$

and

$$U = R(A) \oplus K(A^*) \tag{6.4.10}$$

where

$$R(\boldsymbol{A}^*) = K(\boldsymbol{A})^{\perp} \tag{6.4.11}$$

and

$$R(\boldsymbol{A}) = K(\boldsymbol{A}^*)^{\perp} \tag{6.4.12}$$

Proof: We shall prove eq. (6.4.11) and eq. (6.4.12) and then apply Theorem 5.6.4 to obtain eq. (6.4.9) and eq. (6.4.10). Let $\boldsymbol{u}$ be an arbitrary element in $K(\boldsymbol{A}^*)$. Then for every $\boldsymbol{v} \in V$, $\boldsymbol{u} \cdot (\boldsymbol{A}\boldsymbol{v}) = (\boldsymbol{A}^*\boldsymbol{u}) \cdot \boldsymbol{v} = \boldsymbol{0}$. Thus $K(\boldsymbol{A}^*)$ is contained in $R(\boldsymbol{A})^{\perp}$. Conversely, take $\boldsymbol{u} \in R(\boldsymbol{A})^{\perp}$; then for every $\boldsymbol{v} \in V$, $(\boldsymbol{A}^*\boldsymbol{u}) \cdot \boldsymbol{v} = \boldsymbol{u} \cdot (\boldsymbol{A}\boldsymbol{v}) = \boldsymbol{0}$, and thus $(\boldsymbol{A}^*\boldsymbol{u}) = \boldsymbol{0}$, which implies that $\boldsymbol{u} \in K(\boldsymbol{A}^*)$ and that $R(\boldsymbol{A})^{\perp}$ is in $K(\boldsymbol{A}^*)$. Therefore, $R(\boldsymbol{A})^{\perp} = K(\boldsymbol{A}^*)$. Eq. (6.4.11) follows by an identical argument with $\boldsymbol{A}$ replaced by $\boldsymbol{A}^*$. As mentioned above, eq. (6.4.9) and eq. (6.4.10) now follow Theorem 5.6.4.

Theorem 6.4.4: Given a linear transformation $\boldsymbol{A}: V \to U$, then $\boldsymbol{A}$ and $\boldsymbol{A}^*$ have the same rank.

Proof: By application of Theorems 5.3.9 and 6.4.3,

$$\dim V = \dim R(\boldsymbol{A}^*) + \dim K(\boldsymbol{A})$$

However, by eq. (6.1.6)

$$\dim V = \dim R(\boldsymbol{A}) + \dim K(\boldsymbol{A})$$

Therefore,

$$\dim R(\boldsymbol{A}) = \dim R(\boldsymbol{A}^*) \tag{6.4.13}$$

which is the desired result.

An endomorphism $\boldsymbol{A} \in L(V, V)$ is called Hermitian if $\boldsymbol{A} = \boldsymbol{A}^*$ and skew-Hermitian if $\boldsymbol{A} = -\boldsymbol{A}^*$.

The following theorem, which follows directly the above definitions and from eq. (6.4.1), characterizes Hermitian and skew-Hermitian endomorphisms.

Theorem 6.4.5: An endomorphism $\boldsymbol{A}$ is Hermitian if and only if

$$\boldsymbol{v}_1 \cdot (\boldsymbol{A}\boldsymbol{v}_2) = (\boldsymbol{A}\boldsymbol{v}_1) \cdot \boldsymbol{v}_2 \tag{6.4.14}$$

for all $\boldsymbol{v}_1, \boldsymbol{v}_2 \in V$, and it is skew-Hermitian iff

$$\boldsymbol{v}_1 \cdot (\boldsymbol{A}\boldsymbol{v}_2) = -(\boldsymbol{A}\boldsymbol{v}_1) \cdot \boldsymbol{v}_2 \tag{6.4.15}$$

for all $\boldsymbol{v}_1, \boldsymbol{v}_2 \in V$.

We shall denote by $P(V, V)$ and $F(V, V)$ the subsets of $L(V, V)$ defined by

$$P(V, V) = \{\boldsymbol{A} \mid \boldsymbol{A} \in L(V, V) \text{ and } \boldsymbol{A} = \boldsymbol{A}^*\}$$

and

$$F(V, V) = \{\boldsymbol{A} \mid \boldsymbol{A} \in L(V, V) \text{ and } \boldsymbol{A} = -\boldsymbol{A}^*\}$$

In the special case of a real inner product space, it is easy to show that $P(V,V)$ and $F(V,V)$ are both subspaces of $L(V,V)$. In particular, $L(V,V)$ has the following decomposition:

Theorem 6.4.6: For a real inner product space,

$$L(V,V) = P(V,V) \oplus F(V,V) \tag{6.4.16}$$

Proof: An arbitrary element $\boldsymbol{L} \in L(V,V)$ can always be written as

$$\boldsymbol{L} = \boldsymbol{S} + \boldsymbol{A} \tag{6.4.17}$$

where

$$\boldsymbol{S} = \frac{1}{2}(\boldsymbol{L} + \boldsymbol{L}^{\mathrm{T}}) = \boldsymbol{S}^{\mathrm{T}} \quad \text{and} \quad \boldsymbol{A} = \frac{1}{2}(\boldsymbol{L} - \boldsymbol{L}^{\mathrm{T}}) = -\boldsymbol{A}^{\mathrm{T}}$$

Here the superscript T denotes that transpose, which is the specialization of the adjoint for a real inner product space. Since $\boldsymbol{S} \in P(V,V)$ and $\boldsymbol{A} \in F(V,V)$, eq. (6.4.17) shows that

$$L(V,V) = P(V,V) + F(V,V)$$

Now, let $\boldsymbol{B} \in P(V,V) \cap F(V,V)$. Then $\boldsymbol{B}$ must satisfy the conditions

$$\boldsymbol{B} = \boldsymbol{B}^{\mathrm{T}} \quad \text{and} \quad \boldsymbol{B} = -\boldsymbol{B}^{\mathrm{T}}$$

Thus, $\boldsymbol{B} = 0$ and the proof is complete.

A linear transformation $\boldsymbol{A} \in L(V,U)$ is unitary if

$$\boldsymbol{A}\boldsymbol{v}_2 \cdot \boldsymbol{A}\boldsymbol{v}_1 = \boldsymbol{v}_2 \cdot \boldsymbol{v}_1 \tag{6.4.18}$$

for all $\boldsymbol{v}_1, \boldsymbol{v}_2 \in V$. However, for a real inner product space the above condition defines $\boldsymbol{A}$ to be orthogonal. Essentially, eq. (6.4.18) asserts that unitary (or orthogonal) linear transformations preserve the inner products.

Theorem 6.4.7: If $\boldsymbol{A}$ is unitary, then it is regular.

Proof: Take $\boldsymbol{v}_1 = \boldsymbol{v}_2 = \boldsymbol{v}$ in eq. (6.4.18), and by eq. (5.5.1) we find

$$\| \boldsymbol{A}\boldsymbol{v} \| = \| \boldsymbol{v} \| \tag{6.4.19}$$

Thus, if $\boldsymbol{A}\boldsymbol{v} = \boldsymbol{0}$, then $\boldsymbol{v} = \boldsymbol{0}$, which proves the theorem.

Theorem 6.4.8: $\boldsymbol{A} \in L(V,U)$ is unitary iff $\| \boldsymbol{A}\boldsymbol{v} \| = \| \boldsymbol{v} \|$ for all $\boldsymbol{v} \in V$.

Proof: If $\boldsymbol{A}$ is unitary, we saw in the proof of Theorem 6.4.7 that $\| \boldsymbol{A}\boldsymbol{v} \| = \| \boldsymbol{v} \|$. Thus, we shall assume $\| \boldsymbol{A}\boldsymbol{v} \| = \| \boldsymbol{v} \|$ for all $\boldsymbol{v} \in V$ and attempt to derive eq. (6.4.18). By the example,

$$2\boldsymbol{A}\boldsymbol{v}_1 \cdot \boldsymbol{A}\boldsymbol{v}_2 = \| \boldsymbol{A}(\boldsymbol{v}_1 + \boldsymbol{v}_2) \|^2 + i \| \boldsymbol{A}(\boldsymbol{v}_1 + i\boldsymbol{v}_2) \|^2 - (1+i)(\| \boldsymbol{A}\boldsymbol{v}_1 \|^2 + \| \boldsymbol{A}\boldsymbol{v}_2 \|^2)$$

Therefore, by eq. (6.4.19)

$$2\boldsymbol{A}\boldsymbol{v}_1 \cdot \boldsymbol{A}\boldsymbol{v}_2 = \| \boldsymbol{v}_1 + \boldsymbol{v}_2 \|^2 + i \| \boldsymbol{v}_1 + i\boldsymbol{v}_2 \|^2 - (1+i)(\| \boldsymbol{v}_1 \|^2 + \| \boldsymbol{v}_2 \|^2) = 2\boldsymbol{v}_1 \cdot \boldsymbol{v}_2$$

This proof cannot be specialized directly to a real inner product space since the polar identity is valid for a complex inner product space only. We leave the proof for the real case as an exercise.

If we require V and U to have the same dimension, then Theorem 6.1.10 ensures that a unitary transformation $\boldsymbol{A}$ is an isomorphism. In the case we can use eq. (6.4.1) and eq. (6.4.18) and conclude the following.

Theorem 6.4.9: Given a linear transformation $\boldsymbol{A} \in L(V, U)$, where $\dim V = \dim U$; then $\boldsymbol{A}$ is unitary iff it is an isomorphism whose inverse satisfies

$$\boldsymbol{A}^{-1} = \boldsymbol{A}^{*} \tag{6.4.20}$$

We know that a regular linear transformation maps linearly independent vectors into linearly independent vectors. Therefore if $\{\boldsymbol{e}_1, \ldots, \boldsymbol{e}_N\}$ is a basis for V and $\boldsymbol{A} \in L(V, U)$ is regular, then $\{\boldsymbol{Ae}_1, \ldots, \boldsymbol{Ae}_N\}$ is a basis for $R(\boldsymbol{A})$ which is a subspace in U. If $\{\boldsymbol{e}_1, \ldots, \boldsymbol{e}_N\}$ is orthonormal and $\boldsymbol{A}$ is unitary, it easily follows that $\{\boldsymbol{Ae}_1, \ldots, \boldsymbol{Ae}_N\}$ is also orthonormal. Thus the image of an orthonormal basis under a unitary transformation is also an orthonormal set. Conversely, a linear transformation which sends an orthonormal basis of V into an orthonormal basis of $R(\boldsymbol{A})$ must be unitary. To prove this assertion, let $\{\boldsymbol{e}_1, \ldots, \boldsymbol{e}_N\}$ be orthonormal, and let $\{\boldsymbol{b}_1, \ldots, \boldsymbol{b}_N\}$ be an orthonormal set in U, where $\boldsymbol{b}_k = \boldsymbol{Ae}_k, k = 1, \ldots, N$. Then, if $\boldsymbol{v}_1$ and $\boldsymbol{v}_2$ are arbitrary elements of V,

$$\boldsymbol{v}_1 = \sum_{k=1}^{N} v_1^k \boldsymbol{e}_k \quad \text{and} \quad \boldsymbol{v}_2 = \sum_{l=1}^{N} v_2^l \boldsymbol{e}_l$$

we have

$$\boldsymbol{Av}_1 = \sum_{k=1}^{N} v_1^k \boldsymbol{b}_k \quad \text{and} \quad \boldsymbol{Av}_2 = \sum_{l=1}^{N} v_2^l \boldsymbol{b}_1$$

and, thus,

$$\boldsymbol{Av}_1 \cdot \boldsymbol{Av}_2 = \sum_{k=1}^{N} \sum_{l=1}^{N} v_1^k v_2^{-1} \boldsymbol{b}_k \cdot \boldsymbol{b}_l = \sum_{k=1}^{N} v_1^k v_2^1 = \boldsymbol{v}_1 \cdot \boldsymbol{v}_2 \tag{6.4.21}$$

eq. (6.4.21) establishes the desired result.

We define a subset $U(V)$ of $CL(V)$ by

$$U(V) = \{\boldsymbol{A} \mid \boldsymbol{A} \in CL(V) \text{ and } \boldsymbol{A}^{-1} = \boldsymbol{A}^{*}\}$$

which is easily shown to be a subgroup. This subgroup is called the unitary group of V.

We have defined projections $\boldsymbol{P}: V \to V$ before by the characteristic property

$$\boldsymbol{P}^2 = \boldsymbol{P} \tag{6.4.22}$$

In particular, we have shown in Theorem 6.3.3 that V has the decomposition

$$V = R(\boldsymbol{P}) \oplus K(\boldsymbol{P}) \tag{6.4.23}$$

There are several additional properties of projections which are worthy of discussion here.

Theorem 6.4.10: If $\boldsymbol{P}$ is a projection, then $\boldsymbol{P} = \boldsymbol{P}^* \Leftrightarrow R(\boldsymbol{P}) = K(\boldsymbol{P})^{\perp}$.

Proof: First take $\boldsymbol{P} = \boldsymbol{P}^*$ and let $\boldsymbol{v}$ be an arbitrary element of V. Then by eq. (6.4.23)

$$\boldsymbol{v} = \boldsymbol{u} + \boldsymbol{w}$$

where $\boldsymbol{u} = \boldsymbol{P}\boldsymbol{v}$ and $\boldsymbol{w} = \boldsymbol{v} - \boldsymbol{P}\boldsymbol{v}$, then

$$\begin{aligned} \boldsymbol{u} \cdot \boldsymbol{w} &= \boldsymbol{P}\boldsymbol{v} \cdot (\boldsymbol{v} - \boldsymbol{P}\boldsymbol{v}) \\ &= (\boldsymbol{P}\boldsymbol{v}) \cdot \boldsymbol{v} - (\boldsymbol{P}\boldsymbol{v}) \cdot \boldsymbol{P}\boldsymbol{v} \\ &= (\boldsymbol{P}\boldsymbol{v}) \cdot \boldsymbol{v} - (\boldsymbol{P}^*\boldsymbol{P}\boldsymbol{v}) \cdot \boldsymbol{v} \\ &= (\boldsymbol{P}\boldsymbol{v}) \cdot \boldsymbol{v} - (\boldsymbol{P}^2\boldsymbol{v}) \cdot \boldsymbol{v} \\ &= \boldsymbol{0} \end{aligned}$$

where eq. (6.4.22), eq. (6.4.1), and the assumption that $\boldsymbol{P}$ is Hermitian have been used. Conversely, assume $\boldsymbol{u} \cdot \boldsymbol{w} = \boldsymbol{0}$ for all $\boldsymbol{u} \in R(\boldsymbol{P})$ and all $\boldsymbol{w} \in K(\boldsymbol{P})$, then if $\boldsymbol{v}_1$ and $\boldsymbol{v}_2$ are arbitrary vectors in V,

$$\boldsymbol{P}\boldsymbol{v}_1 \cdot \boldsymbol{v}_2 = \boldsymbol{P}\boldsymbol{v}_1 \cdot (\boldsymbol{P}\boldsymbol{v}_2 + \boldsymbol{v}_2 - \boldsymbol{P}\boldsymbol{v}_2) = \boldsymbol{P}\boldsymbol{v}_1 \cdot \boldsymbol{P}\boldsymbol{v}_2$$

and, by interchanging $\boldsymbol{v}_1$ and $\boldsymbol{v}_2$

$$\boldsymbol{P}\boldsymbol{v}_2 \cdot \boldsymbol{v}_1 = \boldsymbol{P}\boldsymbol{v}_2 \cdot \boldsymbol{P}\boldsymbol{v}_1$$

Therefore,

$$\boldsymbol{P}\boldsymbol{v}_1 \cdot \boldsymbol{v}_2 = \overline{\boldsymbol{P}\boldsymbol{v}_2 \cdot \boldsymbol{P}\boldsymbol{v}_1} = \boldsymbol{v}_1 \cdot (\boldsymbol{P}\boldsymbol{v}_2)$$

This last result and Theorem 6.4.5 show that P is Hermitian.

Because of Theorem 6.4.10, Hermitian projections are called perpendicular projections. If V_1 and V_2 are subspaces of V, they are orthogonal, written as $V_1 \perp V_2$, if $\boldsymbol{v}_1 \cdot \boldsymbol{v}_2 = 0$ for all $\boldsymbol{v}_1 \in V_1$ and $\boldsymbol{v}_2 \in V_2$.

Theorem 6.4.11: If V_1 and V_2 are subspaces of V, $\boldsymbol{P}_1$ is the perpendicular projection of V onto V_1, and $\boldsymbol{P}_2$ is the perpendicular projection of $\boldsymbol{P}$ onto V_2, then $V_1 \perp V_2$ iff $\boldsymbol{P}_2\boldsymbol{P}_1 = \boldsymbol{0}$.

Proof: Assume that $V_1 \perp V_2$; then $\boldsymbol{P}_1\boldsymbol{v} \in V_2^{\perp}$ for all $\boldsymbol{v} \in V$ and thus $\boldsymbol{P}_2\boldsymbol{P}_1\boldsymbol{v} = \boldsymbol{0}$ for all $\boldsymbol{v} \in V$ which yields $\boldsymbol{P}_2\boldsymbol{P}_1 = \boldsymbol{0}$. Next assume $\boldsymbol{P}_2\boldsymbol{P}_1 = \boldsymbol{0}$; this implies that $\boldsymbol{P}_1\boldsymbol{v} \in V_2^{\perp}$ for every $\boldsymbol{v} \in V$. Therefore V_1 is contained in $V_2^{\perp}$ and, as a result, $V_1 \perp V_2$.

6.5 Component Formulas

In this section we shall introduce the components of a linear transformation and several related ideas. Let $\boldsymbol{A} \in L(V, U)$, $\{\boldsymbol{e}_1, \ldots, \boldsymbol{e}_N\}$ a basis for V, and $\{\boldsymbol{b}_1, \ldots, \boldsymbol{b}_M\}$ a basis for U. The vector $\boldsymbol{A}\boldsymbol{e}_k$ is in U and, as a result, can be expanded in a basis of U in the form

$$\boldsymbol{A}\boldsymbol{e}_k = \sum_{\alpha=1}^{M} A_k^\alpha \boldsymbol{b}_\alpha \tag{6.5.1}$$

The MN scalars $A_k^\alpha (\alpha = 1, \ldots, M; k = 1, \ldots, N)$ are called the components of $\boldsymbol{A}$ with respect to the bases. If $\{\boldsymbol{b}^1, \ldots, \boldsymbol{b}^M\}$ is a basis of U which is reciprocal to $\{\boldsymbol{b}_1, \ldots, \boldsymbol{b}_M\}$, eq. (6.5.1) yields

$$A_k^\alpha = (\boldsymbol{A}\boldsymbol{e}_k) \cdot \boldsymbol{b}^\alpha \tag{6.5.2}$$

Under change of bases in V and U defined by

$$\boldsymbol{e}_k = \sum_{j=1}^{N} \hat{T}_k^j \hat{\boldsymbol{e}}_j \tag{6.5.3}$$

and

$$\boldsymbol{b}^\alpha = \sum_{\beta=1}^{M} \overline{S_\beta^\alpha}\, \hat{\boldsymbol{b}}^\beta \tag{6.5.4}$$

eq. (6.5.2) can be used to derive the following transformation rule for the components of $\boldsymbol{A}$:

$$A_k^\alpha = \sum_{\beta=1}^{M} \sum_{j=1}^{N} S_\beta^\alpha \hat{A}_j^\beta \hat{T}_k^j \tag{6.5.5}$$

where

$$\hat{A}_j^\beta = (\boldsymbol{A}\hat{\boldsymbol{e}}_j) \cdot \hat{\boldsymbol{b}}^\beta$$

If $\boldsymbol{A} \in L(V, V)$, eq. (6.5.1) and eq. (6.5.5) specialize

$$\boldsymbol{A}\boldsymbol{e}_k = \sum_{q=1}^{N} A_k^q \boldsymbol{e}_q \tag{6.5.6}$$

and

$$A_k^q = \sum_{s,j=1}^{N} T_s^q \hat{A}_j^s \hat{T}_k^j \tag{6.5.7}$$

The trace of an endomorphism is a function $\mathrm{tr}: L(V, V) \to P$ defined by

$$\mathrm{tr}\boldsymbol{A} = \sum_{k=1}^{N} A_k^k \tag{6.5.8}$$

It easily follows eq. (6.5.7) and eq. (5.7.21) that tr$\boldsymbol{A}$ is independent of the choice of basis of V. Later we shall give a definition of the trace which does not employ the use of a basis.

If $\boldsymbol{A} \in L(V,V)$, then $\boldsymbol{A}^* \in L(U,V)$. The components of $\boldsymbol{A}^*$ are obtained by the same logic as was used in obtaining eq. (6.5.1). For example,

$$\boldsymbol{A}^* \boldsymbol{b}^\alpha = \sum_{k=1}^{N} A^{*\alpha}_{\ k} \boldsymbol{e}^k \tag{6.5.9}$$

where the MN scalars $A^{*\alpha}_{\ k}(k=1,\ldots,N;\alpha=1,\ldots,M)$ are the components of $\boldsymbol{A}^*$ with respect to $\{\boldsymbol{b}^1,\ldots,\boldsymbol{b}^M\}$ and $\{\boldsymbol{e}^1,\ldots,\boldsymbol{e}^N\}$. From the proof of Theorem 6.5.1 we can relate these components of $\boldsymbol{A}^*$ to those of $\boldsymbol{A}$ in eq. (6.5.1); namely,

$$A^{*\alpha}_{\ k} = \overline{A^\alpha_s} \tag{6.5.10}$$

If the inner product spaces U and V are real, eq. (6.5.10) reduces to

$$A^{\mathrm{T}\alpha}_{\ \ s} = A^\alpha_s \tag{6.5.11}$$

By the same logic which produced eq. (6.5.1), we can also write

$$\boldsymbol{A}\boldsymbol{e}_k = \sum_{\alpha=1}^{M} A_{\alpha k} \boldsymbol{b}^\alpha \tag{6.5.12}$$

and

$$\boldsymbol{A}\boldsymbol{e}_k = \sum_{\alpha=1}^{M} A^{\alpha k} \boldsymbol{b}_\alpha = \sum_{\alpha=1}^{M} A_\alpha^k \boldsymbol{b}^\alpha \tag{6.5.13}$$

If we use eq. (5.7.6) and eq. (5.7.9), the various components of $\boldsymbol{A}$ are related by

$$A^\alpha_k = \sum_{s=1}^{N} A^{\alpha s} e_{ks} = \sum_{\beta=1}^{M}\sum_{s=1}^{N} b^{\beta\alpha} A_\beta^s e_{ks} = \sum_{\beta=1}^{M} A_{\beta k} b^{\beta\alpha} \tag{6.5.14}$$

where

$$b_{\alpha\beta} = \boldsymbol{b}_\alpha \cdot \boldsymbol{b}_\beta = \overline{b_{\beta\alpha}} \quad \text{and} \quad b^{\alpha\beta} = \boldsymbol{b}^\alpha \cdot \boldsymbol{b}^\beta = \overline{b^{\beta\alpha}} \tag{6.5.15}$$

A similar set of formulas holds for the components of $\boldsymbol{A}^*$. Eq. (6.5.14) and eq. (6.5.10) can be used to obtain these formulas. The transformation rules for the components defined by eq. (6.5.12) and eq. (6.5.13) are easily established to be

$$A_{\alpha k} = (\boldsymbol{A}\boldsymbol{e}_k) \cdot \boldsymbol{b}_\alpha = \sum_{\beta=1}^{M}\sum_{j=1}^{N} \overline{\hat{S}^\beta_\alpha \hat{A}_{\beta j} \hat{T}^j_k} \tag{6.5.16}$$

$$A^{\alpha k} = (\boldsymbol{A}\boldsymbol{e}^k) \cdot \boldsymbol{b}^\alpha = \sum_{\beta=1}^{M}\sum_{j=1}^{N} S^\alpha_\beta \hat{A}^{\beta j}\, \overline{T^k_j} \tag{6.5.17}$$

and

$$A_{\alpha}^{k} = (\boldsymbol{A}\boldsymbol{e}^{k}) \cdot \boldsymbol{b}_{\alpha} = \sum_{\beta=1}^{M} \sum_{j=1}^{N} \overline{\hat{S}_{\beta}^{\alpha} \hat{A}_{\beta}^{j} \, T_{j}^{k}} \tag{6.5.18}$$

where

$$\hat{A}_{\beta j} = (\boldsymbol{A}\hat{\boldsymbol{e}}_{j}) \cdot \hat{\boldsymbol{b}}_{\beta} \tag{6.5.19}$$

$$\hat{A}^{\beta j} = (\boldsymbol{A}\hat{\boldsymbol{e}}^{j}) \cdot \hat{\boldsymbol{b}}^{\beta} \tag{6.5.20}$$

and

$$\hat{A}_{\beta}^{j} = (\boldsymbol{A}\hat{\boldsymbol{e}}^{j}) \cdot \hat{\boldsymbol{b}}_{\beta} \tag{6.5.21}$$

The quantities $\hat{S}_{\alpha}^{\beta}$ introduced in eq. (6.5.16) and eq. (6.5.18) above are related to the quantities S_{β}^{α} by formulas like eq. (5.7.20) and eq. (5.7.21).

Chapter 7

Determinants and Matrices

In chapter 4 we have learned some basis knowledge about matrices and matrix rings. Here in this chapter we shall consider further the concept of a matrix and a few of new notions such as "permutation" and "Kronecker delta" are presented. We also introduce some special type matrices which are very useful in the last section.

7.1 The Generalized Kronecker Deltas and the Summation Convention

Recall that an $M \times N$ matrix is an array written in any one of the forms:

$$\boldsymbol{A} = \begin{bmatrix} A_1^1 & \cdots & A_N^1 \\ \vdots & \ddots & \vdots \\ A_1^M & \cdots & A_N^M \end{bmatrix} = \left[A_j^\alpha \right]_{\substack{\alpha=1,\ldots,M \\ j=1,\ldots,N}}, \boldsymbol{A} = \begin{bmatrix} A_{11} & \cdots & A_{1N} \\ \vdots & \ddots & \vdots \\ A_{M1} & \cdots & A_{MN} \end{bmatrix} = \left[A_{\alpha j} \right]_{\substack{j=1,\ldots,N \\ \alpha=1,\ldots,M}}$$

$$\boldsymbol{A} = \begin{bmatrix} A_1^1 & \cdots & A_1^N \\ \vdots & \ddots & \vdots \\ A_M^1 & \cdots & A_M^N \end{bmatrix} = \left[A_\alpha^j \right]_{\substack{j=1,\ldots,N \\ \alpha=1,\ldots,M}}, \boldsymbol{A} = \begin{bmatrix} A^{11} & \cdots & A^{1N} \\ \vdots & \ddots & \vdots \\ A^{M1} & \cdots & A^{MN} \end{bmatrix} = \left[A^{\alpha j} \right]_{\substack{j=1,\ldots,N \\ \alpha=1,\ldots,M}}$$

$$(7.1.1)$$

Note: The placement of the indices is of no consequence throughout this section.

The components of the matrix are allowed to be complex numbers. The set of $M \times N$ matrices shall be denoted by $U^{M \times N}$. We know that the rules of addition and scalar multiplication insure that $U^{M \times N}$ is a vector space.

Definition: Consider a set of K elements $\{\alpha_1, \dots, \alpha_K\}$. A permutation is a one-to-one function from $\{\alpha_1, \dots, \alpha_K\}$ to $\{\alpha_1, \dots, \alpha_K\}$. If σ is a permutation, it is customary to write

$$\boldsymbol{\sigma} = \begin{pmatrix} \alpha_1 & \alpha_2 & \dots & \alpha_K \\ \sigma(\alpha_1) & \sigma(\alpha_2) & \dots & \sigma(\alpha_K) \end{pmatrix} \tag{7.1.2}$$

It is a known result in algebra that permutations can be classified into even and odd ones: The permutation $\boldsymbol{\sigma}$ in eq. (7.1.2) is even if an even number of pairwise interchanges of the bottom row is required to order the bottom row exactly like the top row, and $\boldsymbol{\sigma}$ is odd if that number is odd. For a given permutation $\boldsymbol{\sigma}$, the number of pairwise interchanges required to order the bottom row the same as the top row is not unique, but it is proved in algebra that those numbers are either all even or all odd. Therefore the definition for $\boldsymbol{\sigma}$ to be even or odd is meaningful. For example, the permutation

$$\boldsymbol{\sigma} = \begin{pmatrix} 1 & 2 & 3 \\ 2 & 1 & 3 \end{pmatrix}$$

is odd, while the permutation

$$\boldsymbol{\sigma} = \begin{pmatrix} 1 & 2 & 3 \\ 2 & 3 & 1 \end{pmatrix}$$

is even.

The parity of $\boldsymbol{\sigma}$, denoted by $\varepsilon_{\boldsymbol{\sigma}}$, is defined by

$$\varepsilon_{\boldsymbol{\sigma}} = \begin{cases} +1 \text{ if } \boldsymbol{\sigma} \text{ is an even permutation} \\ -1 \text{ if } \boldsymbol{\sigma} \text{ is an odd permutation} \end{cases}$$

All of the applications of permutations we shall have are for permutations defined on $K(K \leqslant N)$ positive integers selected from the set of N positive integers $\{1, 2, 3, \dots, N\}$. Let $\{i_1, \dots, i_K\}$ and $\{j_1, \dots, j_K\}$ be two subsets of $\{1, 2, 3, \dots, N\}$. If we order these two subsets and construct the two K-tuples $(i_1, \dots, i_K)$ and $(j_1, \dots j_K)$, we can define the generalized Kronecker delta as follows:

Definition: The generalized Kronecker delta, denoted by

$$\delta^{i_1 i_2 \dots i_K}_{j_1 j_2 \dots j_K}$$

is defined by

$$\delta^{i_1 i_2 \cdots i_K}_{j_1 j_2 \cdots j_K} = \begin{cases} 0, \text{if the integers } (i_1, \ldots, i_K) \text{ or } (j_1, \ldots, j_K) \text{ are not distinct} \\ 0, \text{if the integers } (i_1, \ldots, i_K) \text{ and } (j_1, \ldots, j_K) \text{ are distinct} \\ \quad \text{but the sets } (i_1, \ldots, i_K) \text{ and } (j_1, \ldots, j_K) \text{ are not equal} \\ \varepsilon_\sigma, \text{if the integers } (i_1, \ldots, i_K) \text{ and } (j_1, \ldots, j_K) \text{ are distinct} \\ \quad \text{and the sets } (i_1, \ldots, i_K) \text{ and } (j_1, \ldots, j_K) \text{ are equal, where} \\ \quad \sigma = \begin{pmatrix} i_1 & i_2 & \cdots & i_K \\ j_1 & j_2 & \cdots & j_K \end{pmatrix} \end{cases}$$

It follows this definition that the generalized Kronecker delta is zero whenever the superscripts are not the same set of integers as the subscripts, or when the superscripts are not distinct. Naturally when $K = 1$ the generalized Kronecker delta reduces to the usual one. As an example, $\delta^{i_1 j_2}_{j_1 j_2}$ has the values

$$\delta^{12}_{12} = 1, \delta^{12}_{21} = -1, \delta^{13}_{12} = 0, \delta^{13}_{21} = 0, \delta^{11}_{12} = 0, \text{etc.}$$

It can be shown that there are $N!\ K!\ /(N-K)!$ nonzero generalized Kronecker deltas for given positive integers K and N.

An ε symbol is one of a pair of quantities

$$\varepsilon^{i_1 i_2 \cdots i_N} = \delta^{i_1 i_2 \cdots i_N}_{12 \cdots N} \quad \text{or} \quad \varepsilon_{j_1 j_2 \cdots j_N} = \delta^{12 \cdots N}_{j_1 j_2 \cdots j_N} \tag{7.1.3}$$

For example, take $N = 3$; then

$$\varepsilon^{123} = \varepsilon^{312} = \varepsilon^{231} = 1$$

$$\varepsilon^{132} = \varepsilon^{321} = \varepsilon^{213} = -1$$

$$\varepsilon^{112} = \varepsilon^{221} = \varepsilon^{222} = \varepsilon^{233} = 0, \text{etc.}$$

As an exercise, the reader is asked to confirm that

$$\varepsilon^{i_1 \cdots i_N} \varepsilon_{j_1 \cdots j_N} = \delta^{i_1 \cdots i_N}_{j_1 \cdots j_N} \tag{7.1.4}$$

An identity involving the quantity δ^{ijq}_{lms} is

$$\sum_{q=1}^{N} \delta^{ijq}_{lms} = (N-2)\delta^{ij}_{lm} \tag{7.1.5}$$

To establish this identity, expand the left side of eq. (7.1.5) in the form

$$\sum_{q=1}^{N} \delta^{ijq}_{lms} = \delta^{ij1}_{lm1} + \ldots + \delta^{ijN}_{lmN} \tag{7.1.6}$$

Clearly we need to verify eq. (7.1.5) for the cases $i \neq j$ and $\{i, j\} = \{l, m\}$ only, since in the remaining cases eq. (7.1.5) reduces to the trivial equation $0 = 0$. In the nontrivial cases, exactly two terms on the right-hand side of eq. (7.1.6) are equal to zero: one for $i=q$ and one for $j=q$. For i, j, l and m not equal to q, δ^{ijq}_{lms} has the same value as δ^{ij}_{lm}. Therefore,

$$\sum_{q=1}^{N} \delta^{ijq}_{lms} = (N-2)\delta^{ij}_{lm}$$

which is the desired result eq. (7.1.5). By the same procedure used above, it is clear that

$$\sum_{j=1}^{N} \delta^{ij}_{lj} = (N-1)\delta^{i}_{l} \tag{7.1.7}$$

Combining eq. (7.1.5) and eq. (7.1.7), we have

$$\sum_{j=1}^{N}\sum_{q=1}^{N} \delta^{ijq}_{ljq} = (N-2)(N-1)\delta^{i}_{j} \tag{7.1.8}$$

Since $\sum_{j=1}^{N} \delta^{j}_{i} = N$, we have from eq. (7.1.8)

$$\sum_{i=1}^{N}\sum_{j=1}^{N}\sum_{q=1}^{N} \delta^{ijq}_{ljq} = (N-2)(N-1)(N) = \frac{N!}{(N-3)!} \tag{7.1.9}$$

Eq. (7.1.9) is a special case of

$$\sum_{i_1, i_2, \cdots, i_K=1}^{N} \delta^{i_1 i_2 \cdots i_K}_{i_1 i_2 \cdots i_K} = \frac{N!}{(N-K)!} \tag{7.1.10}$$

Several other numerical relationships are

$$\sum_{i_{R+1}, i_{R+2}, \cdots, i_K=1}^{N} \delta^{i_1 \cdots i_R i_{R+1} \cdots i_K}_{j_1 \cdots j_R j_{R+1} \cdots j_K} = \frac{(N-R)!}{(N-K)!}\delta^{i_1 \cdots i_R}_{j_1 \cdots j_R} \tag{7.1.11}$$

$$\sum_{i_{K+1}, \cdots, i_N=1}^{N} \varepsilon^{i_1 \cdots i_R i_{R+1} \cdots i_K}\varepsilon_{j_1 \cdots j_K j_{K+1} \cdots j_N} = (N-K)!\,\delta^{i_1 \cdots i_K}_{j_1 \cdots j_K} \tag{7.1.12}$$

$$\sum_{i_{K+1}, \cdots, i_N=1}^{N} \varepsilon^{i_1 \cdots i_K i_{K+1} \cdots i_N}\delta^{j_{K+1} \cdots j_N}_{i_{K+1} \cdots i_N} = (N-K)!\,\varepsilon^{i_1 \cdots i_K j_{K+1} \cdots j_N} \tag{7.1.13}$$

$$\sum_{j_{k+1}, \cdots, j_R=1}^{N} \delta^{i_1 \cdots i_K i_{K+1} \cdots i_R}_{j_1 \cdots j_K j_{K+1} \cdots j_R}\delta^{j_{K+1} \cdots j_R}_{l_{K+1} \cdots l_R} = (R-K)!\,\delta^{i_1 \cdots i_K i_{K+1} \cdots i_R}_{j_1 \cdots j_K l_{K+1} \cdots l_R} \tag{7.1.14}$$

$$\sum_{j_{K+1}, \cdots, j_R=1}^{N}\ \sum_{i_{K+1}, \cdots, i_R=1}^{N} \delta^{i_1 \cdots i_K i_{K+1} \cdots i_R}_{j_1 \cdots j_K j_{K+1} \cdots j_R}\delta^{j_{K+1} \cdots j_R}_{i_{K+1} \cdots i_R} = \frac{(N-K)!}{(N-R)!}(R-K)!\,\delta^{i_1 \cdots i_K}_{j_1 \cdots j_K} \tag{7.1.15}$$

We can simplify our equations if we adopt a summation convention: We automatically sum every repeated index without writing the summation sign. For example, eq. (7.1.5) is written as

$$\delta^{ijq}_{lmq} = (N-1)\delta^{ij}_{lm} \tag{7.1.16}$$

The occurrence of the subscript q and the superscript q implies the summation indicated in eq. (7.1.5). We shall try to arrange things so that we always sum a superscript on a subscript. It is important to know the range of a given summation, so we shall use the summation convention only when the range of summation is understood. Also, observe that the repeated indices are dummy indices in the sense that it is unimportant which symbol is used for them. For example,

$$\delta^{ijs}_{lms} = \delta^{ijt}_{lmt} = \delta^{ijq}_{lmq}$$

Naturally there is no meaning to the occurrence of the same index more than twice in a given term. Other than the summation or dummy indices, many equations have free indices whose values in the given range $\{1, \ldots, N\}$ are arbitrary. For example, the indices j, k, l and m are free indices in eq. (7.1.16).

Note that every term in a given equation must have the same free indices; otherwise, the equation is meaningless.

7.2 Determinants

We have studied the determinant of a matrix in chapter 4, but in this section we shall use the generalized Kronecker deltas and the ε symbols to redefine the determinant of a square matrix.

The determinant of the $N \times N$ matrix $\boldsymbol{A} = [A_{ij}]$, written as $\det \boldsymbol{A}$, is a complex number defined by

$$\det \boldsymbol{A} = \begin{vmatrix} A_{11} & A_{12} & \cdots & A_{1N} \\ A_{21} & A_{22} & \cdots & A_{2N} \\ A_{31} & A_{32} & \cdots & A_{3N} \\ \vdots & \vdots & \ddots & \vdots \\ A_{N1} & A_{N2} & \cdots & A_{NN} \end{vmatrix} = \varepsilon^{i_1 \ldots i_N} A_{i_1 1} A_{i_2 2} \ldots A_{i_N N} \tag{7.2.1}$$

where all summations are from 1 to N. If the elements of the matrix are written as $[A^{ij}]$, its determinant is defined to be

$$\det \boldsymbol{A} = \begin{vmatrix} A^{11} & A^{12} & \cdots & A^{1N} \\ A^{21} & A^{22} & \cdots & A^{2N} \\ A^{31} & A^{32} & \cdots & A^{3N} \\ \vdots & \vdots & \ddots & \vdots \\ A^{N1} & A^{N2} & \cdots & A^{NN} \end{vmatrix} = \varepsilon_{i_1 \ldots j_N} A^{i_1 1} A^{i_2 2} \ldots A^{i_N N} \tag{7.2.2}$$

Likewise, if the elements of the matrix are written as $[A_j^i]$, its determinant is defined by

$$\det \boldsymbol{A} = \begin{vmatrix} A_1^1 & A_2^1 & \cdots & A_N^1 \\ A_1^2 & A_2^2 & \cdots & A_N^2 \\ A_1^3 & A_2^3 & \cdots & A_N^3 \\ \vdots & \vdots & \ddots & \vdots \\ A_1^N & A_2^N & \cdots & A_N^N \end{vmatrix} = \varepsilon_{i_1 \cdots j_N} A_1^{i_1} A_2^{i_2} \cdots A_N^{i_N} \tag{7.2.3}$$

A similar formula holds when the matrix is written $\boldsymbol{A} = [A_i^j]$. The generalized Kronecker delta can be written as the determinant of ordinary Kronecker deltas. For example, one can show, using eq. (7.2.3), that

$$\delta_{kl}^{ij} = \begin{vmatrix} \delta_k^i & \delta_l^i \\ \delta_k^j & \delta_l^j \end{vmatrix} = \delta_k^i \delta_l^j - \delta_l^i \delta_k^j \tag{7.2.4}$$

Eq. (7.2.4) is a special case of the general result

$$\delta_{j_1 \cdots j_K}^{i_1 \cdots i_K} = \begin{vmatrix} \delta_{j_1}^{i_1} & \delta_{j_2}^{i_1} & \cdots & \delta_{j_K}^{i_1} \\ \delta_{j_1}^{i_2} & \delta_{j_2}^{i_2} & \cdots & \delta_{j_K}^{i_2} \\ \vdots & \vdots & \ddots & \vdots \\ \delta_{j_1}^{i_K} & \delta_{j_2}^{i_K} & \cdots & \delta_{j_K}^{i_K} \end{vmatrix} \tag{7.2.5}$$

It is possible to use eq. (7.2.1) – eq. (7.2.3) to show that

$$\varepsilon_{j_1 \cdots j_N} \det \boldsymbol{A} = \varepsilon^{i_1 \cdots i_N} A_{i_1 j_1} \cdots A_{i_N j_N} \tag{7.2.6}$$

$$\varepsilon^{j_1 \cdots j_N} \det \boldsymbol{A} = \varepsilon_{i_1 \cdots i_N} A^{i_1 j_1} \cdots A^{i_N j_N} \tag{7.2.7}$$

and

$$\varepsilon_{j_1 \cdots j_N} \det \boldsymbol{A} = \varepsilon_{i_1 \cdots i_N} A_{j_1}^{i_1} \cdots A_{j_N}^{i_N} \tag{7.2.8}$$

It is possible to use eq. (7.2.6) – eq. (7.2.8) and eq. (7.1.10) with $N = K$ to show that

$$\det \boldsymbol{A} = \frac{1}{N!} \varepsilon^{j_1 \cdots j_N} \varepsilon^{i_1 \cdots i_N} A_{i_1 j_1} \cdots A_{i_N j_N} \tag{7.2.9}$$

$$\det \boldsymbol{A} = \frac{1}{N!} \varepsilon_{j_1 \cdots j_N} \varepsilon_{i_1 \cdots i_N} A^{i_1 j_1} \cdots A^{i_N j_N} \tag{7.2.10}$$

and

$$\det \boldsymbol{A} = \frac{1}{N!} \delta_{i_1 \cdots i_N}^{j_1 \cdots j_N} A_{j_1}^{i_1} \cdots A_{j_N}^{i_N} \tag{7.2.11}$$

eq. (7.2.9), eq. (7.2.10) and eq. (7.2.11) confirm the well-known result that a matrix and its transpose have the same determinant. We shall know that the transpose mentioned here is that of the matrix and not of a linear transformation. By use of this fact it follows that eq. (7.2.1) – eq. (7.2.3) could have been written as

$$\det \boldsymbol{A} = \varepsilon^{i_1 \cdots i_N} A_{1 i_1} \cdots A_{N i_N} \tag{7.2.12}$$

$$\det \boldsymbol{A} = \varepsilon_{i_1 \cdots i_N} A^{1 i_1} \cdots A^{N i_N} \tag{7.2.13}$$

and

$$\det \boldsymbol{A} = \varepsilon^{i_1 \cdots i_N} A^1_{i_1} \cdots A^N_{i_N} \tag{7.2.14}$$

A similar logic also yields

$$\varepsilon_{j_1 \cdots j_N} \det \boldsymbol{A} = \varepsilon^{i_1 \cdots i_N} A_{j_1 i_1} \cdots A_{j_N i_N} \tag{7.2.15}$$

$$\varepsilon^{j_1 \cdots j_N} \det \boldsymbol{A} = \varepsilon_{i_1 \cdots i_N} A^{j_1 i_1} \cdots A^{j_N i_N} \tag{7.2.16}$$

and

$$\varepsilon^{j_1 \cdots j_N} \det \boldsymbol{A} = \varepsilon^{i_1 \cdots i_N} A^{j_1}_{i_1} \cdots A^{j_N}_{i_N} \tag{7.2.17}$$

The cofactor of the element A^s_t in the matrix $\boldsymbol{A} = [A^i_j]$ is defined by

$$\mathrm{cof} A^s_t = \varepsilon_{i_1 i_2 \cdots i_N} A^{i_1}_1 \cdots A^{i_{t-1}}_{t-1} \delta^{i_t}_s A^{i_{t+1}}_{t+1} \cdots A^{i_N}_N \tag{7.2.18}$$

As an illustration of the application of eq. (7.2.18), let $N = 3$ and $s = t = 1$, then

$$\begin{aligned} \mathrm{cof} A^1_1 &= \varepsilon_{ijk} \delta^i_1 A^j_2 A^k_3 \\ &= \varepsilon_{1jk} A^j_2 A^k_3 \\ &= A^2_2 A^3_3 - A^3_2 A^2_3 \end{aligned}$$

Theorem 7.2.1:

$$\sum_{s=1}^{N} A^s_q \, \mathrm{cof} A^s_t = \delta^t_q \det \boldsymbol{A} \quad \text{and} \quad \sum_{t=1}^{N} A^q_t \, \mathrm{cof} A^s_t = \delta^q_s \det \boldsymbol{A} \tag{7.2.19}$$

Proof: It follows eq. (7.2.18) that

$$\begin{aligned} \sum_{s=1}^{N} A^s_q \, \mathrm{cof} A^s_t &= \varepsilon_{i_1 i_2 \cdots i_N} A^{i_1}_1 \cdots A^{i_{t-1}}_{t-1} A^s_q \delta^{i_t}_s A^{i_{t+1}}_{t+1} \cdots A^{i_N}_N \\ &= \varepsilon_{i_1 i_2 \cdots i_N} A^{i_1}_1 \cdots A^{i_{t-1}}_{t-1} A^{i_t}_q A^{i_{t+1}}_{t+1} \cdots A^{i_N}_N \\ &= \varepsilon_{i_1 i_2 \cdots i_N} A^{i_1}_1 \cdots A^{i_{t-1}}_{t-1} A^{i_t}_t A^{i_{t+1}}_{t+1} \cdots A^{i_N}_N \delta^t_q \\ &= \delta^t_q \det \boldsymbol{A} \end{aligned}$$

where the definition eq. (7.2.3) has been used. Eq. (7.2.19) follows by a similar argument.

Eq. (7.2.19) represents the classical Laplace expansion of a determinant.

7.3 The Matrix of a Linear Transformation

In this section we introduce the matrix of a linear transformation with respect to a basis and some of its properties are investigated.

If $\boldsymbol{A} \in L(V,U)$, $\{\boldsymbol{e}_1,\dots,\boldsymbol{e}_N\}$ is a basis for V, and $\{\boldsymbol{b}_1,\dots,\boldsymbol{b}_M\}$ is a basis for U, then we can characterize $\boldsymbol{A}$ by the system eq. (6.5.1):

$$\boldsymbol{A}\boldsymbol{e}_k = A_k^{\alpha}\boldsymbol{b}_{\alpha} \tag{7.3.1}$$

where the summation is in force with the Greek indices ranging from 1 to M. The matrix of $\boldsymbol{A}$ with respect to the bases $\{\boldsymbol{e}_1,\dots,\boldsymbol{e}_N\}$ and $\{\boldsymbol{b}_1,\dots,\boldsymbol{b}_M\}$, denoted by $\boldsymbol{M}(\boldsymbol{A},\boldsymbol{e}_k,\boldsymbol{b}_{\alpha})$ is

$$\boldsymbol{M}(\boldsymbol{A},\boldsymbol{e}_k,\boldsymbol{b}_{\alpha}) = \begin{bmatrix} A_1^1 & A_2^1 & \dots & A_N^1 \\ A_1^2 & A_2^2 & \dots & A_N^2 \\ \vdots & \vdots & \ddots & \vdots \\ A_1^M & A_2^M & \dots & A_N^M \end{bmatrix} \tag{7.3.2}$$

As the above argument indicates, the matrix of $\boldsymbol{A}$ depends upon the choice of basis for V and U. However, unless this point needs to be stressed, we shall often write $\boldsymbol{M}(\boldsymbol{A})$ for the matrix of $\boldsymbol{A}$ and the basis dependence is understood. We can always regard $\boldsymbol{M}$ as a function $M: L(U,V) \to U^{M\times N}$. It is a simple exercise to confirm that

$$\boldsymbol{M}(\lambda\boldsymbol{A}+\mu\boldsymbol{B}) = \lambda\boldsymbol{M}(\boldsymbol{A})+\mu\boldsymbol{M}(\boldsymbol{B}) \tag{7.3.3}$$

for all $\lambda,\mu \in C$ and $\boldsymbol{A},\boldsymbol{B} \in L(U,V)$. Thus $\boldsymbol{M}$ is a linear transformation. Since

$$\boldsymbol{M}(\boldsymbol{A}) = \boldsymbol{0}$$

Implies $\boldsymbol{A}=\boldsymbol{0}$, $\boldsymbol{M}$ is one-to-one. Since $L(U,V)$ and $U^{M\times N}$ have the same dimension, Theorem 6.1.10 tells us that $\boldsymbol{M}$ is an automorphism and, thus, $L(U,V)$ and $U^{M\times N}$ are isomorphic. The dependence of $\boldsymbol{M}(\boldsymbol{A},\boldsymbol{e}_k,\boldsymbol{b}_{\alpha})$ on the bases can be exhibited by use of the transformation rule eq. (6.5.5). In matrix form eq. (6.5.5) is

$$\boldsymbol{M}(\boldsymbol{A},\boldsymbol{e}_k,\boldsymbol{b}_{\alpha}) = \boldsymbol{S}\boldsymbol{M}(\boldsymbol{A},\hat{\boldsymbol{e}}_k,\hat{\boldsymbol{b}}_{\alpha})\boldsymbol{T}^{-1} \tag{7.3.4}$$

where $\boldsymbol{S}$ is the $M\times M$ matrix:

$$S=\begin{bmatrix} S_1^1 & S_2^1 & \dots & S_M^1 \\ S_1^2 & S_2^2 & \dots & S_M^2 \\ \vdots & \vdots & \ddots & \vdots \\ S_1^M & S_2^M & \dots & S_M^M \end{bmatrix} \tag{7.3.5}$$

and T is the $N \times N$ matrix:

$$T=\begin{bmatrix} T_1^1 & T_2^1 & \dots & T_N^1 \\ T_1^2 & T_2^2 & \dots & T_N^2 \\ \vdots & \vdots & \ddots & \vdots \\ T_1^N & T_2^N & \dots & T_N^N \end{bmatrix} \tag{7.3.6}$$

Of course, in constructing eq. (7.3.4) from eq. (6.5.5) we have used the fact expressed by eq. (5.7.20) and eq. (5.7.21) that the matrix T^{-1} has components $\hat{T}_k^j, j,k=1,\dots,N$.

If A is an endomorphism, the transformation formula eq. (7.3.4) becomes

$$M(A,e_k,e_q) = TM(A,\hat{e}_k,\hat{e}_q)T^{\ 1} \tag{7.3.7}$$

We shall use eq. (7.3.7) to motivate the concept of the determinant of an endomorphism. The determinant of $M(A,e_k,e_q)$, written as det $M(A,e_k,e_q)$ can be computed by eq. (7.2.3). It follows eq. (7.3.7) that

$$\begin{aligned}\det M(A,e_k,e_q) &= (\det T)\det(M(A,\hat{e}_k,\hat{e}_q))\det(T^{-1}) \\ &= \det M(A,\hat{e}_k,\hat{e}_q)\end{aligned}$$

Thus, we obtain the important result that det $M(A,e_k,e_q)$ is independent of the choice of basis for V. With this fact we define the determinant of an endomorphism $A \in L(V,V)$, written as det A, by

$$\det A = \det M(A,e_k,e_q) \tag{7.3.8}$$

From the above argument, we are assured that det A is a property of A alone.

Given a linear transformation $A \in L(V,U)$, the adjoint $A^* \in L(U,V)^*$ is defined by the component formula eq. (6.5.9). Consistent with eq. (7.3.1) and eq. (7.3.2), eq. (6.5.9) implies that

$$M(A^*,b^\alpha,e^k)=\begin{bmatrix} A^{*1}_{\ 1} & A^{*2}_{\ 1} & \dots & A^{*M}_{\ 1} \\ A^{*1}_{\ 2} & A^{*2}_{\ 2} & \dots & A^{*M}_{\ 2} \\ \vdots & \vdots & \ddots & \vdots \\ A^{*1}_{\ N} & A^{*2}_{\ N} & \dots & A^{*M}_{\ N} \end{bmatrix} \tag{7.3.9}$$

Notice that the matrix of A^* refers to the reciprocal bases $\{e^k\}$ and $\{b^\alpha\}$. If we now use eq. (6.5.10) and the definition eq. (7.3.2) we see that

$$\boldsymbol{M}(\boldsymbol{A}^*,\boldsymbol{b}^\alpha,\boldsymbol{e}^k)=\overline{\boldsymbol{M}(\boldsymbol{A},\boldsymbol{e}_k,\boldsymbol{b}_\alpha)}^{\mathrm{T}} \tag{7.3.10}$$

where the complex conjugate of a matrix is the matrix formed by taking the complex conjugate of each component of the given matrix. Eq. (7.3.10) gives a simple comparison of the component matrices of a linear transformation and its adjoint. If the vector spaces are real, eq. (7.3.10) reduces to

$$\boldsymbol{M}(\boldsymbol{A}^{\mathrm{T}})=\boldsymbol{M}(\boldsymbol{A})^{\mathrm{T}} \tag{7.3.11}$$

where the basis dependence is understood. For an endomorphism $\boldsymbol{A}\in L(V,V)$, we can use eq. (7.3.10) and eq. (7.3.8) to show that

$$\det \boldsymbol{A}^*=\overline{\det \boldsymbol{A}} \tag{7.3.12}$$

Given a $M\times N$ matrix $\boldsymbol{A}=[A_k^\alpha]$, there correspond M $1\times N$ row matrices and N $M\times 1$ column matrices. The row rank of the matrix $\boldsymbol{A}$ is equal to the number of linearly independent row matrices and the column rank of $\boldsymbol{A}$ is equal to the number of linearly independent column matrices. The following theorem shows a property of matrices that the row rank equals to the column rank. This common rank in turn is equal to the rank of the linear transformation whose matrix is $\boldsymbol{A}$.

Theorem 7.3.1: $\boldsymbol{A}=[A_k^\alpha]$ is an $M\times N$ matrix, the row rank of $\boldsymbol{A}$ equals the column rank of $\boldsymbol{A}$.

Proof: Let $\boldsymbol{A}\in L(V,U)$ and let $\boldsymbol{M}(\boldsymbol{A})=\boldsymbol{A}=[A_k^\alpha]$ with respect to bases $\{\boldsymbol{e}_1,\ldots,\boldsymbol{e}_N\}$ for V and $\{\boldsymbol{b}_1,\ldots,\boldsymbol{b}_M\}$ for U. We can define a linear transformation $B:U\to F^M$ by

$$B\boldsymbol{u}=(u^1,u^2,\ldots,u^M)$$

where $\boldsymbol{u}=u^\alpha\boldsymbol{b}_\alpha$. Observe that B is an isomorphism. The product BA is a linear transformation $V\to F^M$; further

$$B\boldsymbol{A}\boldsymbol{e}_k=B(A_k^\alpha\boldsymbol{b}_\alpha)=A_k^\alpha B\boldsymbol{b}_\alpha=(A_k^1,A_k^2,\ldots,A_k^M)$$

Therefore $B\boldsymbol{A}\boldsymbol{e}_k$ is an M-tuple whose elements are those of the k-th column matrix of $\boldsymbol{A}$. This means $\dim R(B\boldsymbol{A})=$ column rank of $\boldsymbol{A}$. Since B is an isomorphism, $B\boldsymbol{A}$ and $\boldsymbol{A}$ have the same rank and thus the column rank of $\boldsymbol{A}=\dim R(\boldsymbol{A})$. A similar argument applied to the adjoint mapping $\boldsymbol{A}^*$ shows that the row rank of $\boldsymbol{A}=\dim R(\boldsymbol{A}^*)$. If we now apply Theorem 6.4.4 we find the desired result.

Theorem 7.3.2: An endomorphism $\boldsymbol{A}$ is regular iff $\det \boldsymbol{A}\neq 0$.

Proof: If $\det \boldsymbol{A}\neq 0$, eq. (7.2.19) provides us with a formula for the direct calculation of $\boldsymbol{A}^{-1}$, so $\boldsymbol{A}$ is regular. If $\boldsymbol{A}$ is regular, then $\boldsymbol{A}^{-1}$ exists.

Note: We have defined $[A_k^\alpha]$, $[A_{\alpha k}]$, $[A^{\alpha k}]$ and $[A_\alpha^k]$ in chapter 6, but we mean the component matrix of the first kind unless otherwise specified.

7.4 Solution of Systems of Linear Equation

In this section we shall examine the problem system of M equations in N unknowns of the form

$$A_k^\alpha \boldsymbol{v}^k = \boldsymbol{u}^\alpha, \quad \alpha = 1, \ldots, M, \quad k = 1, \ldots, N \tag{7.4.1}$$

where the MN coefficient A_k^α and the M data $\boldsymbol{u}^\alpha$ are given. If we introduce bases $\{\boldsymbol{e}_1, \ldots, \boldsymbol{e}_N\}$ for a vector space V and $\{\boldsymbol{b}_1, \ldots, \boldsymbol{b}_M\}$ for a vector space U, then eq. (7.4.1) can be viewed as the component formula of a certain vector equation

$$\boldsymbol{A}\boldsymbol{v} = \boldsymbol{n} \tag{7.4.2}$$

which immediately yields the following theorem.

Theorem 7.4.1: Eq. (7.4.1) has a solution iff $\boldsymbol{u}$ is in $R(\boldsymbol{A})$.

Theorem 7.4.2: If eq. (7.4.1) has a solution, the solution is unique iff $\boldsymbol{A}$ is regular.

Given the system of eq. (7.4.1), the associated homogeneous system is the set of equations

$$A_k^\alpha \boldsymbol{v}^k = \boldsymbol{0} \tag{7.4.3}$$

Theorem 7.4.3: The set of solutions of the homogeneous system eq. (7.4.3) whose coefficient matrix is of rank R forms a vector space of dimension $N - R$.

Proof: Eq. (7.4.3) can be regarded as the component formula of the vector equation

$$\boldsymbol{A}\boldsymbol{v} = \boldsymbol{0}$$

which implies that $\boldsymbol{v}^k$ solves eq. (7.4.3) iff $\boldsymbol{v} \in K(\boldsymbol{A})$. By eq. (6.1.6) $\dim K(\boldsymbol{A}) = \dim V - \dim R(\boldsymbol{A}) = N - R$.

From Theorems 6.1.7 and eq. (7.4.3) we see that if there are fewer equations than unknowns (i. e. $M \times N$), the system eq. (7.4.3) always has a nonzero solution. This assertion is clear because $N - R \geqslant N - M > 0$, since $R(\boldsymbol{A})$ is a subspace of U and $M = \dim U$.

If in eq. (7.4.1) and eq. (7.4.3) $M = N$, the system eq. (7.4.1) has a solution for all $\boldsymbol{u}^k$ iff eq. (7.4.3) has the trivial solution $\boldsymbol{v}^1 = \boldsymbol{v}^2 = \ldots = \boldsymbol{v}^N = \boldsymbol{0}$

only. For, in this circumstance, $\boldsymbol{A}$ is regular and thus invertible. This means $\det[A_k^\alpha] \neq \mathbf{0}$, and we can use eq. (7.1.19) and write the solution of eq. (7.4.1) in the form

$$v^j = \frac{1}{\det[A_k^\alpha]_{\alpha=1}} \sum_{\alpha=1}^{N} (\mathrm{cof}A_j^\alpha)\, \boldsymbol{u}^\alpha \tag{7.4.4}$$

which is the classical Cramer's rule.

7.5 Special Matrices

In this section we will introduce some special kind of matrices. All of these matrices are very important in studies although they are just shown simply.

Nonnegative Definite Matrices and Positive Definite Matrices

Nonnegative Definite Matrices

A symmetric matrix $\boldsymbol{A}$ such that any quadratic form involving the matrix $\boldsymbol{x}^{\mathrm{T}}\boldsymbol{A}\boldsymbol{x}$ is nonnegative is called a *Nonnegative Definite Matrix*.

Positive Definite Matrices

An important class of nonnegative definite matrices are those that satisfy strict inequalities $\boldsymbol{x}^{\mathrm{T}}\boldsymbol{A}\boldsymbol{x}$ which are called *Positive Definite Matrix*.

A symmetric matrix $\boldsymbol{A}$ is called a nonnegative definite matrix (positive definite matrix) if, for any vector $\boldsymbol{x} = \mathbf{0}$, the quadratic form is nonnegative (positive); that is,

$$\boldsymbol{x}^{\mathrm{T}}\boldsymbol{A}\boldsymbol{x} \geqslant \mathbf{0}\,(\boldsymbol{x}^{\mathrm{T}}\boldsymbol{A}\boldsymbol{x} > \mathbf{0}) \tag{7.5.1}$$

We denote the fact that $\boldsymbol{A}$ is positive definite by

$$\boldsymbol{A} \geqslant \mathbf{0}\,(\boldsymbol{A} > \mathbf{0}) \tag{7.5.2}$$

The properties of nonnegative definite matrices hold also for positive definite matrices.

Gramian Matrices

A (real) matrix $\boldsymbol{A}$ such that for some (real) matrix $\boldsymbol{B}$, $\boldsymbol{A} = \boldsymbol{B}^{\mathrm{T}}\boldsymbol{B}$, is called a *Gramian Matrix*. Any nonnegative definite matrix is Gramian.

Some interesting properties of a Gramian matrix $\boldsymbol{X}^{\mathrm{T}}\boldsymbol{X}$ are:

- $\boldsymbol{X}^{\mathrm{T}}\boldsymbol{X}$ is symmetric.
- $\boldsymbol{X}^{\mathrm{T}}\boldsymbol{X}$ is of full rank iff $\boldsymbol{X}$ is of full column rank or, more generally, rank $(\boldsymbol{X}^{\mathrm{T}}\boldsymbol{X}) = \mathrm{rank}(\boldsymbol{X})$.
- $\boldsymbol{X}^{\mathrm{T}}\boldsymbol{X}$ is nonnegative definite and is positive definite iff $\boldsymbol{X}$ is of full column

rank.

- $X^{\mathrm{T}}X = \mathbf{0} \Rightarrow X = \mathbf{0}$.

Idempotent and Projection Matrices

Idempotent matrix: An important class of matrices are those that, like the identity, have the property that raising them to a power leaves them unchanged. A matrix $\boldsymbol{A}$ such that

$$\boldsymbol{A}\boldsymbol{A} = \boldsymbol{A}$$

is called an *Idempotent Matrix*. An idempotent matrix is square, and it is either singular or the identity matrix.

Projection matrix: An idempotent matrix that is symmetric is called a *Projection Matrix*.

For a given vector space V, a symmetric idempotent matrix $\boldsymbol{A}$ whose columns span V is said to be a projection matrix onto V; in other words, a matrix $\boldsymbol{A}$ is a projection matrix onto span($\boldsymbol{A}$) iff $\boldsymbol{A}$ is symmetric and idempotent. Because a projection matrix is idempotent, the matrix projects any of its columns onto itself, and of course it projects the full matrix onto itself: $\boldsymbol{A}\boldsymbol{A} = \boldsymbol{A}$.

Smoothing Matrices

The hat matrix, either from a full rank $\boldsymbol{X}$ or formed by a generalized inverse, smooths the vector $\boldsymbol{y}$ onto the hyperplane defined by the column space of $\boldsymbol{X}$. It is therefore a *Smoothing Matrix*.

Stochastic Matrices

Stochastic matrices are particularly interesting because of their use in defining a discrete homogeneous Markov chain. In that application, a stochastic matrix plays key roles. In Markov chain models, the stochastic matrix is a probability transition matrix from a distribution at time t, to the distribution at time $t+1$,

$$\pi_{t+1} = \boldsymbol{P}_{\pi_t}$$

A nonnegative matrix $\boldsymbol{P}$ such that

$$\boldsymbol{P}\mathbf{1} = \mathbf{1}$$

is called a *Stochastic Matrix*. It is also clear that if $\boldsymbol{P}$ is a stochastic matrix, then $\|\boldsymbol{P}\|_\infty = 1$.

If $\boldsymbol{P}$ is a stochastic matrix such that

$$\mathbf{1}^{\mathrm{T}}\boldsymbol{P} = \mathbf{1}^{\mathrm{T}}$$

it is called a *Doubly Stochastic Matrix*, and then $\|\boldsymbol{P}\|_1 = 1$ and $\|\boldsymbol{P}\|_\infty = 1$.

Leslie Matrices

Leslie Matrix is often used in population studies, after P. H. Leslie, who used it in models in demography. A Leslie matrix is a matrix of the form

$$\begin{bmatrix} \alpha_1 & \alpha_2 & \cdots & \alpha_{m-1} & \alpha_m \\ \sigma_1 & 0 & \cdots & 0 & 0 \\ 0 & \sigma_2 & \cdots & 0 & 0 \\ \vdots & \vdots & \vdots & \vdots & \vdots \\ 0 & 0 & \cdots & \sigma_{m-1} & 0 \end{bmatrix} \tag{7.5.3}$$

where all elements are nonnegative, and additionally $\sigma_i \leqslant 1$. Furthermore, a Leslie matrix has a single unique positive eigenvalue.

Helmert Matrices

The main use of *Helmert Matrices* in statistics is in defining contrasts in general linear models to compare the second level of a factor with the first level, the third level with the average of the first two, and so on.

A *Helmert Matrix* is a square orthogonal matrix that partitions sums of squares. For example, a partition of the sum $\sum_{i=1}^{n} y_i^2$ into orthogonal sums each involving $\bar{y}_k^2$ and $\sum_{i=1}^{k} (y_i - \bar{y}_k)^2$ is

$$\begin{cases} \bar{y}_i = (i(i+1))^{-1/2} \left(\sum\limits_{j=1}^{i+1} y_j - (i+1) y_{i+1} \right) & \textit{for } i = 1, \ldots n-1, \\ \bar{y}_n = n^{-1/2} \sum\limits_{j=1}^{n} y_j \end{cases}$$

Hadamard Matrices

Hadamard Matrices are in a wide range of applications, including experimental design, cryptology, and other areas of combinatorics.

An $n \times n$ matrix with $-1, 1$ entries whose determinant is $n^{n/2}$ is called a *Hadamard Matrix*. One row and one column of an $n \times n$ Hadamard matrix consist of all 1*s*; all $n-1$ other rows and columns consist of $n/2$ 1*s* and $n/2$ -1*s*. We often denote an $n \times n$ Hadamard matrix by $\boldsymbol{H}_n$, which is the same notation often used for a Helmert matrix, but in the case of Hadamard matrices, the matrix is not unique. All rows are orthogonal and so are all columns. The norm of each row or column is n, so $\boldsymbol{H}_n^{\mathrm{T}} \boldsymbol{H}_n = n\boldsymbol{I}$.

Toeplitz Matrices

Banded Toeplitz matrices arise frequently in time series studies.

If the elements of the matrix $\boldsymbol{A}$ are such that $a_{i,i+c_k} = d_{c_k}$, where d_{c_k} is constant for fixed c_k, then $\boldsymbol{A}$ is called a *Toeplitz Matrix*,

$$\begin{bmatrix} d_0 & d_1 & d_2 & \cdots & d_{n-1} \\ d_{-1} & d_0 & d_1 & \cdots & d_{n-2} \\ \vdots & \vdots & \vdots & \vdots & \vdots \\ d_{-n+2} & d_{-n+3} & d_{-n+4} & \cdots & d_1 \\ d_{-n+1} & d_{-n+2} & d_{-n+3} & \cdots & d_0 \end{bmatrix}$$

that is, a Toeplitz matrix is a matrix with constant codiagonals. A Toeplitz matrix may or may not be a band matrix and it may or may not be symmetric.

Hankel Matrices

One kind of Hankel Matrix occurs in the spectral analysis of time series.

A *Hankel Matrix* is an $n \times m$ matrix $\boldsymbol{H}(c,r)$ generated by an n-vector $\boldsymbol{c}$ and an *m-vector* $\boldsymbol{r}$ such that the (i,j) element is $n \times n$

$$\begin{cases} \boldsymbol{c}_{i+j-1} & \text{if } i+j-1 \leqslant n, \\ \boldsymbol{r}_{i+j-n} & \text{ortherwise} \end{cases}$$

Cauchy Matrices

Cauchy-type matrices often arise in the numerical solution of partial differential equations. For Cauchy matrices, the order of the number of computations for factorization or solutions of linear systems can be reduced from a power of three to a power of two. This is a very significant improvement for large matrices.

A *Cauchy Matrix* is an $n \times m$ matrix $\boldsymbol{C}(x,y,v,w)$ generated by *n-vectors* $\boldsymbol{x}$ and $\boldsymbol{v}$, *m-vectors* $\boldsymbol{y}$ and $\boldsymbol{w}$ of the form

$$\boldsymbol{C}(x,y,v,w) = \begin{bmatrix} \dfrac{v_1 w_1}{x_1 - y_1} & \cdots & \dfrac{v_1 w_m}{x_1 - y_m} \\ \vdots & \cdots & \vdots \\ \dfrac{v_n w_1}{x_n - y_1} & \cdots & \dfrac{v_n w_m}{x_n - y_m} \end{bmatrix} \tag{7.5.4}$$

***M*-Matrices**

M-matrices are very important in applications in physics and in the solution of systems of nonlinear differential equations.

A square matrix all of whose off-diagonal elements are nonpositive is called a Z-matrix. A Z-matrix that is positively stable is called an *M-matrix*. A real symmetric M-matrix is positive definite.

If $\boldsymbol{A}$ is a real M-matrix, then

- all principal minors of $\boldsymbol{A}$ are positive;
- all diagonal elements of $\boldsymbol{A}$ are positive;
- all diagonal elements of $\boldsymbol{L}$ and $\boldsymbol{U}$ in the $\boldsymbol{LU}$ decomposition of $\boldsymbol{A}$ are positive;
- for any i, $\sum_j a_{ij} \geqslant 0$; and
- $\boldsymbol{A}$ is nonsingular and $\boldsymbol{A} - \mathbf{1} \geqslant \mathbf{0}$.

Chapter 8

Spectral Decompositions

In this chapter we will introduce the decomposition which is one of the more advanced topics in the study of linear transformations. Essentially we shall consider the problem of analyzing an endomorphism by decomposing it into elementary parts.

8.1 Direct Sum of Endomorphisms

If $\boldsymbol{A}$ is an endomorphism of a vector space V, a subspace V_1 of V is said to be $\boldsymbol{A}$-invariant if $\boldsymbol{A}$ maps V_1 to V_1. The most obvious example of an $\boldsymbol{A}$-invariant subspace is the null space $K(\boldsymbol{A})$. Let $\boldsymbol{A}_1, \boldsymbol{A}_2, \ldots, \boldsymbol{A}_L$ be endomorphisms of V; then an endomorphism $\boldsymbol{A}$ is the direct sum of $\boldsymbol{A}_1, \boldsymbol{A}_2, \ldots, \boldsymbol{A}_L$ if

$$\boldsymbol{A} = \boldsymbol{A}_1 + \boldsymbol{A}_2 + \ldots + \boldsymbol{A}_L \tag{8.1.1}$$

and

$$\boldsymbol{A}_i\boldsymbol{A}_j = \boldsymbol{0} \quad i \neq j \tag{8.1.2}$$

Theorem 8.1.1: If $\boldsymbol{A} \in L(V, U)$ and $V = V_1 \oplus V_2 \oplus \ldots \oplus V_L$, where each subspace V_j is $\boldsymbol{A}$-invariant, then $\boldsymbol{A}$ has the direct sum decomposition eq. (8.1.1), where each $\boldsymbol{A}_i$ is given by

$$\boldsymbol{A}_i\boldsymbol{v}_i = \boldsymbol{A}\boldsymbol{v}_i \qquad (8.1.3)_1$$

for all $\boldsymbol{v}_i \in V_i$ and

$$\boldsymbol{A}_i\boldsymbol{v}_j = \boldsymbol{0} \qquad (8.1.3)_2$$

for all $v_j \in V_j, i \neq j$, for $i = 1, \ldots, L$. Thus the restriction of A to V_j coincides with that of A_j; further, each V_i is A_j-invariant, for all $j = 1, \ldots, L$.

Proof: Given the decomposition $V = V_1 \oplus V_2 \oplus \ldots \oplus V_L$, then $v \in V$ has the unique representation $v = v_1 + \ldots + v_L$, where each $v_i \in V_i$. By the converse of Theorem 6.3.4, there exist L projections $P_1, \ldots, P_L$ [eq. 6.5.18] and also $v_j \in P_j(V)$. Let $A_j = AP_j$, V_j be A-invariant, and $AP_j = P_jA$, therefore, if $i \neq j$,

$$A_iA_j = AP_iAP_j = AAP_iP_j = 0$$

where eq. (6.3.17) has been used. Also

$$\begin{aligned} A_1 + A_2 + \ldots + A_L &= AP_1 + AP_2 + \ldots + AP_L \\ &= A(P_1 + P_2 + \ldots + P_L) \\ &= A \end{aligned}$$

where eq. (6.3.18) has been used.

When the assumptions of the preceding theorem are satisfied, the endomorphism A is said to be reduced by the subspaces $V_1, \ldots, V_L$. An important result of this circumstance is contained in the following theorem.

Theorem 8.1.2: Under the conditions of Theorem 8.1.1, the determinant of the endomorphism A is given by

$$\det A = \det A_1 \det A_2 \ldots \det A_L \tag{8.1.4}$$

where A_k denotes the restriction of A to V_k for all $k = 1, \ldots, L$.

The proof of this theorem is left as an exercise to the reader.

8.2 Eigenvectors and Eigenvalues

Given an endomorphism $A \in L(V, V)$, the problem of finding a direct sum decomposition of A is closely related to the study of the spectral properties of A. This concept is central in the discussion of eigenvalue problems.

Definition: A scalar λ is an eigenvalue of $A \in L(V, V)$ if there exists a nonzero vector $v \in V$ such that

$$Av = \lambda v \tag{8.2.1}$$

The vector v in eq. (8.2.1) is called an eigenvector of A corresponding to the eigenvalue λ.

The set of all eigenvalues of A is the spectrum of A, denoted by $\sigma(A)$. For any $\lambda \in \sigma(A)$, the set

$$V(\lambda) = \{\boldsymbol{v} \in V | \boldsymbol{A}\boldsymbol{v} = \lambda \boldsymbol{v}\}$$

is a subspace of V, called the eigenspace or characteristic subspace corresponding to λ. The geometric multiplicity of λ is the dimension of $V(\lambda)$.

Theorem 8.2.1: Given any $\lambda \in \sigma(\boldsymbol{A})$, the corresponding eigenspace $V(\lambda)$ is $\boldsymbol{A}$-invariant.

The proof of this theorem involves an elementary use of eq. (8.2.1). Given any $\lambda \in \sigma(\boldsymbol{A})$, the restriction of $\boldsymbol{v}$ to $V(\lambda)$, denoted by $\boldsymbol{A}_{V(\lambda)}$, has the property that

$$\boldsymbol{A}_{V(\lambda)}\boldsymbol{u} = \lambda \boldsymbol{u} \tag{8.2.2}$$

for all $\boldsymbol{u}$ in $V(\lambda)$. Geometrically, $\boldsymbol{A}_{V(\lambda)}$ simply amplifies $\boldsymbol{u}$ by the factor λ. Such linear transformations are often called dilatations.

Theorem 8.2.2: $\boldsymbol{A} \in L(V,V)$ is not regular iff 0 is an eigenvalue of $\boldsymbol{A}$; further, the corresponding eigenspace is $K(\boldsymbol{A})$.

Note that eq. (8.2.1) can be written as $(\boldsymbol{A} - \lambda \boldsymbol{I})\boldsymbol{v} = \boldsymbol{0}$, which shows that

$$V(\lambda) = K(\boldsymbol{A} - \lambda \boldsymbol{I}) \tag{8.2.3}$$

Eq. (8.2.3) implies that λ is an eigenvalue of $\boldsymbol{A}$ iff $\boldsymbol{A} - \lambda \boldsymbol{I}$ is singular. Therefore, by Theorem 7.3.2,

$$\lambda \in \sigma(\boldsymbol{A}) \Leftrightarrow \det(\boldsymbol{A} - \lambda \boldsymbol{I}) = 0 \tag{8.2.4}$$

The polynomial $f(\lambda)$ of degree $N = \dim V$ defined by

$$f(\lambda) = \det(\boldsymbol{A} - \lambda \boldsymbol{I}) \tag{8.2.5}$$

is called the characteristic polynomial of $\boldsymbol{A}$. Eq. (8.2.5) shows that the eigenvalues of $\boldsymbol{A}$ are roots of the characteristic equation

$$f(\lambda) = 0 \tag{8.2.6}$$

8.3 The Characteristic Polynomial

In the last section we found that the eigenvalues of $\boldsymbol{A} \in L(V,V)$ are the roots of the characteristic polynomial

$$f(\lambda) = 0 \tag{8.3.1}$$

where

$$f(\lambda) = \det(\boldsymbol{A} - \lambda \boldsymbol{I}) \tag{8.3.2}$$

If $\{\boldsymbol{e}_1, \ldots, \boldsymbol{e}_N\}$ is a basis for V, then by eq. (6.5.6)

$$\boldsymbol{A}\boldsymbol{e}_k = A^j_k \boldsymbol{e}_j \tag{8.3.3}$$

Therefore, by eq. (8.3.2) and eq. (7.2.11)

$$f(\lambda) = \frac{1}{N!}\delta^{j_1 \cdots j_N}_{i_1 \cdots i_N}(A^{i_1}_{j_1} - \lambda\delta^{i_1}_{j_1}) \ldots (A^{i_N}_{j_N} - \lambda\delta^{i_N}_{j_N}) \tag{8.3.4}$$

If eq. (8.3.4) is expanded and we use eq. (7.1.11), the result is

$$f(\lambda) = (-\lambda)^N + \mu_1(-\lambda)^{N-1} + \ldots + \mu_{N-1}(-\lambda) + \mu_N \tag{8.3.5}$$

where

$$\mu_j = \frac{1}{j!}\delta^{q_1, \ldots, q_j}_{i_1, \ldots, i_j} A^{i_1}_{q_1} \ldots A^{i_j}_{q_j} \tag{8.3.6}$$

Since $f(\lambda)$ is defined by eq. (8.3.2), the coefficients $\mu_j, j = 1, \ldots, N$ are independent of the choice of basis for V. These coefficients are called the fundamental invariants of $\boldsymbol{A}$. Eq. (8.3.6) specializes

$$\mu_1 = \mathrm{tr}\boldsymbol{A}, \mu_2 = \frac{1}{2}\{(\mathrm{tr}\boldsymbol{A})^2 - \mathrm{tr}\boldsymbol{A}^2\}, \text{and } \mu_N = \det\boldsymbol{A} \tag{8.3.7}$$

where eq. (7.2.4) has been used to obtain eq. (8.3.7). Since $f(\lambda)$ is a N-th degree polynomial, it can be factored into the form

$$f(\lambda) = (\lambda_1 - \lambda)^{d_1}(\lambda_2 - \lambda)^{d_2} \ldots (\lambda_L - \lambda)^{d_L} \tag{8.3.8}$$

where $\lambda_1, \ldots, \lambda_L$ are the distinct foots of $f(\lambda) = 0$ and $d_1, \ldots, d_L$ are positive integers which must satisfy $\sum_{j=1}^{L} d_j = N$. It is in writing eq. (8.3.8) that we have made use of the assumption that the scalar field is complex. If the scalar field is real, the polynomial eq. (8.3.5), generally, cannot be factored.

Definition: In general, a scalar field is said to be algebraically closed if every polynomial equation has at least one root in the field, or equivalently, if every polynomial, such as $f(\lambda)$, can be factored into the form eq. (8.3.8).

Theorem 8.3.1: The complex field is algebraically closed.

Note: The real field is not algebraically closed. For example, if λ is real the polynomial equation $f(\lambda) = \lambda^2 + 1 = 0$ has no real roots. By allowing the scalar fields to be complex, we are assured that every endomorphism has at least one eigenvector. In the expression eq. (8.3.8), the integer d_j is called the algebraic multiplicity of the eigenvalue λ_j. It is possible to prove that the algebraic multiplicity of an eigenvalue is not less than the geometric multiplicity of the same eigenvalue.

An expression for the invariant μ_j can be obtained in terms of the eigenvalues if eq. (8.3.8) is expanded and the results are compared to eq. (8.3.5). For example,

$$\mu_1 = \mathrm{tr}\boldsymbol{A} = d_1\lambda_1 + d_2\lambda_2 + \ldots + d_L\lambda_L$$
$$\mu_N = \det\boldsymbol{A} = \lambda_1^{d_1}\lambda_2^{d_2}\ldots\lambda_L^{d_L} \tag{8.3.9}$$

Here we introduce certain ideas associated with polynomials of endomorphisms. If $\boldsymbol{A} \in L(V, V)$, $\boldsymbol{A}^2$ is defined by

$$\boldsymbol{A}^2 = \boldsymbol{A}\boldsymbol{A}$$

Similarly, we define by induction, starting from

$$\boldsymbol{A}^0 = \boldsymbol{I}$$

and in general

$$\boldsymbol{A}^k = \boldsymbol{A}\boldsymbol{A}^{k-1} = \boldsymbol{A}^{k-1}\boldsymbol{A}$$

where k is any integer greater than one. If k and l are positive integers, it is easily established by induction that

$$\boldsymbol{A}^k\boldsymbol{A}^l = \boldsymbol{A}^l\boldsymbol{A}^k = \boldsymbol{A}^{l+k} \tag{8.3.10}$$

thus $\boldsymbol{A}^k$ and $\boldsymbol{A}^l$ commute. A polynomial in $\boldsymbol{A}$ is an endomorphism of the form

$$g(\boldsymbol{A}) = \alpha_0\boldsymbol{A}^M + \alpha_1\boldsymbol{A}^{M-1} + \ldots + \alpha_{M-1}\boldsymbol{A} + \alpha_M\boldsymbol{I} \tag{8.3.11}$$

where M is a positive integer and $\alpha_0, \ldots, \alpha_M$ are scalars. Such polynomials have certain of the properties of polynomials of scalars. For example, if a scalar polynomial

$$g(t) = \alpha_0 t^M + \alpha_1 t^{M-1} + \ldots + \alpha_{M-1}t + \alpha_M$$

can be factored into

$$g(t) = \alpha_0(t - \eta_1)(t - \eta_2)\ldots(t - \eta_M)$$

then the polynomial eq. (8.3.11) can be factored into

$$g(\boldsymbol{A}) = \alpha_0(\boldsymbol{A} - \eta_1\boldsymbol{I})(\boldsymbol{A} - \eta_2\boldsymbol{I})\ldots(\boldsymbol{A} - \eta_M\boldsymbol{I}) \tag{8.3.12}$$

The order of the factors in eq. (8.3.12) is not important since, as a result of eq. (8.3.10), the factors commute. Notice, however, that the product of two nonzero endomorphisms can be zero. Thus the formula

$$g_1(\boldsymbol{A})g_2(\boldsymbol{A}) = 0 \tag{8.3.13}$$

for two polynomials g_1 and g_2 generally does not imply one of the factors is zero. For example, any projection $\boldsymbol{P}$ satisfies the equation $\boldsymbol{P}(\boldsymbol{P} - \boldsymbol{I}) = \boldsymbol{0}$, but generally $\boldsymbol{P}$ and $\boldsymbol{P} - \boldsymbol{I}$ are both nonzero.

Theorem 8.3.2 (Cayley-Hamilton): If $f(\lambda) = (-\lambda)^N + \mu_1(-\lambda)^{N-1} + \ldots + \mu_{N-1}(-\lambda) + \mu_N$ is the characteristic polynomial for an endomorphism $\boldsymbol{A}$, then

$$f(\boldsymbol{A}) = (-\boldsymbol{A})^N + \mu_1(-\boldsymbol{A})^{N-1} + \ldots + \mu_{N-1}(-\boldsymbol{A}) + \mu_N\boldsymbol{I} = \boldsymbol{0} \tag{8.3.14}$$

Proof: If $\mathrm{adj}(\boldsymbol{A}-\lambda\boldsymbol{I})$ is the endomorphism whose matrix is $\mathrm{adj}[A_q^p-\lambda\delta_q^p]$, where $[A_q^p]=\boldsymbol{M}(\boldsymbol{A})$, then by eq. (8.3.2)

$$(\mathrm{adj}(\boldsymbol{A}-\lambda\boldsymbol{I}))(\boldsymbol{A}-\lambda\boldsymbol{I})=f(\lambda)\boldsymbol{I} \tag{8.3.15}$$

by eq. (7.2.18) it follows that $\mathrm{adj}(\boldsymbol{A}-\lambda\boldsymbol{I})$ is a polynomial of degree $N-1$ in λ. Therefore

$$\mathrm{adj}(\boldsymbol{A}-\lambda\boldsymbol{I})=\boldsymbol{B}_0(-\lambda)^{N-1}+\boldsymbol{B}_1(-\lambda)^{N-2}+\ldots+\boldsymbol{B}_{N-2}(-\lambda)+\boldsymbol{B}_{N-1} \tag{8.3.16}$$

where $\boldsymbol{B}_0,\ldots,\boldsymbol{B}_{N-1}$ are endomorphisms determined by $\boldsymbol{A}$. If we now substitute eq. (8.3.16) and eq. (8.3.15) into eq. (8.3.14) and require the result to hold for all λ, we find

$$\begin{aligned}
&\boldsymbol{B}_0=\boldsymbol{I}\\
&\boldsymbol{B}_0\boldsymbol{A}+\boldsymbol{B}_1=\mu_1\boldsymbol{I}\\
&\boldsymbol{B}_1\boldsymbol{A}+\boldsymbol{B}_2=\mu_2\boldsymbol{I}\\
&\quad\vdots\\
&\boldsymbol{B}_{N-2}\boldsymbol{A}+\boldsymbol{B}_{N-1}=\mu_{N-1}\boldsymbol{I}\\
&\boldsymbol{B}_{N-1}\boldsymbol{A}=\mu_N\boldsymbol{I}
\end{aligned} \tag{8.3.17}$$

Now we multiply eq. $(8.3.17)_1$ by $(-\boldsymbol{A})^N$, eq. $(8.3.17)_2$ by $(-\boldsymbol{A})^{N-1}$, eq. $(8.3.17)_3$ by $(-\boldsymbol{A})^{N-2}$, ..., eq. $(8.3.17)_k$ by $(-\boldsymbol{A})^{N-k+1}$, etc., and add the resulting N equations, to find

$$(-\boldsymbol{A})^N+\mu_1(-\boldsymbol{A})^{N-1}+\ldots+\mu_{N-1}(-\boldsymbol{A})+\mu_N\boldsymbol{I}=\boldsymbol{0} \tag{8.3.18}$$

which is the desired result.

8.4 Spectral Decomposition for Hermitian Endomorphisms

In this section we will consider an important problem in mathematics to find a basis for a vector space V. Rather than consider this problem in general, we specialize here the linear transformation $\boldsymbol{A}$ where $\boldsymbol{M}(\boldsymbol{A})$ is Hermitian and show that every Hermitian endomorphism has a matrix which takes on the diagonal form.

First, we prove a general theorem for arbitrary endomorphisms about the linear independence of eigenvectors.

Theorem 8.4.1: If $\{\lambda_1, \ldots, \lambda_L\}$ are distinct eigenvalues of $\boldsymbol{A} \in L(V, V)$ and if $\boldsymbol{u}_1, \ldots, \boldsymbol{u}_L$ are eigenvectors corresponding to them, then $\{\boldsymbol{u}_1, \ldots, \boldsymbol{u}_L\}$ form a linearly independent set.

Proof: If $\{\boldsymbol{u}_1, \ldots, \boldsymbol{u}_L\}$ is not linearly independent, we choose a maximal, linearly independent subset, say $\{\boldsymbol{u}_1, \ldots, \boldsymbol{u}_s\}$, from the set $\{\boldsymbol{u}_1, \ldots, \boldsymbol{u}_L\}$; then the remaining vectors can be expressed uniquely as linear combinations of $\{\boldsymbol{u}_1, \ldots, \boldsymbol{u}_S\}$, say

$$\boldsymbol{u}_{S+1} = \alpha_1 \boldsymbol{u}_1 + \ldots + \alpha_S \boldsymbol{u}_S \tag{8.4.1}$$

where $\alpha_1, \ldots, \alpha_S$ are not all zero and unique, because $\{\boldsymbol{u}_1, \ldots, \boldsymbol{u}_S\}$ is linearly independent. Applying $\boldsymbol{A}$ to eq. (8.4.1) yields

$$\lambda_{S+1} \boldsymbol{u}_{S+1} = (\alpha_1 \lambda_1) \boldsymbol{u}_1 + \ldots + (\alpha_S \lambda_S) \boldsymbol{u}_S \tag{8.4.2}$$

Now, if $\lambda_{S+1} = 0$, then $\lambda_1, \ldots, \lambda_L$ are nonzero because the eigenvalues are distinct and eq. (8.4.2) contradicts the linear independence of $\{\boldsymbol{u}_1, \ldots, \boldsymbol{u}_S\}$; on the other hand, if $\lambda_{S+1} \neq 0$, then we can divide eq. (8.4.2) by λ_{S+1}, obtaining another expression of $\boldsymbol{u}_{S+1}$ as a linear combination of $\{\boldsymbol{u}_1, \ldots, \boldsymbol{u}_S\}$ contradicting the uniqueness of the coefficients $\alpha_1, \ldots, \alpha_S$. Hence in any case the maximal linearly independent subset cannot be a proper subset of $\{\boldsymbol{u}_1, \ldots, \boldsymbol{u}_L\}$; thus $\{\boldsymbol{u}_1, \ldots, \boldsymbol{u}_L\}$ is linearly independent.

We see that if the geometric multiplicity is equal to the algebraic multiplicity for each eigenvalue of $\boldsymbol{A}$, then the vector space V admits the direct sum representation

$$V = V(\lambda_1) \oplus V(\lambda_2) \oplus \ldots \oplus V(\lambda_L)$$

where $\lambda_1, \ldots, \lambda_L$ are the distinct eigenvalues of $\boldsymbol{A}$. The reason for this representation is obvious, since the right-hand side of the above equation is a subspace having the same dimension as V; thus that subspace is equal to V. Whenever the representation holds, we can always choose a basis of V formed by bases of the subspaces $V(\lambda_1), \ldots, V(\lambda_L)$. Then this basis consisting entirely of eigenvectors of V becomes a diagonal matrix, namely,

$$
M(A) = \begin{bmatrix}
\lambda_1 & & & & & & & & & & & & \\
& \cdot & & & & & & & & & & & \\
& & \cdot & & & & & & & & & & \\
& & & \cdot & & & & & & & & & \\
& & & & \lambda_1 & & & & & & 0 & & \\
& & & & & \lambda_2 & & & & & & & \\
& & & & & & \cdot & & & & & & \\
& & & & & & & \cdot & & & & & \\
& & & & & & & & \cdot & & & & \\
& & & & & & & & & \lambda_2 & & & \\
& & & & & & & & & & \cdot & & \\
& & & & & & & & & & & \cdot & \\
& & & & & & & & & & & & \cdot \\
& & & & & & & & & & & & & \lambda_L \\
& & & & 0 & & & & & & & & & & \cdot \\
& & & & & & & & & & & & & & & \cdot \\
& & & & & & & & & & & & & & & & \cdot \\
& & & & & & & & & & & & & & & & & \lambda_L
\end{bmatrix}
\tag{8.4.3$_1$}
$$

where each λ_K is repeated d_K times, d_K being the algebraic as well as the geometric multiplicity of λ_K. Of course, the representation of V by direct sum of eigenspaces of $\boldsymbol{A}$ is possible if $\boldsymbol{A}$ has $N = \dim V$ distinct eigenvalues. In this case the matrix of $\boldsymbol{A}$ taken with respect to a basis of eigenvectors has the diagonal form

$$
M(A) = \begin{bmatrix}
\lambda_1 & & & & & & \\
& \lambda_2 & & & & & \\
& & \lambda_3 & & & & \\
& & & \cdot & & & \\
& & & & \cdot & & \\
& & & & & \cdot & \\
& & & & & & \lambda_N
\end{bmatrix}
\tag{8.4.3$_2$}
$$

If the eigenvalues of $\boldsymbol{v}$ are not all distinct, then in general the geometric multiplicity of an eigenvalue may be less than the algebraic multiplicity. Whenever the two multiplicities are different for at least one eigenvalue of $\boldsymbol{A}$, it is no longer possible to find any basis in which the matrix of $\boldsymbol{A}$ is diagonal. However, if V is an inner product space and if $\boldsymbol{A}$ is Hermitian, then a diagonal matrix of $\boldsymbol{A}$ can always be found; we shall now investigate this problem.

If $\boldsymbol{u}$ and $\boldsymbol{v}$ are arbitrary vectors in V, the adjoint $\boldsymbol{A}^*$ of $\boldsymbol{A} \in L(V,V)$ is defined by

$$\boldsymbol{Au} \cdot \boldsymbol{v} = \boldsymbol{u} \cdot \boldsymbol{A}^* \boldsymbol{v} \tag{8.4.4}$$

As usual, if the matrix of $\boldsymbol{A}$ refers to a basis $\{\boldsymbol{e}_k\}$, then the matrix of $\boldsymbol{A}^*$ refers to the reciprocal basis $\{\boldsymbol{e}^k\}$

$$\boldsymbol{M}(\boldsymbol{A}^*)\overline{\boldsymbol{M}(\boldsymbol{A})}^{\mathrm{T}} \tag{8.4.5}$$

where the over bar indicates the complex conjugate as usual. If $\boldsymbol{A}$ Hermitian, i. e. if $\boldsymbol{A} = \boldsymbol{A}^*$, then eq. (8.4.4) reduces to

$$\boldsymbol{Au} \cdot \boldsymbol{v} = \boldsymbol{u} \cdot \boldsymbol{Av}$$

for all $\boldsymbol{u}, \boldsymbol{v} \in V$.

Theorem 8.4.2: The eigenvalues of a Hermitian endomorphism are all real.

Proof: Let $\boldsymbol{A} \in L(V,V)$ be Hermitian. Since $\boldsymbol{Au} = \lambda \boldsymbol{u}$ for any eigenvalue λ, we have

$$\lambda = \frac{\boldsymbol{Au} \cdot \boldsymbol{u}}{\boldsymbol{u} \cdot \boldsymbol{u}} \tag{8.4.6}$$

Therefore we must show that $\boldsymbol{Au} \cdot \boldsymbol{u}$ is real or, equivalently, we must show $\boldsymbol{Au} \cdot \boldsymbol{u} = \overline{\boldsymbol{Au} \cdot \boldsymbol{u}}$. By eq. (8.4.6)

$$\boldsymbol{Au} \cdot \boldsymbol{u} = \boldsymbol{u} \cdot \boldsymbol{Au} = \overline{\boldsymbol{Au} \cdot \boldsymbol{u}} \tag{8.4.7}$$

where the rule $\boldsymbol{u} \cdot \boldsymbol{v} = \overline{\boldsymbol{v} \cdot \boldsymbol{u}}$ has been used. Eq. (8.4.7) yields the desired result.

Theorem 8.4.3: If $\boldsymbol{A}$ is Hermitian and if V_1 is an $\boldsymbol{A}$-invariant subspace of V, then $V_1^{\perp}$ is also $\boldsymbol{A}$-invariant.

Proof: If $\boldsymbol{v} \in V_1$ and $\boldsymbol{u} \in V_1^{\perp}$, then $\boldsymbol{Av} \cdot \boldsymbol{u} = \boldsymbol{0}$ because $\boldsymbol{Av} \in V_1$. But since $\boldsymbol{A}$ is Hermitian, $\boldsymbol{Av} \cdot \boldsymbol{u} = \boldsymbol{v} \cdot \boldsymbol{Au} = \boldsymbol{0}$. Therefore, $\boldsymbol{Au} \in V_1^{\perp}$, which proves the theorem.

Theorem 8.4.4: If $\boldsymbol{A}$ is Hermitian, the algebraic multiplicity of each eigenvalue equals the geometric multiplicity.

Proof: Let $V(\lambda_0)$ be the characteristic subspace associated with an eigenvalue λ_0. Then the geometric multiplicity of λ_0 is $M = \dim V(\lambda_0)$. By Theorems 5.6.4 and 8.4.3

$$V = V(\lambda_0) \oplus V(\lambda_0)^{\perp} \tag{8.4.8}$$

where $V(\lambda_0)$ and $V(\lambda_0)^{\perp}$ are $\boldsymbol{A}$-invariant. By Theorem 8.1.1

$$A = A_1 + A_2$$

and by Theorem 6.3.4

$$I = P_1 + P_2$$

where P_1 projects V onto $V(\lambda_0)$, P_2 projects V onto $V(\lambda_0)^{\perp}$, $A_1 = AP_1$, and $A_2 = AP_2$. By Theorem 6.4.10, P_1 and P_2 are Hermitian and they also commute with A. Indeed, for any $v \in V$, $P_1 v \in V(\lambda_0)$ and $P_2 v \in V(\lambda_0)^{\perp}$, and thus $AP_1 v \in V(\lambda_0)$ and $AP_2 v \in V(\lambda_0)^{\perp}$. But since

$$Av = A(P_1 + P_2)v = AP_1 v + AP_2 v$$

we see that $AP_1 v$ is the $V(\lambda_0)$ component of Av and $AP_2 v$ is the $V(\lambda_0)^{\perp}$ component of Av. Therefore

$$P_1 Av = AP_1 v, \quad P_2 Av = AP_2 v$$

for all $v \in V$, or, equivalently

$$P_1 A = AP_1, \quad P_2 A = AP_2$$

Together with the fact that P_1, and P_2 are Hermitian, these equations imply that A_1 and A_2 are also Hermitian. Further, A is reduced by the subspaces $V(\lambda_0)$ and $V(\lambda_0)^{\perp}$, since

$$A_1 A_2 = AP_1 AP_2 = A^2 P_1 P_2$$

Thus if we select a basis $\{e_1, \ldots, e_N\}$ such that $\{e_1, \ldots, e_N\}$ span $V(\lambda_0)$ and $\{e_{M+1}, \ldots, e_N\}$ span $V(\lambda_0)^{\perp}$, then the matrix of A to $\{e_k\}$ takes the form

$$M(A) = \begin{bmatrix} A_1^1 & \cdots & A_M^1 & & & \\ \vdots & & \vdots & & \mathbf{0} & \\ A_1^M & \cdots & A_M^M & & & \\ & & & A_{M+1}^{M+1} & \cdots & A_N^{M+1} \\ & \mathbf{0} & & \vdots & & \vdots \\ & & & A_{M+1}^N & \cdots & A_N^N \end{bmatrix}$$

and the matrices of A_1 and A_2 are

$$M(A_1) = \begin{bmatrix} A_1^1 & \cdots & A_M^1 & \\ \vdots & & \vdots & \mathbf{0} \\ A_1^M & \cdots & A_M^M & \\ & \mathbf{0} & & \mathbf{0} \end{bmatrix}$$

$$M(A)_2 = \begin{bmatrix} \mathbf{0} & & \mathbf{0} & \\ & A_{M+1}^{M+1} & \cdots & A_N^{M+1} \\ \mathbf{0} & \vdots & & \vdots \\ & A_{M+1}^N & \cdots & A_N^N \end{bmatrix} \tag{8.4.9}$$

which imply

$$\det(\boldsymbol{A}-\lambda\boldsymbol{I})=\det(\boldsymbol{A}_1-\lambda\boldsymbol{I}_{V(\lambda_0)})\det(\boldsymbol{A}_2-\lambda\boldsymbol{I}_{V(\lambda_0)^\perp})$$

By eq. (8.2.2), $\boldsymbol{A}_1=\lambda_0\boldsymbol{I}_{V(\lambda_0)}$; thus by eq. (7.1.21)

$$\det(\boldsymbol{A}-\lambda\boldsymbol{I})=(\lambda_0-\lambda)^M\det(\boldsymbol{A}_2-\lambda\boldsymbol{I}_{V(\lambda_0)^\perp})$$

On the other hand, λ_0 is not an eigenvalue of $\boldsymbol{A}_2$. Therefore

$$\det(\boldsymbol{A}_2-\lambda\boldsymbol{I}_{V(\lambda_0)^\perp})\neq 0$$

Hence the algebraic multiplicity of λ_0 equals M, the geometric multiplicity.

Theorem 8.4.5: If $\boldsymbol{A}$ is Hermitian, the eigenspaces corresponding to distinct eigenvalues λ_1 and λ_2 are orthogonal.

Proof: Let $\boldsymbol{A}\boldsymbol{u}_1=\lambda_1\boldsymbol{u}_1$ and $\boldsymbol{A}\boldsymbol{u}_2=\lambda_2\boldsymbol{u}_2$, then

$$\lambda_1\boldsymbol{u}_1\cdot\boldsymbol{u}_2=\boldsymbol{A}\boldsymbol{u}_1\cdot\boldsymbol{u}_2=\boldsymbol{u}_1\cdot\boldsymbol{A}\boldsymbol{u}_2=\lambda_2\boldsymbol{u}_1\cdot\boldsymbol{u}_2$$

Since, $\lambda_1\neq\lambda_2$, $\boldsymbol{u}_1\cdot\boldsymbol{u}_2=\boldsymbol{0}$ which proves the theorem.

Theorem 8.4.6: If $\boldsymbol{A}$ is a Hermitian endomorphism with (distinct) eigenvalues $\lambda_1,\lambda_2,\ldots\lambda_N$, then V has the representation

$$V=V_1(\lambda_1)\oplus V_2(\lambda_2)\oplus\ldots\oplus V_L(\lambda_L) \qquad (8.4.10)$$

where the eigenspaces $V(\lambda_k)$ are mutually orthogonal.

The proof of this theorem is just a use of the theorems we have learned so here we leave it as an exercise.

Corollary: (Spectral Theorem). If $\boldsymbol{A}$ is a Hermitian endomorphism with (distinct) eigenvalues $\lambda_1,\ldots,\lambda_L$, then

$$\boldsymbol{A}=\sum_{j=1}^{L}\lambda_j\boldsymbol{P}_j \qquad (8.4.11)$$

where $\boldsymbol{P}_j$ is the perpendicular projection of V onto $V(\lambda_j)$, for $j=1,\ldots,L$.

Proof: By Theorem 8.4.6, $\boldsymbol{A}$ has a representation of the form (8.2.1). Let $\boldsymbol{u}$ be an arbitrary element of V, then by eq. (8.4.10),

$$\boldsymbol{u}=\boldsymbol{u}_1+\ldots+\boldsymbol{u}_L \qquad (8.4.12)$$

where $\boldsymbol{u}_j\in V(\lambda_j)$. By eq. (8.1.3), eq. (8.4.12), and eq. (8.2.1)

$$\boldsymbol{A}_j\boldsymbol{u}=\boldsymbol{A}_j\boldsymbol{u}_j=\boldsymbol{A}\boldsymbol{u}_j=\lambda_j\boldsymbol{u}_j$$

But $\boldsymbol{u}_j=\boldsymbol{P}_j\boldsymbol{u}$; therefore

$$\boldsymbol{A}_j=\lambda_j\boldsymbol{P}_j \qquad \text{(no sum)}$$

which, with eq. (8.1.1) proves the corollary.

The reader is reminded that the L perpendicular projections satisfy the equations

$$\sum_{j=1}^{L} \boldsymbol{P}_j = \boldsymbol{I} \tag{8.4.13}$$

$$\boldsymbol{P}_j^2 = \boldsymbol{P}_j \tag{8.4.14}$$

$$\boldsymbol{P}_j = \boldsymbol{P}_j^* \tag{8.4.15}$$

and

$$\boldsymbol{P}_j \boldsymbol{P}_i = \boldsymbol{0} \qquad i \neq j \tag{8.4.16}$$

Certain other endomorphisms also have a spectral representation of the form eq. (8.4.11); however, the projections are not perpendicular ones and do not obey the condition eq. (8.4.16).

Corollary: If $\boldsymbol{A}$ is Hermitian, there exists an orthogonal basis for V consisting entirely of eigenvectors of $\boldsymbol{A}$. With respect to this basis of eigenvectors, the matrix of $\boldsymbol{A}$ is clearly diagonal. Thus the problem of finding a basis for V such that $\boldsymbol{M}(\boldsymbol{A})$ is diagonal is solved for Hermitian endomorphisms.

If $f(\boldsymbol{A})$ is any polynomial in the Hermitian endomorphism, then eq. (8.4.11), eq. (8.4.14) and eq. (8.4.16) can be used to show

$$f(\boldsymbol{A}) = \sum_{j=1}^{L} f(\lambda_j) \boldsymbol{P}_j \tag{8.4.17}$$

where $f(\lambda)$ is the same polynomial except that the variable $\boldsymbol{A}$ is replaced by the scalar λ. For example, the polynomial $\boldsymbol{P}^2$ has the representation

$$\boldsymbol{A}^2 = \sum_{j=1}^{L} \lambda_j^2 \boldsymbol{P}_j$$

In general, $f(\boldsymbol{A})$ is Hermitian iff $f(\lambda_j)$ is real for all $j = 1, \ldots, L$. If the eigenvalues of $\boldsymbol{A}$ are all nonnegative, then we can extract Hermitian roots of $\boldsymbol{A}$ by the following rule:

$$\boldsymbol{A}^{1/k} = \sum_{j=1}^{L} \lambda_j^{1/k} \boldsymbol{P}_j \tag{8.4.18}$$

where $\lambda_j^{1/k} \geqslant 0$. Then we can verify easily that $(\boldsymbol{A}^{1/k})^k = \boldsymbol{A}$. If $\boldsymbol{A}$ has no zero eigenvalues, then

$$\boldsymbol{A}^{-1} = \sum_{j=1}^{L} \frac{1}{\lambda_j} \boldsymbol{P}_j \tag{8.4.19}$$

which is easily confirmed.

A Hermitian endomorphism $\boldsymbol{A}$ is defined to be

$$\left.\begin{cases}\text{positive definite}\\ \text{positive semidefinite}\\ \text{negative semidefinite}\\ \text{negative definite}\end{cases}\right\}\text{if } \boldsymbol{v}\cdot\boldsymbol{A}\boldsymbol{v}\begin{cases}>0\\ \geqslant 0\\ \leqslant 0\\ <0\end{cases}$$

all nonzero $\boldsymbol{v}$, it follows eq. (8.4.11) that

$$\boldsymbol{v}\cdot\boldsymbol{A}\boldsymbol{v}=\sum_{j=1}^{L}\lambda_j\boldsymbol{v}_j\cdot\boldsymbol{v}_j \tag{8.4.20}$$

where

$$\boldsymbol{v}=\sum_{j=1}^{L}\boldsymbol{v}_j$$

Eq. (8.4.20) implies the following important theorem.

Theorem 8.4.7: A Hermitian endomorphism $\boldsymbol{A}$ is

$$\begin{Bmatrix}\text{positive definite}\\ \text{positive semidefinite}\\ \text{negeative semidefinite}\\ \text{negative definite}\end{Bmatrix}$$

iff every eigenvalue of $\boldsymbol{A}$ is

$$\begin{Bmatrix}>0\\ \geqslant 0\\ \leqslant 0\\ <0\end{Bmatrix}$$

As corollaries to Theorem 8.4.7 it follows that positive-definite and negative-definite Hermitian endomorphisms are regular[see eq. (8.4.20)], and positive-definite and positive-semidefinite endomorphisms possess Hermitian roots.

Singular Value Decomposition

$\boldsymbol{A}$ is a linear transformation from n-dimensional vector space V to m-dimensional vector space U, $\boldsymbol{A}^*$ is the adjoint transformation of $\boldsymbol{A}$, that is to say $\boldsymbol{A}: V\to U$ and $\boldsymbol{A}^*: U\to V$.

Theorem 8.4.8: λ_i, $i=1,2,\ldots,m$ is the eigenvalues of transformation $\boldsymbol{A}\boldsymbol{A}^*\in L(U)$ and μ_i, $i=1,2,\ldots,n$ is the eigenvalues of transformation $\boldsymbol{A}^*\boldsymbol{A}\in L(V)$, and then

(1) $\lambda_1\geqslant\lambda_2\geqslant\ldots\geqslant\lambda_r>\lambda_{r+1}=\lambda_{r+2}=\ldots=\lambda_m=0$

(2) $\mu_1\geqslant\mu_2\geqslant\ldots\geqslant\mu_r>\mu_{r+1}=\mu_{r+2}=\ldots=\mu_n=0$

(3) $\lambda_i = \mu_i, \quad i = 1, 2, \ldots, r$

Proof: Assume that $\boldsymbol{u}_i$ is the standard orthogonal basis of $\boldsymbol{AA}^*$ in U, and then $\boldsymbol{AA}^*(\boldsymbol{u}_i) = \lambda_i \boldsymbol{u}_i$. From

$$\langle \boldsymbol{u}_i, \boldsymbol{AA}^*(\boldsymbol{u}_i)\rangle_U = \lambda_i \langle \boldsymbol{u}_i, \boldsymbol{u}_i\rangle_U = \lambda_i \|\boldsymbol{u}_i\|^2$$
$$= \langle \boldsymbol{A}^*(\boldsymbol{u}_i), \boldsymbol{A}^*(\boldsymbol{u}_i)\rangle_V = \|\boldsymbol{A}^*(\boldsymbol{u}_i)\|^2$$

we know that, $\lambda_i = \|\boldsymbol{A}^*(\boldsymbol{u}_i)\|^2 \geqslant 0$. Because that rank$(\boldsymbol{AA}^*) = r$, arrange the eigenvalues we can get (1). Similarity, we can prove (2). Assume $\boldsymbol{AA}^*(\boldsymbol{u}_i) = \lambda_i \boldsymbol{u}_i$, then we have

$$\boldsymbol{A}^*\boldsymbol{A}[\boldsymbol{A}^*(\boldsymbol{u}_i)] = \lambda_i \boldsymbol{A}^*(\boldsymbol{u}_i)$$

Show that the eigenvalues λ_i of transformation $\boldsymbol{AA}^*$ are equal to the eigenvalues μ_i of transformation $\boldsymbol{A}^*\boldsymbol{A}$ which proves (3).

Definition: V and U, $\boldsymbol{A}$ and $\boldsymbol{A}^*$, λ_i and μ_i are the same as above. We define

$$\sigma_i = \sqrt{\lambda_i} = \sqrt{\mu_i}$$

as the singular value of $\boldsymbol{A}$ and $\boldsymbol{A}^*$.

Theorem 8.4.9: $\boldsymbol{A}$ is the matrix representation of $\boldsymbol{A}$ with respect to the orthonormal bases in V and U. $\boldsymbol{A} \in C_r^{m \times n}$, $\sigma_1, \sigma_2, \ldots, \sigma_r$ be the r positive singular values of $\boldsymbol{A}$. Then there exists $U \in u^{m \times m}$ and $V \in u^{n \times n}$ such that

$$\boldsymbol{A} = \boldsymbol{UDV}^* = \boldsymbol{U}\begin{bmatrix} \sum & \boldsymbol{0}_{r \times (n-r)} \\ \boldsymbol{0}_{(m-r) \times r} & \boldsymbol{0}_{(m-r) \times (n-r)} \end{bmatrix} \boldsymbol{V}^*$$

where $\sum = \text{diag}\{\sigma_1, \sigma_2, \ldots, \sigma_r\}$ with $\sigma_1 \geqslant \sigma_2 \geqslant \ldots \geqslant \sigma_r > 0$. Furthermore, let $\boldsymbol{U} = [\boldsymbol{U}_1, \boldsymbol{U}_2]$ with $\boldsymbol{U}_1 \in C^{m \times r}$, then $\boldsymbol{V} = [\boldsymbol{A}^*\boldsymbol{U}_1 \sum^{-1} \ \boldsymbol{V}_2]$ with $\boldsymbol{V}_2 \in C^{n \times (n-r)}$ being any matrix that $\boldsymbol{V}_2^*\boldsymbol{A}^*\boldsymbol{U}_1 \sum^{-1} = \boldsymbol{0}$ and $\boldsymbol{V}_2^*\boldsymbol{V}_2 = \boldsymbol{E}_{n-r}$.

Proof: The decomposition of $\boldsymbol{A}$ follows directly

$$\boldsymbol{A}([\boldsymbol{v}_1 \boldsymbol{v}_2 \ldots \boldsymbol{v}_n]) = [\boldsymbol{u}_1 \boldsymbol{u}_2 \ldots \boldsymbol{u}_m]\boldsymbol{D}, \Leftrightarrow \boldsymbol{AV} = \boldsymbol{UD}$$

$[\boldsymbol{v}_1 \boldsymbol{v}_2 \ldots \boldsymbol{v}_n]$ and $[\boldsymbol{u}_1 \boldsymbol{u}_2 \ldots \boldsymbol{u}_m]$ are basis of $\boldsymbol{V}$ and $\boldsymbol{U}$. Thus $\boldsymbol{A} = \boldsymbol{UDV}^*$. To show that $\boldsymbol{V}$ has the required structure, we note that

$$\boldsymbol{A}^*[\boldsymbol{u}_1 \boldsymbol{u}_2 \ldots \boldsymbol{u}_m] = [\boldsymbol{v}_1 \boldsymbol{v}_2 \ldots \boldsymbol{v}_n]\boldsymbol{D}^*, \Leftrightarrow \boldsymbol{A}^*\boldsymbol{U} = \boldsymbol{VD}^*$$

The first r columns in last equation are

$$\boldsymbol{A}^*\boldsymbol{U}_1 = \boldsymbol{V}_1 \sum \Leftrightarrow \boldsymbol{V}_1 = \boldsymbol{A}^*\boldsymbol{U}_1 \sum^{-1}$$

Let $\boldsymbol{V}_2$ satisfy $\boldsymbol{V}_2^*\boldsymbol{A}^*\boldsymbol{U}_1 \sum^{-1} = \boldsymbol{0}$ and $\boldsymbol{V}_2^*\boldsymbol{V}_2 = \boldsymbol{E}_{n-r}$, then

$$\begin{bmatrix} U_1^* \\ U_2^* \end{bmatrix} A[A^* U_1 \Sigma^{-1} \ V_2] = \begin{bmatrix} U_1^* AA^* U_1 \Sigma^{-1} & U_1^* AV_2 \\ U_2^* AA^* U_1 \Sigma^{-1} & U_2^* AV_2 \end{bmatrix}$$

$$U_1^* AA^* U_1 = \Sigma^2 \Leftrightarrow U_1^* AA^* U_1 \Sigma^{-1} = \Sigma$$

$U_1^* AV_2 = \mathbf{0}$ follows

$$\mathbf{0} = (A^* U_1 \Sigma^{-1})^* V_2 = \Sigma^{-1} U_1 AV_2$$

$$U_2^* AA^* U_1 = \mathbf{0} \Leftrightarrow U_2^* AA^* U_1 \Sigma^{-1} = \mathbf{0}$$

$$U_2^* AA^* U_2 = \mathbf{0} \Leftrightarrow U_2^* A = \mathbf{0}$$

Thus $U_2^* AV_2 = \mathbf{0}$ and

$$U^* AV = \begin{bmatrix} \Sigma & \mathbf{0}_{r \times (n-r)} \\ \mathbf{0}_{(m-r) \times r} & \mathbf{0}_{(m-r) \times (n-r)} \end{bmatrix}$$

Full Rank Decomposition

Theorem 8.4.10: Let $A \in C_r^{m \times n}$, then there exist $B \in C_r^{m \times r}$ and $C \in C_r^{r \times n}$ such that

$$A = BC$$

Furthermore, one of such factorizations is given by

$$B = P^{-1} \begin{bmatrix} E_r \\ \mathbf{0} \end{bmatrix} \in C_r^{m \times r}, \ C = [E_r \ D] Q^{-1} \in C_r^{r \times n}$$

where $P \in C_m^{m \times m}$ and $Q \in C_n^{n \times n}$ are nonsingular matrices such that

$$PAQ = \begin{bmatrix} E_r & D \\ \mathbf{0} & \mathbf{0} \end{bmatrix}$$

Proof: Suppose that the first r columns of A are linearly independent. Then there exists a nonsingular matrix $P \in C_m^{m \times m}$ such that

$$PA = \begin{bmatrix} E_r & D \\ \mathbf{0} & \mathbf{0} \end{bmatrix}$$

or equivalently

$$A = P^{-1} \begin{bmatrix} E_r & D \\ \mathbf{0} & \mathbf{0} \end{bmatrix}$$

$$= P^{-1} \begin{bmatrix} E_r \\ \mathbf{0} \end{bmatrix} [E_r \ D] = BC$$

where

$$B = P^{-1}\begin{bmatrix} E_r \\ 0 \end{bmatrix} \in C_r^{m\times r}, C = [E_r D] \in C_r^{r\times n}$$

If the first r columns of A are linearly dependent, by interchanging the columns we can get a matrix whose first r columns are linearly independent. Since interchanging the columns of a matrix is equivalent to post-multiplying it by a permutation matrix, there exists a nonsingular matrix $Q \in C_n^{n\times n}$ (the permutation matrix) such that the first r columns of AQ are linearly independent and the previous procedure can be applied to AQ. Hence there exist $P \in C_m^{m\times m}$ and $Q \in C_n^{n\times n}$ such that

$$\begin{aligned} A &= P^{-1}\begin{bmatrix} E_r & D \\ 0 & 0 \end{bmatrix} Q^{-1} \\ &= P^{-1}\begin{bmatrix} E_r \\ 0 \end{bmatrix} [E_r \quad D] Q^{-1} = BC \end{aligned}$$

which completes the proof.

***UR* and *QR* Decomposition**

Theorem 8.4.11: Let $A \in C_r^{m\times r}$ be of full column rank, then A can be uniquely factorized as

$$A = UR$$

where $U \in U_r^{m\times r}$ is a sub-unitary matrix, R is an $r\times r$ upper triangular matrix with positive diagonal entries.

Proof: Let A be partitioned into the column form

$$A = [\alpha_1 \quad \alpha_2 \quad \cdots \quad \alpha_r]$$

Applying the Gram-Schmidt orthonormalization process to the linearly independent vectors $\alpha_1, \alpha_2, \cdots, \alpha_r$ we get r orthonormal vectors $\gamma_1, \gamma_2, \cdots, \gamma_r$ satisfying

$$\begin{aligned} \alpha_1 &= k_{11}\gamma_1 \\ \alpha_2 &= k_{21}\gamma_1 + k_{22}\gamma_2 \\ \alpha_3 &= k_{31}\gamma_1 + k_{32}\gamma_2 + k_{33}\gamma_3 \\ &\vdots \\ \alpha_r &= k_{r1}\gamma_1 + k_{r2}\gamma_2 + \cdots + k_{rr}\gamma_r \end{aligned}$$

from the above equation we get

$$\begin{aligned} A &= [\alpha_1\alpha_2\cdots\alpha_r] \\ &= [\gamma_1\gamma_2\cdots\gamma_r]\begin{bmatrix} k_{11} & k_{12} & \cdots & k_{11} \\ 0 & k_{22} & \cdots & k_{2l} \\ \vdots & \vdots & \ddots & \vdots \\ 0 & k_{r2} & \cdots & k_{rl} \end{bmatrix} = UR \end{aligned}$$

where $\boldsymbol{U} = [\boldsymbol{\gamma}_1\boldsymbol{\gamma}_2 \ldots \boldsymbol{\gamma}_r] \in U_r^{m\times r}$ is dub-unitary, $\boldsymbol{R} = (k_{ij})$ is an upper triangular matrix with positive diagonal entries.

We show the uniqueness of the factorization. Let

$$\boldsymbol{A} = \boldsymbol{U}_1\boldsymbol{R}_1 = \boldsymbol{U}_2\boldsymbol{R}_2$$

be two factorizations, then

$$\boldsymbol{A}^*\boldsymbol{A} = \boldsymbol{R}_1^*\boldsymbol{R}_1 = \boldsymbol{R}_2^*\boldsymbol{R}_2$$

Since $\boldsymbol{A}^*\boldsymbol{A}$ is positive definite matrix, the Cholesky factorization is unique and $\boldsymbol{R}_1 = \boldsymbol{R}_2$. Thus $\boldsymbol{U}_1 = \boldsymbol{U}_2$.

It is clear that $\boldsymbol{U}$ can be expanded to a unitary matrix $\boldsymbol{U}_1$, then $\boldsymbol{UR} = \boldsymbol{U}_1\boldsymbol{R}_1$, where

$$\boldsymbol{R}_1 = \begin{bmatrix} \boldsymbol{R} \\ \boldsymbol{0}_{(m-r)\times r} \end{bmatrix}$$

For a matrix $\boldsymbol{A}$ of full row rank, performing the factorization for $\boldsymbol{A}^{\mathrm{T}}$ we get the following result.

Theorem 8.4.12: Let $\boldsymbol{A} \in C_r^{r\times n}$ be of full row rank, then $\boldsymbol{A}$ can be factorized uniquely as

$$\boldsymbol{A} = \boldsymbol{LU}$$

where $\boldsymbol{L}$ is an $r\times r$ lower triangular matrix with positive diagonal entries, $\boldsymbol{U} \in U_r^{m\times r}$ is a sub-unitary.

Theorem 8.4.13: If $\boldsymbol{A} \in C_r^{m\times n}$, then $\boldsymbol{A}$ can be factorized uniquely as

$$\boldsymbol{A} = \boldsymbol{U}_1\boldsymbol{R}_1\boldsymbol{L}_2\boldsymbol{U}_2$$

where $\boldsymbol{U}_1 \in U_r^{m\times r}$, $\boldsymbol{U}_2 \in U_r^{r\times n}$ are sub-unitary matrices, $\boldsymbol{R}_1$ is $r\times r$ upper triangular matrix with positive diagonal entries, $\boldsymbol{L}_2$ is $r\times r$ lower triangular matrix with positive diagonal entries.

Proof: From theorem 8.4.10, $\boldsymbol{A}$ can be factorized as $\boldsymbol{A} = \boldsymbol{BC}$, where $\boldsymbol{B} \in C_r^{m\times r}$, $\boldsymbol{C} \in C_r^{r\times n}$. Applying Theorem 8.4.11 to $\boldsymbol{B}$ and Theorem 8.4.12 to $\boldsymbol{C}$, we get the factorization.

8.5 Illustrative Examples

In this section we shall illustrate certain of the results of the preceding section by working selected numerical examples. For simplicity, the basis of V shall be taken to be the orthonormal basis $\{\boldsymbol{i}_1, \ldots \boldsymbol{i}_N\}$ introduced in eq. (5.6.1). The vector eq. (8.2.1) takes the component form

$$\boldsymbol{A}_{kj}\boldsymbol{v}_j = \lambda \boldsymbol{v}_k \tag{8.5.1}$$

where

$$\boldsymbol{v} = \boldsymbol{v}_j \boldsymbol{i}_j \tag{8.5.2}$$

and

$$\boldsymbol{A}\boldsymbol{i}_j = \boldsymbol{A}_{kj}\boldsymbol{i}_k \tag{8.5.3}$$

Since the basis is orthonormal, we have written all indices as subscripts, and the summation convention is applied in the usual way.

Example 8.5.1: Consider a real three-dimensional vector space V. Let the matrix of an endomorphism $\boldsymbol{A} \in L(V, V)$ be

$$\boldsymbol{M}(\boldsymbol{A}) = \begin{bmatrix} 1 & 1 & 0 \\ 1 & 2 & 1 \\ 0 & 1 & 1 \end{bmatrix} \tag{8.5.4}$$

Clearly $\boldsymbol{A}$ is symmetric and, thus, the theorems of the preceding section can be applied. By direct expansion of the determinant of $\boldsymbol{M}(\boldsymbol{A} - \lambda \boldsymbol{I})$ the characteristic polynomial is

$$f(\lambda) = (-\lambda)(1-\lambda)(3-\lambda) \tag{8.5.5}$$

Therefore the three eigenvalues of $\boldsymbol{A}$ are distinct and are given by

$$\lambda_1 = 0, \lambda_2 = 1, \lambda_3 = 3 \tag{8.5.6}$$

The ordering of the eigenvalues is not important. Since the eigenvalues are distinct, their corresponding characteristic subspaces are one-dimensional. For definiteness, let $\boldsymbol{v}^{(p)}$ be an eigenvector associated with λ_p. As usual, we can represent $\boldsymbol{v}^{(p)}$ by

$$\boldsymbol{v}^{(p)} = v_k^{(p)} \boldsymbol{i}_k \tag{8.5.7}$$

Then eq. (8.5.1), for $p = 1$, reduces to

$$v_1^{(1)} + v_2^{(1)} = 0, v_1^{(1)} + 2v_2^{(1)} + v_3^{(1)} = 0, v_2^{(1)} + v_3^{(1)} = 0 \tag{8.5.8}$$

The general solution of this linear system is

$$v_1^{(1)} = t, v_2^{(1)} = -t, v_3^{(1)} = t \tag{8.5.9}$$

For all $t \in R$. In particular, if $\boldsymbol{v}^{(1)}$ is required to be a unit vector, then we can choose $t = \pm 1/\sqrt{3}$, where the choice of sign is arbitrary, say

$$v_1^{(1)} = 1/\sqrt{3}, v_2^{(1)} = -1/\sqrt{3}, v_3^{(1)} = 1/\sqrt{3} \tag{8.5.10}$$

So

$$\boldsymbol{v}^{(1)} = (1/\sqrt{3})\boldsymbol{i}_1 - (1/\sqrt{3})\boldsymbol{i}_2 + (1/\sqrt{3})\boldsymbol{i}_3 \tag{8.5.11}$$

Likewise we find for $p=2$

$$\mathbf{v}^{(2)} = (1/\sqrt{2})\mathbf{i}_1 - (1/\sqrt{2})\mathbf{i}_3 \tag{8.5.12}$$

and for $p=3$

$$\mathbf{v}^{(3)} = (1/\sqrt{6})\mathbf{i}_1 + (2/\sqrt{6})\mathbf{i}_2 + (1/\sqrt{6})\mathbf{i}_3 \tag{8.5.13}$$

It is easy to check that $\{\mathbf{v}^{(1)}, \mathbf{v}^{(2)}, \mathbf{v}^{(3)}\}$ is an orthonormal basis.

By eq. (8.5.6) and eq. (8.4.12), $\mathbf{A}$ has the spectral decomposition

$$\mathbf{A} = 0\mathbf{P}_1 + 1\mathbf{P}_2 + 3\mathbf{P}_3 \tag{8.5.14}$$

where $\mathbf{P}_k$ is the perpendicular projection defined by

$$\mathbf{P}_k\mathbf{v}^{(k)} = \mathbf{v}^{(k)}, \mathbf{P}_k\mathbf{v}^{(j)} = \mathbf{0}, \quad j \neq k \tag{8.5.15}$$

for $k = 1, 2, 3$. In component form relative to the original orthonormal basis $\{\mathbf{i}_1, \mathbf{i}_2, \mathbf{i}_3\}$ these projections are given by

$$\mathbf{P}_k\mathbf{i}_j = v_j^{(k)} v_l^{(k)} \mathbf{i}_i \quad (\text{no sum on } k) \tag{8.5.16}$$

This result follows eq. (8.5.15) and the transformation law eq. (7.3.7) or directly the representations

$$\begin{aligned}
\mathbf{i}_1 &= (1/\sqrt{3})\mathbf{v}^{(1)} + (1/\sqrt{2})\mathbf{v}^{(2)} + (1/\sqrt{2})\mathbf{v}^{(3)} \\
\mathbf{i}_2 &= -(1/\sqrt{3})\mathbf{v}^{(1)} + (2/\sqrt{6})\mathbf{v}^{(3)} \\
\mathbf{i}_3 &= (1/\sqrt{3})\mathbf{v}^{(1)} - (1/\sqrt{2})\mathbf{v}^{(2)} + (1/\sqrt{6})\mathbf{v}^{(3)}
\end{aligned} \tag{8.5.17}$$

since the coefficient matrix of $\{\mathbf{i}_1, \mathbf{i}_2, \mathbf{i}_3\}$ relative to $\{\mathbf{v}^{(1)}, \mathbf{v}^{(2)}, \mathbf{v}^{(3)}\}$ is the transpose of that of $\{\mathbf{v}^{(1)}, \mathbf{v}^{(2)}, \mathbf{v}^{(3)}\}$ relative to $\{\mathbf{i}_1, \mathbf{i}_2, \mathbf{i}_3\}$.

There is a result, known as Sylvester's Theorem, which enables one to compute the projections directly. We shall not prove this theorem here, but we shall state the formula in the case when the eigenvalues are distinct. The result is

$$\mathbf{P}_j = \frac{\prod_{k=1;j\neq k}^{N} (\lambda_k \mathbf{I} - \mathbf{A})}{\prod_{k=1;j\neq k}^{N} (\lambda_k - \lambda_j)} \tag{8.5.18}$$

The advantage of this formula is that one does not need to know the eigenvectors in order to find the projections. With respect to an arbitrary basis, eq. (8.5.18) yields

$$\mathbf{M}(\mathbf{P}_j) = \frac{\prod_{k=1;j\neq k}^{N} (\lambda_k \mathbf{M}(\mathbf{I}) - \mathbf{M}(\mathbf{A}))}{\prod_{k=1;j\neq k}^{N} (\lambda_k - \lambda_j)} \tag{8.5.19}$$

Example 8.5.2: To illustrate eq. (8.5.19), let

$$\mathbf{M}(\mathbf{A}) = \begin{bmatrix} 2 & 2 \\ 2 & -1 \end{bmatrix} \tag{8.5.20}$$

with respect to an orthonormal basis. The eigenvalues of this matrix are easily found to be $\lambda_1 = -2, \lambda_2 = 3$. Then, eq. (8.5.19) yields

$$M(P_1) = \left(\lambda_2\begin{bmatrix}1 & 0\\0 & 1\end{bmatrix} - \begin{bmatrix}2 & 2\\2 & -1\end{bmatrix}\right)\Big/(\lambda_2 - \lambda_1) = \frac{1}{5}\begin{bmatrix}1 & -2\\-2 & 4\end{bmatrix}$$

and

$$M(P_2) = \left(\lambda_1\begin{bmatrix}1 & 0\\0 & 1\end{bmatrix} - \begin{bmatrix}2 & 2\\2 & -1\end{bmatrix}\right)\Big/(\lambda_2 - \lambda_1) = \frac{1}{5}\begin{bmatrix}1 & 4\\2 & 1\end{bmatrix}$$

The spectral theorem for this linear transformation yields the matrix equation

$$M(A) = -\frac{2}{5}\begin{bmatrix}1 & -2\\-2 & 4\end{bmatrix} + \frac{3}{5}\begin{bmatrix}4 & 2\\2 & 1\end{bmatrix}$$

8.6 The Minimal Polynomial

We remarked that for any given endomorphism $\boldsymbol{A}$ there exist some polynomials $f(t)$ such that

$$f(\boldsymbol{A}) = \boldsymbol{0} \tag{8.6.1}$$

For example, by the Cayley-Hamilton theorem, we can always choose, $f(t)$ to be the characteristic polynomial of $\boldsymbol{A}$. Another obvious choice can be found by observing the fact that since $L(V,V)$ has dimension N^2, the set

$$\{\boldsymbol{A}^0 = \boldsymbol{I}, \boldsymbol{A}^1, \boldsymbol{A}^2, \ldots, \boldsymbol{A}^{N^2}\}$$

is necessarily linearly dependent and, thus there exist scalars $\{\alpha_0, \alpha_1, \ldots, \alpha_{N^2}\}$, not all equal to zero, such that

$$\alpha_0\boldsymbol{I} + \alpha_1\boldsymbol{A}^1 + \ldots + \alpha_{N^2}\boldsymbol{A}^{N^2} = \boldsymbol{0} \tag{8.6.2}$$

For definiteness, let us denote the set of all polynomials f satisfying the condition eq. (8.6.1) for a given $\boldsymbol{A}$ by the symbol $P(\boldsymbol{A})$. We shall now show that $P(\boldsymbol{A})$ has a very simple structure, called a principal ideal.

In general, an ideal ℓ is a subset of an integral domain D such that the following two conditions are satisfied:

1) If f and g belong to ℓ, so is their sum $f + g$.

2) If f belongs to ℓ and h arbitrary element of D, then $fh = hf$ also belong to ℓ.

Of course, D itself and the subset $\{0\}$ consisting in the zero element of D are obvious examples of ideals, and these are called trivial ideals or improper ideals.

Another example of ideal is the subset $\ell \subset D$ consisting in all multiples of a particular element $g \in D$, namely

$$\ell = \{hg, h \in D\} \tag{8.6.3}$$

It is easy to verify that this subset satisfies the two conditions for an ideal. Ideals of the special form eq. (8.6.3) are called principal ideals.

For the set $P(\boldsymbol{A})$, we choose the integral domain D to be the set of all polynomials with complex coefficients. (For real vector space, the coefficients are required to be real, of course.) Then it is obvious that $P(\boldsymbol{A})$ is an ideal in D, since if f and g satisfy the condition eq. (8.6.1), so does their sum $f+g$ and similarly if f satisfies eq. (8.6.1) and h is an arbitrary polynomial, then

$$(hf)(\boldsymbol{A}) = h(\boldsymbol{A})f(\boldsymbol{A}) = h(\boldsymbol{A})\mathbf{0} = \mathbf{0} \tag{8.6.4}$$

Theorem 8.6.1: Every ideal of the polynomial domain is a principal ideal.

Proof: We assume that the reader is similar with the operation of division for polynomials. If f and $g \neq 0$ are polynomials, we can divide f by g and obtain a remainder r having degree less than g, namely

$$r(t) = f(t) - h(t)g(t) \tag{8.6.5}$$

Now, to prove that $P(\boldsymbol{A})$ can be represented by the form eq. (8.6.3), we choose a polynomial $g \neq 0$ having the lowest degree in $P(\boldsymbol{A})$. Then we claim that

$$P(\boldsymbol{A}) = \{hg, h \in D\} \tag{8.6.6}$$

To see this, we must show that every $f \in P(\boldsymbol{A})$ can be devided through by g without a remainder. Suppose that the division of f by g yields a remainder r as shown in eq. (8.6.5). Then since $P(\boldsymbol{A})$ is an ideal and since $f, g \in P(\boldsymbol{A})$, eq. (8.6.5) shows that $r \in P(\boldsymbol{A})$ also. But since the degree of r is less than the degree of g, $r \in P(\boldsymbol{A})$ is possible iff $r = 0$. Thus $f = hg$, so the representation eq. (8.6.6) is valid.

Corollary: The nonzero polynomial g having the lowest degree in $P(\boldsymbol{A})$ is unique to within an arbitrary nonzero multiple of a scalar.

Definition: Define as above. If we require the leading coefficient of g to be 1, then g becomes unique, and we call this particular polynomial g the minimal polynomial of the endomorphism $\boldsymbol{A}$.

We pause here to give some examples of minimal polynomials.

Example 8.6.1: The minimal polynomial of the zero endomorphism $\mathbf{0}$ is the polynomial $f(t) = 1$ of zero degree, since by convention

$$f(\mathbf{0}) = 1\mathbf{0}^0 = \mathbf{0} \tag{8.6.7}$$

In general, if $\boldsymbol{A} \neq \boldsymbol{0}$, then the minimal polynomial of $\boldsymbol{A}$ is at least of degree 1, since in this case

$$1\boldsymbol{A}^0 = 1\boldsymbol{I} \neq \boldsymbol{0} \tag{8.6.8}$$

Example 8.6.2: Let $\boldsymbol{P}$ be a nontrivial projection. Then the minimal polynomial g of $\boldsymbol{P}$ is

$$g(t) = t^2 - t \tag{8.6.9}$$

by the definition of a projection,

$$\boldsymbol{P}^2 - \boldsymbol{P} = \boldsymbol{P}(\boldsymbol{P} - \boldsymbol{I}) = \boldsymbol{0} \tag{8.6.10}$$

and since $\boldsymbol{P}$ is assumed to be non trivial, the two lower degree divisors t and $t-1$ no longer satisfy the condition eq. (8.6.1) for $\boldsymbol{P}$.

Theorem 8.6.2: If f and g are polynomials, then there exists a greatest common divisor d which is a divisor of f and g and is also a multiple of every common divisor of f and g.

Proof: We define the ideal ℓ in the polynomial domain D by

$$\ell \equiv \{hf + kg, h, k \in D\} \tag{8.6.11}$$

By Theorem 8.6.1, ℓ is a principal ideal, and thus it has a representation

$$\ell \equiv \{hd, h \in D\} \tag{8.6.12}$$

We claim that d is a greatest common divisor of f and g. Clearly, d is a common divisor of f and g, since f and g are themselves members of ℓ, so by eq. (8.6.12) there exist h and k in D such that

$$f = hd, \quad g = kd \tag{8.6.13}$$

On the other hand, since d is also a member of ℓ, by eq. (8.6.12) there exist also p and q in D such that

$$d = pf + qg \tag{8.6.14}$$

Therefore if c is any common divisor of f and g, say

$$f = ac, \quad g = bc \tag{8.6.15}$$

then from eq. (8.6.14)

$$d = (pa + qd)c \tag{8.6.16}$$

So d is a multiple of c. Thus d is a greatest common divisor of f and g.

We see that the greatest common divisor d of f and g is unique to within a nonzero scalar factor. So we can render d unique by requiring its leading coefficient to be 1. Also, it is clear that the preceding theorem can be extended in an obvious way to more than two polynomials. If the greatest common divisor of $f_1, \ldots, f_L$ is the zero degree polynomial 1, then $f_1, \ldots, f_L$ are said to be relatively prime. Similarly $f_1, \ldots, f_L$ are pairwise prime if each pair $f_i, f_j, i \neq j$, from $f_1, \ldots, f_L$ is relatively prime.

Here we present another important concept associated with the algebra of polynomials: the least common multiple.

Theorem 8.6.3: If f and g are polynomials, then there exists a least common multiple m which is a multiple of f and g and is a divisor of every common multiple of f and g.

The proof of this theorem is based on the same argument as the proof of the preceding theorem, so it is left as an exercise.

8.7 Spectral Decomposition for Arbitrary Endomorphisms

In this section we shall consider the problem in general and we shall find decompositions which are, in some sense, closest to the simple decomposition eq. (8.4.12) for endomorphisms in general.

Theorem 8.7.1: If f is any polynomial, then the null space $K(f(\boldsymbol{A}))$ is $\boldsymbol{A}$-invariant.

Proof: Since the multiplication of polynomials is a commutative operation, we have

$$\boldsymbol{A}f(\boldsymbol{A}) = f(\boldsymbol{A})\boldsymbol{A} \tag{8.7.1}$$

Hence if $\boldsymbol{v} \in K(f(\boldsymbol{A}))$, then

$$f(\boldsymbol{A})\boldsymbol{A}\boldsymbol{v} = \boldsymbol{A}f(\boldsymbol{A})\boldsymbol{v} = \boldsymbol{A}\boldsymbol{0} = \boldsymbol{0} \tag{8.7.2}$$

which shows that $\boldsymbol{A}\boldsymbol{v} \in K(f(\boldsymbol{A}))$. Therefore $K(f(\boldsymbol{A}))$ is $\boldsymbol{A}$-invariant.

Theorem 8.7.2: If f is a multiple of g, say

$$f = hg \tag{8.7.3}$$

Then

$$K(f(\boldsymbol{A})) \supset K(g(\boldsymbol{A})) \tag{8.7.4}$$

Proof: This result is a general property of the null space. Indeed, for any endomorphisms $\boldsymbol{B}$ and $\boldsymbol{C}$ we always have

$$K(\boldsymbol{B}\boldsymbol{C}) \subset K(\boldsymbol{C}) \tag{8.7.5}$$

So if we set $h(\boldsymbol{A}) = \boldsymbol{B}$ and $g(\boldsymbol{A}) = \boldsymbol{C}$, then eq. (8.7.5) reduces to eq. (8.7.4).

The preceding theorem does not imply that $K(g(\boldsymbol{A}))$ is necessarily a proper subspace of $K(f(\boldsymbol{A}))$, however. It is quite possible that the two subspaces, in fact, coincide. For example, if g and hence f both belong to $P(\boldsymbol{A})$, then $g(\boldsymbol{A}) = f(\boldsymbol{A}) = 0$, and thus

$$K(f(\boldsymbol{A})) = K(g(\boldsymbol{A})) = K(\boldsymbol{0}) = V \tag{8.7.6}$$

However, if m is the minimal polynomial of $\boldsymbol{A}$, and f is a proper divisor of m (i. e. m is not a divisor of f) so that $f \notin P(\boldsymbol{A})$, then $K(f(\boldsymbol{A}))$ is strictly a proper subspace of $V = K(m(\boldsymbol{A}))$.

Theorem 8.7.3: If f is a divisor (proper or improper) of the minimal polynomial m of $\boldsymbol{A}$, and g is a proper divisor of f, then $K(g(\boldsymbol{A}))$ is strictly a proper subspace of $K(f(\boldsymbol{A}))$.

Proof: By assumption there exists a polynomial h such that

$$m = hf \tag{8.7.7}$$

We set

$$k = hg \tag{8.7.8}$$

Then k is a proper divisor of m, since by assumption, g is a proper divisor of f. By the remark preceding the theorem, $K(k(\boldsymbol{A}))$ is strictly a subspace of V, which is equal to $K(m(\boldsymbol{A}))$. Thus there exists a vector $\boldsymbol{v}$ such that

$$k(\boldsymbol{A})\boldsymbol{v} = g(\boldsymbol{A})h(\boldsymbol{A})\boldsymbol{v} \neq 0 \tag{8.7.9}$$

which implies that the vector $\boldsymbol{u} = h(\boldsymbol{A})\boldsymbol{v}$ does not belong to $K(g(\boldsymbol{A}))$. On the other hand, from eq. (8.7.7)

$$f(\boldsymbol{A})\boldsymbol{u} = f(\boldsymbol{A})h(\boldsymbol{A})\boldsymbol{v} = m(\boldsymbol{A})\boldsymbol{v} = 0 \tag{8.7.10}$$

which implies that $\boldsymbol{u}$ belongs to $K(f(\boldsymbol{A}))$. Thus $K(g(\boldsymbol{A}))$ is strictly a proper subspace of $K(f(\boldsymbol{A}))$.

The next theorem shows the role of the greatest common divisor in terms of the null space.

Theorem 8.7.4: Let f and g be any polynomials, and suppose that d is their greatest common divisor, then

$$K(d(\boldsymbol{A})) = K(f(\boldsymbol{A})) \cap K(g(\boldsymbol{A})) \tag{8.7.11}$$

Proof: Since d is a common divisor of f and g, the inclusion

$$K(d(\boldsymbol{A})) \subset K(f(\boldsymbol{A})) \cap K(g(\boldsymbol{A})) \tag{8.7.12}$$

follows readily Theorem 8.7.2. To prove the reversed inclusion

$$K(d(\boldsymbol{A})) \supset K(f(\boldsymbol{A})) \cap K(g(\boldsymbol{A})) \tag{8.7.13}$$

recall that from eq. (8.6.14) there exist polynomials p and q such that

$$d = pf + qg \tag{8.7.14}$$

and thus

$$d(\boldsymbol{A}) = p(\boldsymbol{A})f(\boldsymbol{A}) + q(\boldsymbol{A})g(\boldsymbol{A}) \tag{8.7.15}$$

This equation means that if $\boldsymbol{v} \in K(f(\boldsymbol{A})) \cap K(g(\boldsymbol{A}))$, then

$$\begin{aligned} d(\boldsymbol{A})\boldsymbol{v} &= p(\boldsymbol{A})f(\boldsymbol{A})\boldsymbol{v} + q(\boldsymbol{A})g(\boldsymbol{A})\boldsymbol{v} \\ &= p(\boldsymbol{A})\boldsymbol{0} + q(\boldsymbol{A})\boldsymbol{0} = \boldsymbol{0} \end{aligned} \tag{8.7.16}$$

so that $\boldsymbol{v} \in K(d(\boldsymbol{A}))$. Therefore eq. (8.7.13) is valid and hence eq. (8.7.11).

A corollary of the preceding theorem is the fact that if f and g are relatively prime, then

$$K(f(\boldsymbol{A})) \cap K(g(\boldsymbol{A})) = \{0\} \tag{8.7.17}$$

since in this case the greatest common divisor of f and g is $d(t) = 1$, so that

$$K(d(\boldsymbol{A})) = K(\boldsymbol{A}^0) = K(\boldsymbol{I}) = \{0\} \tag{8.7.18}$$

Here we have assumed $\boldsymbol{A} \neq \boldsymbol{0}$ of course.

Next we consider the role of the least common multiple in terms of the null space.

Theorem 8.7.5: Let f and g be any polynomials, and suppose that l is their least common multiplier, then

$$K(l(\boldsymbol{A})) = K(f(\boldsymbol{A})) + K(g(\boldsymbol{A})) \tag{8.7.19}$$

where the operation on the right-hand side of eq. (8.7.19) is the sum of some subspaces. Like the result eq. (8.7.11), the result eq. (8.7.19) can be generalized in an obvious way for more than two polynomials.

The proof is left as an exercise.

Corollary: If f and g are relatively prime (pairwise prime if there are more than two polynomials), then eq. (8.7.19) can be strengthened to

$$K(l(\boldsymbol{A})) = K(f(\boldsymbol{A})) \oplus K(g(\boldsymbol{A})) \tag{8.7.20}$$

Again, we have assumed that $\boldsymbol{A} \neq \boldsymbol{0}$. We leave the proof of eq. (8.7.20) also as an exercise.

Having summarized the preliminary theorems, we are now ready to state the main theorem of this section.

Theorem 8.7.6: If m is the minimal polynomial of $\boldsymbol{A}$ which is factored into the form

$$m(t) = (t - \lambda_1)^{\alpha_1} \ldots (t - \lambda_1)^{\alpha_L} = m_1(t) \ldots m_L(t) \tag{8.7.21}$$

where $\lambda_1, \ldots, \lambda_L$ are distinct and $\alpha_1, \ldots, \alpha_L$ are positive integers, then V has the representation

$$V = K(m_1(\boldsymbol{A})) \oplus \ldots \oplus K(m_L(\boldsymbol{A})) \tag{8.7.22}$$

Proof: Since $\lambda_1, \ldots, \lambda_L$ are distinct, the polynomials $m_1, \ldots, m_L$ are pairwise prime and their least common multiplier is m. Hence by eq. (8.7.20) we have

$$K(m(\boldsymbol{A})) = K(m_1(\boldsymbol{A})) \oplus \ldots \oplus K(m_L(\boldsymbol{A})) \tag{8.7.23}$$

But since $m(\boldsymbol{A}) = 0, K(m(\boldsymbol{A})) = V$, eq. (8.7.22) holds.

Now from Theorem 8.7.1 we know that each subspace $K(m_i(\boldsymbol{A})), i, \ldots, L$ is $\boldsymbol{A}$-invariant; then from eq. (8.7.22) we see that $\boldsymbol{A}$ is reduced by the subspaces $K(m_1(\boldsymbol{A})), \ldots, K(m_L(\boldsymbol{A}))$. Therefore, the results of Theorem 8.1.1 can be applied.

Theorem 8.7.7: Each factor m_k of m is the minimal polynomial of the restriction of $\boldsymbol{A}$ to the subspace $K(m_k(\boldsymbol{A}))$. More generally, any product of factors, say $m_1, \ldots, m_M$ is the minimal polynomial of the restriction of $\boldsymbol{A}$ to the corresponding subspace $K(m_1(\boldsymbol{A})) \oplus \ldots \oplus K(m_M(\boldsymbol{A}))$.

Proof: We prove the special case of one factor only, say m_1; the proof of the general case of several factors is similar and is left as an exercise. For definiteness, let $\bar{\boldsymbol{A}}$ denote the restriction of $\boldsymbol{A}$ to $K(m_1(\boldsymbol{A}))$, then $m_1(\bar{\boldsymbol{A}})$ is equal to the restriction of $m_1(\boldsymbol{A})$ to $m_1(\bar{\boldsymbol{A}})$, and, thus $m_1(\bar{\boldsymbol{A}}) = 0$, which means that $m_1 \in P(\bar{\boldsymbol{A}})$. Now if g is any proper divisor of m_1, then $g(\bar{\boldsymbol{A}}) \neq 0$; for otherwise, we would have $g(\boldsymbol{A}) m_2(\boldsymbol{A}) \ldots m_L(\boldsymbol{A})$, contradicting the fact that m is minimal for $\boldsymbol{A}$. Therefore m_1 is minimal for $\bar{\boldsymbol{A}}$ and the proof is complete.

Here if the minimal polynomial eq. (8.7.21) has simple roots only, i. e. the powers $\alpha_1, \ldots, \alpha_L$ are all equal to 1, then

$$m_i(\boldsymbol{A}) = \boldsymbol{A} - \lambda_i \boldsymbol{I} \tag{8.7.24}$$

for all i. In this case, the restriction of $\boldsymbol{A}$ to $K(m_i(\boldsymbol{A}))$ coincides with $\lambda_i \boldsymbol{I}$ on that subspace, namely

$$\boldsymbol{A}_{V(\lambda_i)} = \lambda_i \boldsymbol{I} \tag{8.7.25}$$

Here we have used the fact from (8.7.24), that

$$V(\lambda_i) = K(m_i(\boldsymbol{A})) \tag{8.7.26}$$

Then the form eq. (8.1.5) reduces to the diagonal form eq. $(8.4.3)_1$ or eq. $(8.4.3)_2$.

The condition that m has simple roots only turns out to be necessary for the existence of a diagonal matrix for $\boldsymbol{A}$ also, as we shall now see in the following theorem.

Theorem 8.7.8: An endomorphism $\boldsymbol{A}$ has a diagonal matrix iff its minimal polynomial can be factored into distinct factors all of the first degree.

Proof:

1) Sufficiency: Sufficiency has already been proven.

2) Necessity: Assume that the matrix of $\boldsymbol{A}$ relative to some basis $\{\boldsymbol{e}_1, \ldots, \boldsymbol{e}_N\}$ has the form eq. (8.4.3)$_1$, then the polynomial

$$m(t) = (t - \lambda_1) \ldots (t - \lambda_L) \tag{8.7.27}$$

is the minimal polynomial of $\boldsymbol{A}$. Indeed, each basis vector $\boldsymbol{e}_i$ is contained in the null space $K(\boldsymbol{A} - \lambda_k \boldsymbol{I})$ for one particular k. Consequently $\boldsymbol{e}_i \in K(m(\boldsymbol{A}))$ for all $i = 1, \ldots N$ and thus

$$K(m(\boldsymbol{A})) = V \tag{8.7.28}$$

or equivalently

$$m(\boldsymbol{A}) = 0 \tag{8.7.29}$$

which implies that $m \in P(\boldsymbol{A})$. But since $\lambda_1, \ldots, \lambda_L$ are distinct, no proper divisor of m still belongs to $P(\boldsymbol{A})$. Therefore m is the minimal polynomial of $\boldsymbol{A}$.

As before, when the condition of the preceding theorem is satisfied, then we can define projections $\boldsymbol{P}_1, \ldots, \boldsymbol{P}_L$ by

$$R(\boldsymbol{P}_i) = V(\lambda_i), \quad K(\boldsymbol{P}_i) = \bigoplus_{\substack{j=1 \\ j \neq i}}^{L} V(\lambda_j) \tag{8.7.30}$$

for all $i = 1, \ldots, L$, and the diagonal form eq. (8.7.3) shows that $\boldsymbol{A}$ has the representation

$$\boldsymbol{A} = \lambda_1 \boldsymbol{P}_1 + \ldots + \lambda_L \boldsymbol{P}_L = \sum_{i=1}^{L} \lambda_i \boldsymbol{P}_i \tag{8.7.31}$$

Note: Instating this result we have not made use of any inner product, so it is not meaningful to say whether or not the projections $\boldsymbol{P}_1, \ldots, \boldsymbol{P}_L$ are perpendicular; furthermore, the eigenvalues $\lambda_1, \ldots, \lambda_L$ are generally complex numbers. In fact, the factorization eq. (8.7.21) for m in general is possible only if the scalar field is algebraically closed, such as the complex field used here. If the scalar field is the real field, we should define the factors $m_1, \ldots, m_L$ of m to be powers of irreducible polynomials, i. e. polynomials having no proper divisors. Then the decomposition eq. (8.7.22) for V remains valid, since the argument of the proof is based entirely on the fact that the factors of m are pairwise prime.

Theorem 8.7.8 shows that in order to know whether or not $\boldsymbol{A}$ has a diagonal form, we must know the roots and their multiplicities in the minimal polynomial m of $\boldsymbol{A}$. Now since the characteristic polynomial f of $\boldsymbol{A}$ belongs to $P(\boldsymbol{A})$, m is a divisor of f. Hence the roots of m are always roots of f. The next theorem gives the converse of this result.

Theorem 8.7.9: Each eigenvalue of $\boldsymbol{A}$ is a root of the minimal polynomial m of $\boldsymbol{A}$ and vice versa.

Proof: Sufficiency has already been proved. To prove necessity, let λ be an eigenvalue of $\boldsymbol{A}$. Then we wish to show that the polynomial

$$g(t) = t - \lambda \tag{8.7.32}$$

is a divisor of m. Since g is of the first degree, if g is not a divisor of m, then m and g are relatively prime. By eq. (8.7.17) and the fact that $m \in P(\boldsymbol{A})$, we have

$$\{0\} = K(m(\boldsymbol{A})) \cap K(g(\boldsymbol{A})) = V \cap K(g(\boldsymbol{A})) = K(g(\boldsymbol{A})) \tag{8.7.33}$$

But this is impossible, since $K(g(\boldsymbol{A}))$, being the characteristic subspace corresponding to the eigenvalue λ, cannot be of zero dimension.

Note: The root λ generally has a smaller multiplicity in m than in f, because m is a divisor of f. The characteristic polynomial f yields not only the (distinct) roots $\lambda_1, \dots, \lambda_L$ of m, it determines also the dimensions of their corresponding subspaces $K(m_1(\boldsymbol{A})), \dots, K(m_L(\boldsymbol{A}))$ in the decomposition eq. (8.7.22). This result is made explicit in the following theorem.

Theorem 8.7.10: Let d_k denote the algebraic multiplicity of the eigenvalue λ_k as before, i. e. d_k is the multiplicity of λ_k in f[cf. eq. (8.7.8)]. Then we have

$$\dim K(m_k(\boldsymbol{A})) = d_k \tag{8.7.34}$$

Proof: We prove this result by induction. Clearly, it is valid for all $\boldsymbol{A}$ having a diagonal form, since in this case $m_k(\boldsymbol{A}) = \boldsymbol{A} - \lambda_k \boldsymbol{I}$, so that $K(m_k(\boldsymbol{A}))$ is the characteristic subspace corresponding to λ_k and its dimension is the geometric multiplicity as well as the algebraic multiplicity of λ_k. Now assuming that the result is valid for all $\boldsymbol{A}$ whose minimal polynomial has at most M multiple roots, where $M = 0$ is the starting induction hypothesis, we wish to show that the same holds for all $\boldsymbol{A}$ whose minimal polynomial has $M + 1$ multiple roots. To see this, we make use of the decomposition eq. (8.7.23) and, for definiteness, we assume that λ_1 is a multiple root of m. We put

$$U = K(m_2(\boldsymbol{A})) \oplus \ldots \oplus K(m_L(\boldsymbol{A})) \tag{8.7.35}$$

Then U is $\boldsymbol{A}$-invariant. From Theorem 8.7.7 we know that the minimal polynomial m_U of the restriction $\boldsymbol{A}_U$ is

$$m_U = m_2 \ldots m_L \tag{8.7.36}$$

which has at most M multiple roots. Hence by the induction hypothesis we have

$$\dim U = \dim K(m_2(\boldsymbol{A})) + \ldots + \dim K(m_L(\boldsymbol{A})) = d_2 + \ldots + d_L \tag{8.7.37}$$

But from eq. (8.3.8) and eq. (8.7.23) we have also

$$N = d_1 + d_2 + \ldots + d_L$$
$$N = \dim K(m_1(\boldsymbol{A})) + \dim K(m_2(\boldsymbol{A})) + \ldots + \dim K(m_L(\boldsymbol{A})) \tag{8.7.38}$$

Comparing eq. (8.7.38) with eq. (8.7.37), we see that

$$d_1 = \dim K(m_1(\boldsymbol{A})) \tag{8.7.39}$$

Thus the result (8.7.34) is valid for all $\boldsymbol{A}$ whose minimal polynomial has $M+1$ multiple roots.

An immediate consequence of the preceding theorem is the following.

Theorem 8.7.11: Let b_k be the geometric multiplicity of λ_k, namely

$$b_k \equiv \dim V(\lambda_k) \equiv \dim K(g_k(\boldsymbol{A})) \equiv \dim K(\boldsymbol{A} - \lambda_k \boldsymbol{I}) \tag{8.7.40}$$

Then we have

$$1 \leqslant b_k \leqslant d_k - \alpha_k + 1 \tag{8.7.41}$$

where d_k is the algebraic multiplicity of λ_k and α_k is the multiplicity of λ_k in m, as shown in eq. (8.7.21). Furthermore, $b_k = d_k$ if and only if

$$K(g_k(\boldsymbol{A})) = K(g_k^2(\boldsymbol{A})) \tag{8.7.42}$$

Proof: If λ_k is a simple root of m, i. e. $\alpha_k = 1$ and $g_k = m_k$, then from eq. (8.7.34) and eq. (8.7.40) we have $b_k = d_k$. On the other hand, if λ_k is a multiple root of m, i. e. $\alpha_k > 1$ and $m_k = g_k^{\alpha_k}$, then the polynomials $g_k, g_k^2, \ldots, g_k^{(\alpha_k - 1)}$ are proper divisors of m_k. Hence by Theorem 8.7.3

$$V(\lambda_k) = K(g_k(\boldsymbol{A})) \subset K(g_k^2(\boldsymbol{A})) \subset \ldots \subset K(g_k^{(\alpha_k-1)}(\boldsymbol{A}) \subset K(m_k(\boldsymbol{A})) \tag{8.7.43}$$

where the inclusions are strictly proper and the dimensions of the subspaces change by at least one in each inclusion. Thus eq. (8.7.41) holds.

The second part of the theorem can be proved as follows: If λ_k is a simple root of m, then $m_k = g_k$, and, thus $K(g_k(\boldsymbol{A})) = K(g_k^2(\boldsymbol{A})) = V(\lambda_k)$. On the other hand, if λ_k is not a simple root of m, then m_k is at least of second degree. In this case g_k and g_k^2 are both divisors of m. But since g_k is also a proper divisor of g_k^2, by Theorem 8.7.3, $K(g_k(\boldsymbol{A}))$ is strictly a proper subspace of $K(g_k^2(\boldsymbol{A}))$, so that eq. (8.7.42) can not hold, and the proof is complete.

The preceding three theorems show that for each eigenvalue λ_k of $\boldsymbol{A}$, generally there are two nonzero $\boldsymbol{A}$-invariant subspaces, namely, the eigenspace $V(\lambda_k)$ and the subspace $K(m_k(\boldsymbol{A}))$. For definiteness, let us call the latter subspace the characteristic subspace corresponding to λ_k and denote it by the more compact notation $U(\lambda_k)$. Then $V(\lambda_k)$ is a subspace of $U(\lambda_k)$ in general, and the two subspaces coincide iff λ_k is a simple root of m. Since λ_k is the only eigenvalue of the restriction of $\boldsymbol{A}$ to $U(\lambda_k)$, by the Cayley-Hamilton theorem we have also

$$U(\lambda_k) = K((\boldsymbol{A} - \lambda_k \boldsymbol{I})^{d_k}) \tag{8.7.44}$$

where d_k is the algebraic multiplicity of λ_k, which is also the dimension of $U(\lambda_k)$. Thus we can determine the characteristic subspace directly form the characteristic polynomial of $\boldsymbol{A}$ by eq. (8.7.44).

Now if we define $\boldsymbol{P}_k$ to be the projection on $U(\lambda_k)$ in the direction of the remaining $U(\lambda_j)$, $j \neq k$, namely

$$R(\boldsymbol{P}_k) = U(\lambda_k), K(\boldsymbol{P}_k) = \bigoplus_{\substack{i=1 \\ j\neq k}}^{L} U(\lambda_j) \tag{8.7.45}$$

and we define $\boldsymbol{B}_k$ to be $\boldsymbol{A} - \lambda_k \boldsymbol{P}_k$ on $U(\lambda_k)$ and 0 on $U(\lambda_k)$, $j \neq k$, then $\boldsymbol{A}$ has the spectral decomposition by a direct sum

$$\boldsymbol{A} = \sum_{j=1}^{L} (\lambda_j \boldsymbol{P}_j + \boldsymbol{B}_j) \tag{8.7.46}$$

where

$$\left.\begin{aligned} &\boldsymbol{P}_j^2 = \boldsymbol{P}_j \\ &\boldsymbol{P}_j \boldsymbol{B}_j = \boldsymbol{B}_j \boldsymbol{P}_j = \boldsymbol{B}_j \\ &\boldsymbol{B}_j^{\alpha_j} = 0, 1 \leqslant \alpha_j \leqslant d_j \end{aligned}\right\} j = 1, \ldots, L \tag{8.7.47}$$

$$\left.\begin{aligned} &\boldsymbol{P}_j \boldsymbol{P}_k = \boldsymbol{0} \\ &\boldsymbol{P}_j \boldsymbol{B}_k = \boldsymbol{B}_k \boldsymbol{P}_j = \boldsymbol{0} \\ &\boldsymbol{B}_j \boldsymbol{B}_k = \boldsymbol{0}, \end{aligned}\right\} j \neq k, j, k = 1, \ldots, L \tag{8.7.48}$$

In general, an endomorphism $\boldsymbol{B}$ satisfying the condition

$$\boldsymbol{B}^{\alpha} = \boldsymbol{0} \tag{8.7.49}$$

for some power α is called nilpotent. From eq. (8.7.49) or from Theorem 8.7.9

the only eigenvalue of a nilpotent endomorphism is 0, and the lowest power a satisfying (8.7.49) is an integer $\alpha, 1 \leqslant \alpha \leqslant N$, such that t^N is the characteristic polynomial of $\boldsymbol{B}$ and t^α is the minimal polynomial of $\boldsymbol{B}$. In view of eq. (8.7.47) we see that each endomorphism $\boldsymbol{B}_j$ in the decomposition eq. (8.7.46) is nilpotent and can be regarded also as a nilpotent endomorphism on $U(\lambda_j)$. In order to decompose $\boldsymbol{A}$ further from eq. (8.7.46), we must determine a spectral decomposition for each $\boldsymbol{B}_j$. This problem is solved in general as follows.

First, we define a nilcyclic endomorphism $\boldsymbol{C}$ to be a nilpotent endomorphism such that

$$\boldsymbol{C}^N = \boldsymbol{0} \quad \text{but} \quad \boldsymbol{C}^{N-1} \neq \boldsymbol{0} \tag{8.7.50}$$

where N is the dimension of the underlying vector space V. For such an endomorphism we can find a cyclic basis $\{\boldsymbol{e}_1, \ldots, \boldsymbol{e}_N\}$ which satisfies the conditions

$$\boldsymbol{C}^{N-1}\boldsymbol{e}_1 = \boldsymbol{0}, \boldsymbol{C}^{N-2}\boldsymbol{e}_2 = \boldsymbol{e}_1, \ldots, \boldsymbol{C}\boldsymbol{e}_N = \boldsymbol{e}_{N-1} \tag{8.7.51}$$

or equivalently

$$\boldsymbol{C}^{N-1}\boldsymbol{e}_N = \boldsymbol{e}_1, \boldsymbol{C}^{N-2}\boldsymbol{e}_N = \boldsymbol{e}_2, \ldots, \boldsymbol{C}\boldsymbol{e}_N = \boldsymbol{e}_{N-1} \tag{8.7.52}$$

so that the matrix of $\boldsymbol{C}$ takes the simple form eq. (8.3.19). Indeed, we can choose $\boldsymbol{e}_N$ to be any vector such that $\boldsymbol{C}^{N-1}\boldsymbol{e}_N \neq \boldsymbol{0}$; then the set $\{\boldsymbol{e}_1, \ldots, \boldsymbol{e}_N\}$ defined by eq. (8.7.52) is linearly independent and thus forms a cyclic basis for $\boldsymbol{C}$. Nilcyclic endomorphisms constitute only a special class of nilpotent endomorphisms, but in some sense the former can be regarded as the building blocks for the latter. The result is made precise by the following theorem.

Theorem 8.7.12: Let $\boldsymbol{B}$ be a nonzero nilpotent endomorphism of V in general, say $\boldsymbol{B}$ satisfies the conditions

$$\boldsymbol{B}^\alpha = \boldsymbol{0} \quad \text{but} \quad \boldsymbol{B}^{\alpha-1} \neq \boldsymbol{0} \tag{8.7.53}$$

for some integer α between 1 and N. Then there exists a direct sum decomposition for V:

$$V = V_1 \oplus \ldots \oplus V_M \tag{8.7.54}$$

and a corresponding direct sum decomposition for $\boldsymbol{B}$ (in the sense explained in the above chapter):

$$\boldsymbol{B} = \boldsymbol{B}_1 + \ldots + \boldsymbol{B}_M \tag{8.7.55}$$

such that each $\boldsymbol{B}_j$ is nilpotent and its restriction to V_j is nilcyclic. The subspaces $V_1, \ldots, V_M$ in the decomposition eq. (8.7.54) are not unique, but their dimensions are unique and obey the following rules: The maximum of the dimension of

$V_1, \ldots, V_M$ is equal to the integer a in eq. (8.7.53); the number N_α of subspaces among $V_1, \ldots, V_M$ having dimension α is given by

$$N_\alpha = N - \dim K(\boldsymbol{B}^{\alpha-1}) \tag{8.7.56}$$

More generally, the number N_b of subspaces among $V_1, \ldots, V_M$ having dimensions greater than or equal to b is given by

$$N_b = \dim K(\boldsymbol{B}^b) - \dim K(\boldsymbol{B}^{b-1}) \tag{8.7.57}$$

for all $b = 1, \ldots, \alpha$. In particular, when $b = 1$, N_1 is equal to the integer M in eq. (8.7.54), and eq. (8.7.57) reduces to

$$M = \dim K(\boldsymbol{B}) \tag{8.7.58}$$

Proof: We prove the theorem by induction on the dimension of V. Clearly, the theorem is valid for one-dimensional space since a nilpotent endomorphism there is simply the zero endomorphism which is nilcyclic. Assuming now the theorem is valid for vector spaces of dimension less than or equal to $N-1$, we shall prove that the same is valid for vector spaces of dimension N.

Notice first if the integer α in eq. (8.7.53) is equal to N, then $\boldsymbol{B}$ is nilcyclic and the assertion is trivially satisfied with $M=1$, so we can assume that $1 < \alpha < N$. By $(8.7.53)_2$, there exists a vector $\boldsymbol{e}_a \in V$ such that $\boldsymbol{B}^{a-1}\boldsymbol{e}_a \neq \boldsymbol{0}$. As in eq. (8.7.52) we define

$$\boldsymbol{B}^{a-1}\boldsymbol{e}_a = \boldsymbol{e}_1, \ldots, \boldsymbol{B}\boldsymbol{e}_{a-1} \tag{8.7.59}$$

Then the set $\{\boldsymbol{e}_1, \ldots, \boldsymbol{e}_a\}$ is linearly independent. We put V_1 to be the subspace generated by $\{\boldsymbol{e}_1, \ldots, \boldsymbol{e}_a\}$, then by definition $\dim V_1 = \alpha$, and the restriction of $\boldsymbol{B}$ on V_1 is nilcyclic.

We know that for any subspace of a vector space we can define a factor space. As usual we denote the factor space of V over V_1 by V/V_1. We have

$$\dim V/V_1 = N - a < N - 1 \tag{8.7.60}$$

Thus we can apply the theorem to the factor space V/V_1. For definiteness, if $\boldsymbol{v} \in V$, then $\overline{\boldsymbol{v}}$ denotes the equivalence set of $\boldsymbol{v}$ in V/V_1. From eq. (8.7.59) it is easy to see that V_1 is $\boldsymbol{B}$-invariant. Hence if $\boldsymbol{u}$ and $\boldsymbol{v}$ belong to the same equivalence set, so do $\boldsymbol{Bu}$ and $\boldsymbol{Bv}$. Therefore we can define an endomorphism $\overline{\boldsymbol{B}}$ on the factor space V/V_1, by

$$\overline{\boldsymbol{B}}\overline{\boldsymbol{v}} = \overline{\boldsymbol{B}\boldsymbol{v}} \tag{8.7.61}$$

for all $\boldsymbol{v} \in V$ or equivalently for all $\overline{\boldsymbol{v}} \in V/V_1$. Applying eq. (8.7.60) repeatedly, we have also

$$\overline{B}^k\overline{v} = \overline{B^k v} \tag{8.7.62}$$

for all integers k. In particular, $\overline{B}$ is nilpotent and

$$\overline{B}^a = \mathbf{0} \tag{8.7.63}$$

By the induction hypothesis we can then find a direct sum decomposition of the form

$$V/V_1 = U_1 \oplus \ldots \oplus U_P \tag{8.7.64}$$

for the factor space V/V_1 and a corresponding direct sum decomposition

$$\overline{B} = F_1 + \ldots + F_P \tag{8.7.65}$$

for $\overline{B}$. In particular, there are cyclic bases in the subspaces $U_1, \ldots, U_P$ for the nilcyclic endomorphisms which are the restrictions of $F_1, \ldots, F_P$ to the corresponding subspaces. For definiteness, let $\{\overline{f}_1, \ldots, \overline{f}_b\}$ be a cyclic basis in U_1, say

$$\overline{B}^b\overline{f}_b = \mathbf{0}, \quad \overline{B}^{b-1}\overline{f}_b = \overline{f}_1, \ldots, \overline{B}\overline{f}_b = \overline{f}_{b-1} \tag{8.7.66}$$

From (8.7.63), b is necessarily less than or equal to α.

From $(8.7.66)_1$ and (8.7.62) we see that $\overline{B^b f_b}$ belongs to V_1 and, thus can be expressed as a linear combination of $\{e_1, \ldots, e_\alpha\}$, say

$$\overline{B^b f_b} = \alpha_1 e_1 + \ldots + \alpha_\alpha e_\alpha = (\alpha_1 B^{\alpha-1} + \ldots + \alpha_{\alpha-1} B + \alpha_\alpha I) e_\alpha \tag{8.7.67}$$

Now there are two possibilities: (i) $\overline{B^b f_b} = 0$ or (ii) $\overline{B^b f_b} \neq 0$.

In case (i) we define as before

$$B^{b-1} f_b = f_1, \ldots, B f_b = f_{b-1} \tag{8.7.68}$$

Then $\{f_1, \ldots, f_b\}$ is a linearly independent set in V, and we define V_2 to be the subspace generated by $\{f_1, \ldots, f_b\}$.

In case (ii) from eq. (8.7.53) we see that b is strictly less than α; moreover, from eq. (8.7.67) we have

$$\mathbf{0} = B^a f_b = (\alpha_{a-b+1} B^{a-1} + \ldots + \alpha_a B^{a-b}) e_a = \alpha_{a-b+1} e_1 + \ldots + \alpha_a e_b \tag{8.7.69}$$

which implies

$$\alpha_1 = \ldots = \alpha_{a-b+1} = \ldots = \alpha_a = 0 \tag{8.7.70}$$

or equivalently

$$B^a f_b = (\alpha_1 B^{a-1} + \ldots + \alpha_{a-b} B^b) e_a \tag{8.7.71}$$

Hence we can choose another vector f'_b in the same equivalence set of f_b by

$$f'_b = f_b - (\alpha_1 B^{a-b-1} + \ldots + \alpha_{a-b} I) e_a = f_b - \alpha_1 e_{b+1} - \ldots - \alpha_{a-b} e_a \tag{8.7.72}$$

which now obeys the condition $B^b f'_b = \mathbf{0}$, and we can proceed in exactly the same way as in case (i). Thus in any case every cyclic basis $\{\bar{f}_1, \ldots, \bar{f}_b\}$ for U_1 gives rise to a cyclic set $\{f_1, \ldots, f_b\}$ in V.

Applying this result to each one of the subspaces $U_1, \ldots, U_P$ we obtain cyclic sets $\{f_1, \ldots, f_b\}, \{g_1, \ldots, g_c\}, \ldots,$ and subspaces $V_2, \ldots, V_{P+1}$ generated by them in V. Now it is clear that the union of $\{e_1, \ldots, e_a\}, \{f_1, \ldots, f_b\}, \{g_1, \ldots, g_c\}, \ldots,$ form a basis of V since from eq. (8.7.59), eq. (8.7.60), and eq. (8.7.64) there are precisely N vectors in the union; furthermore, if we have

$$\alpha_1 e_1 + \ldots + \alpha_a e_a + \beta_1 f_1 + \ldots + \beta_b f_b + \gamma_1 g_1 + \ldots + \gamma_c g_c = \mathbf{0} \quad (8.7.73)$$

then taking the canonical projection to V/V_1 yields

$$\beta_1 \bar{f}_1 + \ldots + \beta_b \bar{f}_b + \gamma_1 \bar{g}_1 + \ldots + \gamma_c \bar{g}_c + \ldots = \mathbf{0}$$

which implies

$$\beta_1 = \ldots = \beta_b = \gamma_1 = \ldots = \gamma_c = \ldots = 0$$

and substituting this result back into eq. (8.7.73) yields

$$\alpha_1 = \ldots = \alpha_a = 0$$

Thus V has a direct sum decomposition given by eq. (8.7.54) with $M = P + 1$ and $\boldsymbol{B}$ has a corresponding decomposition given by eq. (8.7.55) where $\boldsymbol{B}_1, \ldots, \boldsymbol{B}_M$ have the prescribed properties.

Now the only assertion yet to be proved is eq. (8.7.57). This result follows the general rule that for any nilcyclic endomorphism $\boldsymbol{C}$ on an L-dimensional space we have

$$\dim K(\boldsymbol{C}^k) - \dim K(\boldsymbol{C}^{k-1}) = \begin{cases} 1 & \text{for} \quad 1 \leqslant k \leqslant L \\ 0 & \text{for} \quad k > L \end{cases}$$

Applying this rule to the restriction of $\boldsymbol{B}_j$ to V for all $j = 1, \ldots, M$ and using the fact that the kernel of $\boldsymbol{B}^k$ is equal to the direct sum of the kernel of the restriction of $(\boldsymbol{B}_j)^k$ for all $j = 1, \ldots, M$, prove easily that eq. (8.7.57) holds. Thus the proof is complete.

In general, we cannot expect the subspaces $V_1, \ldots, V_M$ in the decomposition eq. (8.7.54) to be unique. Indeed, if there are two subspaces among $V_1, \ldots, V_M$ having the same dimension, say $\dim V_1 = \dim V_2 = a$, then we can decompose the direct sum $V_1 \oplus V_2$ in many other ways, e. g.

$$V_1 \oplus V_2 = \bar{V}_1 \oplus \bar{V}_2 \quad (8.7.74)$$

and when we substitute eq. (8.7.74) into eq. (8.7.54) the new decomposition

$$V = \bar{V}_1 \oplus \bar{V}_2 \oplus V_3 \oplus \ldots \oplus V_N$$

possesses exactly the same properties as the original decomposition eq. (8.7.54). For instance we can define V_1 and V_2 to be the subspaces generated by the linearly independent cyclic set $\{\tilde{\boldsymbol{e}}_1, \ldots, \tilde{\boldsymbol{e}}_a\}$ and $\{\tilde{\boldsymbol{f}}_1, \ldots, \tilde{\boldsymbol{f}}_a\}$, where we choose the starting vectors $\tilde{\boldsymbol{e}}_a$ and $\tilde{\boldsymbol{f}}_a$ by

$$\tilde{\boldsymbol{e}}_a = \alpha \boldsymbol{e}_a + \beta \boldsymbol{f}_a, \quad \tilde{\boldsymbol{f}}_a = \gamma \boldsymbol{e}_a + \delta \boldsymbol{f}_a$$

provided that the coefficient matrix on the right hand side is nonsingular.

If we apply the preceding theorem to the restriction of $\boldsymbol{A} - \lambda_k \boldsymbol{I}$ on U_k, we see that the inequality eq. (8.7.41) is the best possible one in general. Indeed, eq. (8.7.41) becomes an equality iff U_k has the decomposition

$$U_k = U_{k1} \oplus \ldots \oplus U_{kM} \tag{8.7.75}$$

where the dimensions of the subspaces $U_{k1}, \ldots, U_{kM}$ are

$$\dim U_{k1} = a_k, \quad \dim U_{k2} = \ldots = \dim U_{kM} = 1$$

If there are more than one subspaces among $U_{k1}, \ldots, U_{kM}$ having dimension greater than one, then eq. (8.7.41) is a strict inequality.

The matrix of the restriction of $\boldsymbol{A} - \lambda_k \boldsymbol{I}$ to U_k relative to the union of the cyclic basis for $U_{k1}, \ldots, U_{kM}$ has the form

$$\boldsymbol{A}_k = \begin{bmatrix} \boldsymbol{A}_{k1} & & & & & \boldsymbol{0} \\ & \boldsymbol{A}_{k2} & & & & \\ & & \cdot & & & \\ & & & \cdot & & \\ & & & & \cdot & \\ \boldsymbol{0} & & & & & \boldsymbol{A}_{kM} \end{bmatrix} \tag{8.7.76}$$

where each submatrix $\boldsymbol{A}_{kj}$ in the diagonal of $\boldsymbol{A}_k$ has the form

$$\boldsymbol{A}_{kj} = \begin{bmatrix} \lambda_k & 1 & & & & 0 \\ & \lambda_k & 1 & & & \\ & & \cdot & & & \\ & & & \cdot & & \\ & & & & \cdot & 1 \\ 0 & & & & & \lambda_k \end{bmatrix} \tag{8.7.77}$$

Substituting eq. (8.7.76) into $\boldsymbol{M}(\boldsymbol{A}) = \begin{bmatrix} \boldsymbol{A}_1 & & \boldsymbol{0} \\ & \boldsymbol{A}_2 & \\ & & \ddots & \\ \boldsymbol{0} & & & \boldsymbol{A}_L \end{bmatrix}$ yields the Jordan normal form for $\boldsymbol{A}$:

$$\boldsymbol{M}(\boldsymbol{A}) = \begin{bmatrix} \boldsymbol{A}_{11} & & \boldsymbol{0} & & & & & \\ & \boldsymbol{A}_{12} & & & & & & \\ & & \cdot & & & & \boldsymbol{0} & \\ & & & \cdot & & & & \\ & & & & \cdot & & & \\ \boldsymbol{0} & & & & & & & \\ & & & & & \boldsymbol{A}_2 & & \\ & & & & & & \cdot & \\ & & & & & & & \cdot \\ & & & & & & & & \cdot \\ & & \boldsymbol{0} & & & & & & \boldsymbol{A}_M \end{bmatrix} \tag{8.7.78}$$

The Jordan normal form is an important result since it gives a geometric interpretation of an arbitrary endomorphism of a vector space. In general, we say that two endomorphisms $\boldsymbol{A}$ and $\boldsymbol{A}'$ are similar if the matrix of $\boldsymbol{A}$ relative to a basis is identical to the matrix of $\boldsymbol{A}'$ relative to another basis. From the transformation law eq. (7.3.7), we see that $\boldsymbol{A}$ and $\boldsymbol{A}'$ are similar iff there exists a nonsingular endomorphism $\boldsymbol{T}$ such that

$$\boldsymbol{A}' = \boldsymbol{T}\boldsymbol{A}\boldsymbol{T}^{-1} \tag{8.7.79}$$

Clearly, eq. (8.7.79) defines an equivalence relation on $L(V,V)$. We call the equivalence sets relative to eq. (8.7.79) the conjugate subsets of $L(V,V)$. Now for each $\boldsymbol{A} \in L(V,V)$ the Jordan normal form of $\boldsymbol{A}$ is a particular matrix of $\boldsymbol{A}$ and is unique to within an arbitrary change of ordering of the various square blocks on the diagonal of the matrix. Hence $\boldsymbol{A}$ and $\boldsymbol{A}'$ are similar iff they have the same Jordan normal form. Thus the Jordan normal form characterizes the conjugate subsets of $L(V,V)$.

Chapter 9

Tensor Algebra

The concept of a tensor is of major importance in applied mathematics. Virtually ever discipline in the physical sciences makes some use of tensors. We begin with a brief discussion of linear functions on a vector space. Since in the applications the scalar field is usually the real field, from now on we shall consider real vector spaces only.

9.1 Linear Functions, the Dual Space

Let V be a real vector space of dimension N. We consider the space of linear functions $L(V,R)$ from V into the real numbers R. By Theorem 6.2.1, $L(V,R) = \dim V = N$. Thus V and $L(V,R)$ are isomorphic. We call $L(V,R)$ the dual space of V, and we denote it by the special notation V^*. To distinguish elements of V from those of V^*, we shall call the former elements vectors and the latter elements covectors. However, these two names are strictly relative to each other. Since V^* is an N-dimensional vector space by itself, we can apply any result valid for a vector space in general to V^* as well as to V. In fact, we can even define a dual space (V^*) for V^* just as we define a dual space V^* for V. In order not to introduce too many new concepts at the same time, we shall postpone the second dual space (V^*) until the next section. Hence in this section V shall be a given N-dimensional space and V^* shall denote its dual space. As usual, we denote typical elements of V by $\boldsymbol{u},\boldsymbol{v},\boldsymbol{w},\ldots$. Then the typical elements of V^* are denoted by $\boldsymbol{u}^*$, $\boldsymbol{v}^*,\boldsymbol{w}^*,\ldots$. However, it should be noted that the asterisk here is strictly a convenient notation, not a symbol for a function from V to V^*. Thus $\boldsymbol{u}^*$ is not related in any particular way to $\boldsymbol{u}$. Also, for some covectors, such as those that constitute a dual basis to be defined shortly, this mutation becomes rather cumbersome. In such cases, the notation is simply abandoned. For instance, without fear of ambiguity we denote the null covector in V^* by the same notation as the null vector in V, namely $\mathbf{0}$, instead of $\mathbf{0}^*$.

If $\boldsymbol{v}^* \in V^*$, then $\boldsymbol{v}^*$ is a linear function from V to R, i. e.

$$\boldsymbol{v}^*: V \to R$$

such that for any vectors $\boldsymbol{u},\boldsymbol{v} \in V$ and scalars $\alpha,\beta \in R$

$$\boldsymbol{v}^*(\alpha\boldsymbol{u}+\beta\boldsymbol{v}) = \alpha\boldsymbol{v}^*(\boldsymbol{u}) + \beta\boldsymbol{v}^*(\boldsymbol{v})$$

Of course, the linear operations on the right hand side are those of R while those on the left hand side are linear operations in V. For a reason they will become apparent later, it is more convenient to denote the value of $\boldsymbol{v}^*$ at $\boldsymbol{v}$ by the notation $\langle \boldsymbol{v}^*,\boldsymbol{v}\rangle$. Then the bracket $\langle\ ,\ \rangle$ operation can be viewed as a function

$$\langle\ ,\ \rangle : V \times V \to R \tag{9.1.1}$$

It is easy to verify that this operation has the following properties:

1) $\langle \alpha\boldsymbol{v}^* + \beta\boldsymbol{u}^*, \boldsymbol{v}\rangle = \alpha\langle \boldsymbol{v}^*,\boldsymbol{v}\rangle + \beta\langle \boldsymbol{u}^*,\boldsymbol{v}\rangle$

2) $\langle \boldsymbol{v}^*, \alpha\boldsymbol{u} + \beta\boldsymbol{v}\rangle = \alpha\langle \boldsymbol{v}^*,\boldsymbol{u}\rangle + \beta\langle \boldsymbol{v}^*,\boldsymbol{v}\rangle$

3) For any given $\boldsymbol{v}$, $\langle \boldsymbol{v}^*,\boldsymbol{v}\rangle$ vanishes for all $\boldsymbol{v} \in V$ iff $\boldsymbol{v}^* = \mathbf{0}$.

4) Similarly, for any given $\boldsymbol{v}$, $\langle \boldsymbol{v}^*,\boldsymbol{v}\rangle$ vanishes for all $\boldsymbol{v}^* \in V^*$ iff $\boldsymbol{v} = \mathbf{0}$.

The first two properties define $\langle\ ,\ \rangle$ to be a bilinear operation on $V^* \times V$, and the last two properties define $\langle\ ,\ \rangle$ to be a definite operation. These properties resemble the properties of an inner product, so that we call the operation $\langle\ ,\ \rangle$ the scalar product. As we shall see, we can define many concepts associated with the scalar product similar to corresponding concepts associated with an inner product. The first example is the concept of the dual basis, which is the counterpart of the concept of the reciprocal basis.

If $\{\boldsymbol{e}_1,\ldots,\boldsymbol{e}_N\}$ is a basis for V, we define dual basis to be a basis $\{\boldsymbol{e}^1,\ldots,\boldsymbol{e}^N\}$ for V^* such that

$$\langle \boldsymbol{e}^j, \boldsymbol{e}_i\rangle = \delta_i^j \tag{9.1.2}$$

for all $i,j = 1,\ldots,N$. The reader should compare this condition with the condition eq. (5.7.1) that defines the reciprocal basis. By exactly the same argument as before we can prove the following theorem.

Theorem 9.1.1: The dual basis relative to a given basis exists and it is unique.

Notice that we have dropped the asterisk notation for the covector $\boldsymbol{e}^j$ in a dual basis; the superscript alone is enough to distinguish $\{\boldsymbol{e}^1, \ldots, \boldsymbol{e}^N\}$ from $\{\boldsymbol{e}_1, \ldots, \boldsymbol{e}_N\}$. However, it should be kept in mind that, unlike the reciprocal basis, the dual basis is a basis for V^*, not a basis for V. In particular, it makes no sense to require a basis to be the same as its dual basis. This means the component form of a vector $\boldsymbol{v} \in V$ relative to a basis $\{\boldsymbol{e}_1, \ldots, \boldsymbol{e}_N\}$.

$$\boldsymbol{v} = v^i \boldsymbol{e}_i \tag{9.1.3}$$

must never be confused with the component form of a covector $\boldsymbol{v}^* \in V^*$ relative to the dual basis,

$$\boldsymbol{v}^* = v_i \boldsymbol{e}^i \tag{9.1.4}$$

In order to emphasize the difference of these two component forms, we call v^i the contravariant components of $\boldsymbol{v}$ and v_i the covariant components of $\boldsymbol{v}^*$. A vector has contravariant components only and a covector has covariant components only. The terminology for the components is not inconsistent with the same terminology defined earlier for an inner product space, since we have the following theorem.

Theorem 9.1.2: Given any inner product on V, there exists a unique isomorphism

$$\boldsymbol{G}: V \rightarrow V^* \tag{9.1.5}$$

which is induced by the inner product in such a way that

$$\langle \boldsymbol{G}\boldsymbol{v}, \boldsymbol{w} \rangle = \boldsymbol{v} \cdot \boldsymbol{w}, \quad \boldsymbol{v}, \boldsymbol{w} \in V \tag{9.1.6}$$

Under this isomorphism the image of any orthonormal basis $\{\boldsymbol{i}_1, \ldots, \boldsymbol{i}_N\}$ is the dual basis $\{\boldsymbol{i}^1, \ldots, \boldsymbol{i}^N\}$, namely

$$\boldsymbol{G}\bar{\boldsymbol{e}}^k = \boldsymbol{e}^k, \quad k = 1, \ldots, N \tag{9.1.7}$$

Proof: Since we now consider only real vector spaces and real inner product spaces, the right-hand side of eq. (9.1.6), clearly, is a linear function of $\boldsymbol{w}$ for each $\boldsymbol{v} \in V$. Thus $\boldsymbol{G}$ is well defined by the condition eq. (9.1.6). We must show that $\boldsymbol{G}$ is an isomorphism. The fact that $\boldsymbol{G}$ is a linear transformation is obvious, since the right-hand side of eq. (9.1.6) is linear in $\boldsymbol{v}$ for each $\boldsymbol{w} \in V$. Also, $\boldsymbol{G}$ is one-to-one because, from eq. (9.1.6), if $\boldsymbol{G}\boldsymbol{u} = \boldsymbol{G}\boldsymbol{v}$, then $\boldsymbol{u} \cdot \boldsymbol{w} = \boldsymbol{v} \cdot \boldsymbol{w}$ for all $\boldsymbol{w}$ and thus $\boldsymbol{u} = \boldsymbol{v}$. Now since we already know that $\dim V = \dim V^*$, any one-to-one linear transformation from V to V^* is necessarily surjective and hence an isomorphism. The proof of eq. (9.1.7) is obvious, since by the definition of the reciprocal basis we have

$$\bar{e}^i \cdot e_j = \delta^i_j, \quad i,j = 1,\ldots,N$$

and by the definition of the dual basis we have

$$\langle e^i, e_j \rangle = \delta^i_j, \quad i,j = 1,\ldots,N$$

Comparing these definitions with eq. (9.1.6), we obtain

$$\langle G\bar{e}^i, e_j \rangle = \langle e^i, e_j \rangle \quad i,j = 1,\ldots,N$$

which implies eq. (9.1.7) because $\{e_1,\ldots,e_N\}$ is a basis of V.

Because of this theorem, if a particular inner product is assigned on V, then we can identify V with V^* by suppressing the notation for the isomorphisms G and G^{-1}. In other words, we regard a vector v also as a linear function on V:

$$\langle v, w \rangle \equiv v \cdot w \tag{9.1.8}$$

According to this rule the reciprocal basis is identified with the duel basis and the inner product becomes the scalar product. However, since a vector space can be equipped with many inner products, unless a particular inner product is chosen, we cannot identify V with V^* in general. In this section, we shall not assign any particular inner product in V, so V and V^* are different vector spaces.

We shall now derive some formulas which generalize the results of an inner product space to a vector space in general. First, if $v \in V$ and $v^* \in V^*$ are arbitrary, then their scalar products $\langle v^*, v \rangle$ can be computed in component form as follows: Choose a basis $\{e_i\}$ and its dual basis $\{e^i\}$ for V and V^*, respectively, so that we can express v and v^* in component form eq. (9.1.3) and eq. (9.1.4). Then from eq. (9.1.1) and eq. (9.1.2) we have

$$\langle v^*, v \rangle = \langle v_i e^i, v^j e_j \rangle = v_i v^j \langle e^i, e_j \rangle = v_i v^j \delta^i_j = v_i v^i \tag{9.1.9}$$

Applying eq. (9.1.9) to $v^* = e^i$, we obtain

$$\langle e^i, v \rangle = v^i \tag{9.1.10}$$

similarly applying eq. (9.1.11) to $v = e_j$, we obtain

$$\langle v^*, e_i \rangle = v_i \tag{9.1.11}$$

Next recall that for inner product spaces V and U we define the adjoint A^* of a linear transformation $A: V \to U$ to be a linear transformation $A^*: U \to V$ such that the following condition [cf. eq. (6.4.1)] is satisfied:

$$u \cdot Av = A^* u \cdot v, \quad u \in U, v \in V$$

If we do not make use of any inner product, we simply replace this condition by

$$\langle \boldsymbol{u}^*, \boldsymbol{A}\boldsymbol{v} \rangle = \langle \boldsymbol{A}^* \boldsymbol{u}^*, \boldsymbol{v} \rangle, \quad \boldsymbol{u}^* \in U^*, \boldsymbol{v} \in V \tag{9.1.12}$$

Then $\boldsymbol{A}^*$ is a linear transformation from U^* to V^*,

$$\boldsymbol{A}^*: U^* \to V^*$$

and is called the dual of $\boldsymbol{A}$. By the same argument as before we can prove the following theorem.

Theorem 9.1.3: For every linear transformation $\boldsymbol{A}: V \to U$ there exists a unique dual $\boldsymbol{A}^*: U^* \to V$ satisfying the condition eq. (9.1.12).

If we choose a basis $\{\boldsymbol{e}_1, \ldots, \boldsymbol{e}_N\}$ for V and a basis $\{\boldsymbol{b}_1, \ldots, \boldsymbol{b}_M\}$ for U and express the linear transformation $\boldsymbol{A}$ by eq. (6.4.2) and the linear transformation $\boldsymbol{A}^*$ by eq. (6.4.3), where $\{\boldsymbol{b}^\alpha\}$ and $\{\boldsymbol{e}^k\}$ are now regarded as the dual bases of $\{\boldsymbol{b}_\alpha\}$ and $\{\boldsymbol{e}_k\}$, respectively, then eq. (6.4.4) remains valid in the more general context, except that we now have

$$\boldsymbol{A}^{*\alpha}_{\ k} = \boldsymbol{A}^\alpha_k \tag{9.1.13}$$

since we no longer consider complex spaces. Of course, the formulas eq. (6.4.5) and eq. (6.4.6) are now replaced by

$$\langle \boldsymbol{b}^\alpha, \boldsymbol{A}\boldsymbol{e}_k \rangle = \boldsymbol{A}^\alpha_k \tag{9.1.14}$$

and

$$\langle \boldsymbol{A}^* \boldsymbol{b}^\alpha, \boldsymbol{e}_k \rangle = \boldsymbol{A}^{*\alpha}_{\ k} \tag{9.1.15}$$

respectively.

For an inner product space the orthogonal complement of a subspace U of V is a subspace $U^\perp$,

$$U^\perp = \{\boldsymbol{v} \mid \boldsymbol{u} \cdot \boldsymbol{v} = \boldsymbol{0} \quad \text{for all } \boldsymbol{u} \in U\}$$

By the same token, if V is a vector space in general, then we define the orthogonal complement of U to be the subspace $U^\perp$ of V^* given by

$$^\perp U = \{\boldsymbol{v}^* \mid \langle \boldsymbol{v}^*, \boldsymbol{u} \rangle = 0 \quad \text{for all } \boldsymbol{u} \in U\} \tag{9.1.16}$$

In general if $\boldsymbol{v} \in V$ and $\boldsymbol{v}^* \in V^*$ are arbitrary, then $\boldsymbol{v}$ and $\boldsymbol{v}^*$ are said to be orthogonal to each other if $\langle \boldsymbol{v}^*, \boldsymbol{v} \rangle = 0$. We can prove the following theorem by the same argument used previously for inner product spaces.

Theorem 9.1.4: If U is a subspace of V, then

$$\dim U + \dim U^\perp = \dim V \tag{9.1.17}$$

However, V is no longer the direct sum of U and $U^\perp$ since $U^\perp$ is a subspace of V^*, not a subspace of V.

Theorem 9.1.5: If $\boldsymbol{A}: V \to U$ is a liner transformation, then

$$K(\boldsymbol{A}^*) = R(\boldsymbol{A})^\perp \tag{9.1.18}$$

and

$$K(A) = R(A^*)^{\perp} \tag{9.1.19}$$

Theorem 9.1.6: Any linear transformation A and its dual A^* have the same rank.

Finally, formulas for transferring from one basis to another basis can be generalized from inner product spaces to vector spaces in general. If $\{e_1, \dots, e_N\}$ and $\{\hat{e}, \dots, \hat{e}_N\}$ are bases for V, then as before we can express one basis in component form relative to another, as shown by eq. (5.7.17) and eq. (5.7.18). Now suppose that $\{e^1, \dots, e^N\}$ and $\{\hat{e}^1, \dots, \hat{e}^N\}$ are the dual bases of $\{e_1, \dots, e_N\}$ and $\{\hat{e}, \dots, \hat{e}_N\}$, respectively. Then it can be verified easily that

$$\hat{e}^q = \hat{T}^q_k e^k, \qquad e^q = T^q_k \hat{e}^k \tag{9.1.20}$$

where T^q_k and $\hat{T}^q_k$ are defined by eq. (5.7.17) and eq. (5.7.18), i. e.

$$e_k = \hat{T}^q_k \hat{e}_q, \quad \hat{e}_k = T^q_k e_q \tag{9.1.21}$$

From these relations if $v \in V$ and $v^* \in V^*$ have the component forms eq. (9.1.3) and eq. (9.1.4) relative to $\{e_i\}$ and $\{e^i\}$ and the component forms

$$v = \hat{V}^i \hat{e}_i \tag{9.1.22}$$

and

$$v^* = \hat{V}_i \hat{e}^i \tag{9.1.23}$$

relative to $\{\hat{e}_i\}$ and $\{\hat{e}^i\}$, respectively, then we have the following transformation laws:

$$\hat{V}_q = T^k_q V_K \tag{9.1.24}$$

$$\hat{V}^q = \hat{T}^q_k V^k \tag{9.1.25}$$

$$\hat{V}^k = T^k_q \hat{V}^q \tag{9.1.26}$$

and

$$V_k = \hat{T}^q_k \hat{V}_q \tag{9.1.27}$$

9.2 The Second Dual Space, Canonical Isomorphisms

In the preceding section we defined the dual space V^* of any vector space V

to be the space of liner functions $L(V,R)$ from V to R. By the same procedure we can define the dual space $(V^*)^*$ of V^* by

$$(V^*)^* = L(V^*,R) = L(L(V,R),R) \tag{9.2.1}$$

For simplicity let us denote this space by V^{**}, called the second dual space of V. Of course, the dimension of V^{**} is the same as that of V, namely

$$\dim V = \dim V^* = \dim V^{**} \tag{9.2.2}$$

Using the system of notation introduced in the preceding section, we write a typical element of V^{**} by $\boldsymbol{v}^{**}$. Then $\boldsymbol{v}^{**}$ is a linear function on V^*

$$\boldsymbol{v}^{**}: V^* \to R$$

Further, for each $\boldsymbol{v}^* \in V^*$ we denote the value of V^{**} at $\boldsymbol{v}^*$ by $\langle \boldsymbol{v}^{**}, \boldsymbol{v}^* \rangle$. The $\langle , \rangle$ operation is now a mapping from $V^{**} \times V^*$ to R and possesses the same four properties given in the preceding section.

Unlike the dual space V^*, the second dual space V^{**} can always be identified as V without using any inner product. The isomorphism

$$\boldsymbol{J}: V \to V^{**} \tag{9.2.3}$$

is defined by the condition

$$\langle \boldsymbol{J}\boldsymbol{v}, \boldsymbol{v}^* \rangle = \langle \boldsymbol{v}^*, \boldsymbol{v} \rangle, \quad \boldsymbol{v} \in V, \boldsymbol{v}^* \in V^* \tag{9.2.4}$$

Clearly, $\boldsymbol{J}$ is well defined by eq. (9.2.4) since for each $\boldsymbol{v} \in V$ the right-hand side of eq. (9.2.4) is a linear function of $\boldsymbol{v}^*$. To see that $\boldsymbol{J}$ is an isomorphism, we notice first that $\boldsymbol{J}$ is a linear transformation, because for each $\boldsymbol{v}^* \in V$ the right-hand side is linear in $\boldsymbol{v}$. Now $\boldsymbol{J}$ is also one-to-one, since if $\boldsymbol{J}\boldsymbol{v} = \boldsymbol{J}\boldsymbol{u}$, then eq. (9.2.4) implies that

$$\langle \boldsymbol{v}^*, \boldsymbol{v} \rangle = \langle \boldsymbol{v}^*, \boldsymbol{u} \rangle, \quad \boldsymbol{v}^* \in V^* \tag{9.2.5}$$

which then implies $\boldsymbol{u} = \boldsymbol{v}$. From eq. (9.2.2), we conclude that the one-to-one linear transformation $\boldsymbol{J}$ is surjective, and thus $\boldsymbol{J}$ is an isomorphism. We summarize this result in the following theorem.

Theorem 9.2.1: There exists a unique isomorphism $\boldsymbol{J}$ from V to V^{**} satisfying the condition eq. (9.2.4).

Since the isomorphism $\boldsymbol{J}$ is defined without using any structure in addition to the vector space structure on V, its notation can often be suppressed without any ambiguity. We shall adopt such a convention here and identify any $\boldsymbol{v} \in V$ as a linear function on V^*

$$\boldsymbol{v}: V^* \to R$$

by the condition that defines $\boldsymbol{J}$, namely

$$\langle \boldsymbol{v}, \boldsymbol{v}^* \rangle = \langle \boldsymbol{v}^*, \boldsymbol{v} \rangle, \quad \text{for all } \boldsymbol{v}^* \in V^* \tag{9.2.6}$$

In doing so, we allow the same symbol $\boldsymbol{v}$ to represent two different objects: an element of the vector space V and a linear function on the vector space V^*, and the two objects are related to each other through the condition eq. (9.2.6).

To distinguish an isomorphism such as $\boldsymbol{J}$, whose notation may be suppressed without causing any ambiguity, from an isomorphism such as $\boldsymbol{G}$, defined by eq. (9.1.6), whose notation may not be suppressed, because there are many isomorphisms of similar nature, we call the former isomorphism a canonical or natural isomorphism. Whether or not an isomorphism is canonical is usually determined by a convention, not by any axioms. A general rule for choosing a canonical isomorphism is that the isomorphism must be defined without using any additional structure other than the basic structure already assigned to the underlying spaces; furthermore, by suppressing the notation of the canonical isomorphism no ambiguity is likely to arise. Hence the choice of a canonical isomorphism depends on the basic structure of the vector spaces. If we deal with inner product spaces equipped with particular inner products, the isomorphism $\boldsymbol{G}$ can safely be regarded as canonical, and by choosing $\boldsymbol{G}$ to be canonical, we can achieve much economy in writing. On the other hand, if we consider vector spaces without any pre-assigned inner product, then we cannot make all possible isomorphisms $\boldsymbol{G}$ canonical, otherwise the notation becomes ambiguous.

It should be noticed that not every isomorphism whose definition depends only on the basis structure of the underlying space can be made canonical. For example, the operation of taking the dual:

$$* : L(V, U) \to L(U^*, V^*)$$

is defined by using the vector space structure of V and U only. However, by suppressing the notation $*$, we encounter immediately much ambiguity, especially when U is equal to V. Surely, we do not wish to make every endomorphism $\boldsymbol{A}: V \to V$ self-adjoint! Another example will illustrate the point even clearer. The operation of taking the opposite vector of any vector is an isomorphism

$$- : V \to V$$

which is defined by using the vector space structure alone. Evidently, we cannot suppress the minus sign without any ambiguity.

To test whether the isomorphism $\boldsymbol{J}$ can be made a canonical one without any ambiguity, we consider the effect of this choice on the notations for the dual basis and the dual of a linear transformation. Of course, we wish to have $\{e_i\}$, when considered as a basis for V^{**}, to be the dual basis of $\{e^i\}$, and A, when considered as a linear transformation from V^{**} to U^{**}, to be the dual of A^*. These results are indeed correct and they are contained in the following.

Theorem 9.2.2: Given any basis $\{e_i\}$ for V, then the dual basis of its dual basis $\{e^i\}$ is $\{Je_i\}$.

Proof: This result is more or less obvious. By definition, the basis $\{e_i\}$ and its dual basis $\{e^i\}$ are related by

$$\langle e^i, e_j\rangle = \delta_j^i$$

for $i,j = 1,\ldots,N$. From eq. (9.2.4) we have

$$\langle Je_j, e^i\rangle = \langle e^i, e_j\rangle$$

Comparing the preceding two equations, we see that

$$\langle Je_j, e^i\rangle = \delta_j^i$$

which means that $\{Je_1,\ldots,Je_N\}$ is the dual basis of $\{e^1,\ldots,e^N\}$.

Because of this theorem, after suppressing the notation for J we say that $\{e_i\}$ and $\{e^j\}$ are dual relative to each other. The next theorem shows that the relation holds between A and A^*.

Theorem 9.2.3: Given any linear transformation $A: V \to U$, the dual of its dual A^* is $J_U A J_V^{-1}$. Here J_V denotes the isomorphism from V to V^{**} defined by eq. (9.2.4) and J_U denotes the isomorphism from U to U^{**} defined by a similar condition.

Proof: By definition A and A^* are relayed by eq. (9.1.13)

$$\langle u^*, Av\rangle = \langle A^* u^*, v\rangle$$

for all $u^* \in U^*$, $v \in V$. From eq. (9.2.4) we have

$$\langle u^*, Av\rangle = \langle J_U Av, u^*\rangle, \langle A^* u^*, v\rangle = \langle J_V v, A^* u^*\rangle$$

Comparing the preceding three equations, we see that

$$\langle J_U A J_V^{-1}(J_V v), u^*\rangle = \langle J_V v, A^* u^*\rangle$$

Since J_V is an isomorphism, we can rewrite the last equation as $J_U A J_V^{-1}$ is the dual of

$$\langle J_U A J_V^{-1} v^{**}, u^*\rangle = \langle v^{**}, A^* u^*\rangle$$

Because $v^{**} \in V^{**}$ and $u^* \in U^*$ are arbitrary, it follows that A^*. So if we suppress the notations for J_U and J_V, then A and A^* are the dual relative to each other.

A similar result exists for the operation of taking the orthogonal complement of a subspace; we have the following result.

Theorem 9.2.4: Given any subspace U of V, the orthogonal complement of its orthogonal complement $U^{\perp}$ is $\boldsymbol{J}(U)$.

We leave the proof of this theorem as an exercise. Because of this theorem we say that U and $U^{\perp}$ are orthogonal to each other. As we shall see in the next few sections, the use of canonical isomorphisms, like the summation convention, is an important device to achieve economy in writing. We shall make use of this device whenever possible, so the reader should be prepared to allow one symbol to represent two or more different objects.

The last three theorems show clearly the advantage of making $\boldsymbol{J}$ a canonical isomorphism, so from now on we shall substitute the symbol for $\boldsymbol{J}$. In general, if an isomorphism from a vector space V to a vector space U is chosen to be canonical, then we write

$$V \cong U \tag{9.2.7}$$

In particular, we have

$$V \cong V^{**} \tag{9.2.8}$$

Part Ⅲ

Chapter 10

Linear Programming

10.1 Basic Properties of Linear Programs

A linear program (LP) is an optimization problem in which the objective function is linear in the unknowns and the constraints consist of linear equalities and linear inequalities. The exact form of these constraints may differ from one problem to another, but as shown below, any linear program can be transformed into the following standard form:

$$\begin{aligned} \text{minimize} \quad & \boldsymbol{c}^{\mathrm{T}}\boldsymbol{x} \\ \text{subject to} \quad & \boldsymbol{A}\boldsymbol{x} = \boldsymbol{b} \\ & \boldsymbol{x} \geqslant \boldsymbol{0} \end{aligned} \tag{10.1.1}$$

where $\boldsymbol{x}$ is an n-dimensional column vector, $\boldsymbol{c}^{\mathrm{T}}$ is an n-dimensional row vector, $\boldsymbol{A}$ is an $m \times n$ matrix, and $\boldsymbol{b}$ is an m-dimensional column vector. The vector inequality $\boldsymbol{x} \geqslant \boldsymbol{0}$ means that each component of $\boldsymbol{x}$ is nonnegative.

Definition: Given a linear program in standard form eq. (10.1.1), feasible solution to the constraints that achieves the minimum value of the objective function subject to those constraints is said to be an optimal feasible solution. If this solution is basic, it is an optimal basic feasible solution.

Fundamental theorem of linear programming: Given a linear program in standard form eq. (10.1.1) where $\boldsymbol{A}$ is an $m \times n$ matrix of rank m,

1) if there exists a feasible solution, there is a basic feasible solution;

2) if there exists an optimal feasible solution, there is an optimal basic feasible solution.

Definition: A point $\boldsymbol{x}$ in a convex set C is said to be an extreme point of C if there are no two distinct points x_1 and x_2 in C such that $\boldsymbol{x} = \alpha x_1 + (1-\alpha)x_2$ for some $\alpha, 0<\alpha<1$.

Theorem 10.1.1: (Equivalence of extreme points and basic solutions). Let $\boldsymbol{A}$ be an $m \times n$ matrix of rank m and $\boldsymbol{b}$ an m-vector. Let K be the convex polytope consisting of all n-vectors $\boldsymbol{x}$ satisfying

$$\begin{aligned} \boldsymbol{Ax} &= \boldsymbol{b} \\ \boldsymbol{x} &\geqslant \boldsymbol{0} \end{aligned} \tag{10.1.2}$$

Vector $\boldsymbol{x}$ is an extreme point of K iff $\boldsymbol{x}$ is a basic feasible solution to eq. (10.1.2).

The above theorem enables to prove certain geometric properties of the convex polytope K, which defines the constraint set as eq. (10.1.2).

The Simplex Method: Pivots

Definition: Consider the set of simultaneous linear equations

$$\boldsymbol{Ax} = \boldsymbol{b} \tag{10.1.3}$$

where $\boldsymbol{A}$ is an $m \times n$ matrix and $m \leqslant n$. In the space E^n we interpret this as a collection of m linear relations that must be satisfied by a vector $\boldsymbol{x}$. Thus denoting by a^i the i-th row of $\boldsymbol{A}$, we may express eq. (10.1.2) as:

$$\begin{aligned} \boldsymbol{a}^1\boldsymbol{x} &= b_1 \\ \boldsymbol{a}^2\boldsymbol{x} &= b_2 \\ &\vdots \\ \boldsymbol{a}^m\boldsymbol{x} &= b_m \end{aligned} \tag{10.1.4}$$

This corresponds to the most natural interpretation of eq. (10.1.2) as a set of m equations.

If $m<n$ and the equations are linearly independent, then there is not a unique solution but a whole linear variety of solutions. A unique solution results, however, if $n-m$ additional independent linear equations are adjoined. Different basic solutions are obtained by imposing different additional equations of this special form.

If the eq. (10.1.4) are linearly independent, we may replace a given equation by any non-zero multiple of itself plus any linear combination of the other equations in the system. This leads to the well-known Gaussian reduction schemes, whereby multiples of equations are systematically subtracted from one another to yield either a triangular or canonical form. It is well known, and easily proved, that if the first m columns of $\boldsymbol{A}$ are linearly independent, the system eq. (10.1.2) can, by a sequence of such multiplications and subtractions, be converted to the following canonical form:

$$\begin{aligned} x_1 + y_{1,m+1}x_{m+1} + y_{1,m+2}x_{m+2} + \dots + y_{1,n}x_n &= b_{10} \\ x_2 + y_{2,m+1}x_{m+1} + y_{2,m+2}x_{m+2} + \dots + y_{2,n}x_n &= b_{20} \\ \vdots \qquad\qquad\qquad\qquad & \vdots \\ x_m + y_{m,m+1}x_{m+1} + \quad \dots \quad + y_{m,n}x_n &= b_{m0} \end{aligned} \tag{10.1.5}$$

Corresponding to this canonical representation of the system, the variables $\boldsymbol{x}_1$, $\boldsymbol{x}_2, \dots, \boldsymbol{x}_m$ are called basic and the other variables are nonbasic. The corresponding basic solution is then:

$$x_1 = b_{10}, x_2 = b_{20}, \dots, x_m = b_{m0}, \quad x_{m+1} = 0, \dots, x_n = 0$$

or in vector form: $\boldsymbol{x} = (\boldsymbol{b}_0, \boldsymbol{0})$ where $\boldsymbol{b}_0$ is m-dimensional and $\boldsymbol{0}$ is the $(n-m)$-dimensional zero vector.

Actually, we relax our definition somewhat and consider a system to be in canonical form if, among the n variables, there are m basic ones with the property that each appears in only one equation, its coefficient in that equation is unity, and no two of these m variables appear in any one equation. This is equivalent to saying that a system is in canonical form if by some reordering of the equations and the variables it takes the form eq. (10.1.5).

Also it is customary, from the dictates of economy, to represent the system eq. (10.1.5) by its corresponding array of coefficients or tableau:

$$\begin{array}{cccccccccc} 1 & 0 & \dots & 0 & y_{1,m+1} & y_{1,m+2} & \cdots & y_{1n} & y_{10} \\ 0 & 1 & \dots & 0 & y_{2,m+1} & y_{2,m+2} & \cdots & y_{2n} & y_{20} \\ 0 & 0 & \dots & 0 & \dots & \dots & \dots & \dots & \dots \\ \vdots & \vdots & \vdots & \vdots & \vdots & \vdots & \vdots & \vdots & \vdots \\ 0 & 0 & \dots & 1 & y_{m,m+1} & y_{m,m+2} & \cdots & y_{mn} & y_{m0} \end{array}$$

The question solved by pivoting is this: given a system in canonical form, suppose a basic variable is to be made non-basic and a non-basic variable is to be made basic; what is the new canonical form corresponding to the new set of basic variables? The procedure is quite simple. Suppose in the canonical system eq. (10.1.5), we wish to replace the basic variable $\boldsymbol{x}_p, 1 \leqslant p \leqslant m$, by the non-basic variable $\boldsymbol{x}_q$. This can be done iff y_{pq} is nonzero; it is accomplished by dividing row p by y_{pq} to get a unit coefficient for $\boldsymbol{x}_q$ in the p-th equation, and then subtracting suitable multiples of row p from each of the other rows in order to get a zero coefficient for $\boldsymbol{x}_q$ in all other equations. This transforms the q-th column of the tableau so that it is zero except in its p-th entry (which is unity) and does not affect the columns of the other basic variables. Denoting the coefficients of the new system in canonical form by y'_{ij}, we have explicitly

$$\begin{cases} y'_{ij} = y_{ij} - \dfrac{y_{pj}}{y_{pq}} y_{iq} \\ \qquad\qquad\qquad\qquad , i \neq p \\ y'_{pj} = \dfrac{y_{pj}}{y_{pq}}. \end{cases} \tag{10.1.6}$$

Eq. (10.1.6) are the pivot equations that arise frequently in linear programming. The element y_{pq} in the original system is said to be the pivot element.

Adjacent Extreme Points

It is only necessary to consider basic feasible solutions to the system

$$\begin{aligned} \boldsymbol{Ax} &= \boldsymbol{b} \\ \boldsymbol{x} &\geqslant \boldsymbol{0} \end{aligned} \tag{10.1.7}$$

when solving a linear program, however, that although the pivot operation takes one basic solution into another, the non-negativity of the solution will not in general be preserved. Special conditions must be satisfied in order that a pivot operation maintains feasibility. In this section we show how it is possible to select pivots so that we may transfer from one basic feasible solution to another.

Determining a Minimum Feasible Solution

Suppose we have a basic feasible solution

$$(\boldsymbol{x}_B, \boldsymbol{0}) = (y_{10}, y_{20}, \cdots, y_{m0}, 0, 0, \cdots, 0)$$

together with a tableau having an identity matrix appearing in the first m columns as shown below:

$$\begin{array}{cccccccc} a_1 & a_2 & \cdots & a_m & a_{m+1} & \cdots & a_n & b \\ 1 & 0 & & 0 & y_{1,m+1} & \cdots & y_{1n} & y_{10} \\ 0 & 1 & & 0 & y_{2,m+1} & \cdots & y_{2n} & y_{20} \\ \vdots & \vdots & & \vdots & \vdots & & \vdots & \vdots \\ 0 & 0 & & 1 & y_{m,m+1} & & y_{mn} & y_{m0} \end{array} \tag{10.1.8}$$

The value of the objective function corresponding to any solution $\boldsymbol{x}$ is

$$z = c_1 x_1 + c_2 x_2 + \ldots + c_n x_n \tag{10.1.9}$$

and hence for the basic solution, the corresponding value is

$$z_0 = \boldsymbol{c}_B^{\mathrm{T}} \boldsymbol{x}_B \tag{10.1.10}$$

where $\boldsymbol{c}_B^{\mathrm{T}} = [c_1, c_2, \ldots, c_m]$.

Although it is natural to use the basic solution $(\boldsymbol{x}_B, \boldsymbol{0})$ when we have the tableau eq. (10.1.8), it is clear that if arbitrary values are assigned to $x_{m+1}, x_{m+2}, \ldots, x_n$, we can easily solve for the remaining variables as

$$\begin{aligned} x_1 &= y_{10} - \sum_{j=m+1}^{n} y_{1j} x_j \\ x_2 &= y_{20} - \sum_{j=m+1}^{n} y_{2j} x_j \\ &\vdots \\ x_m &= y_{m0} - \sum_{j=m+1}^{n} y_{mj} x_j \end{aligned} \tag{10.1.11}$$

Using eq. (10.1.11), we may eliminate $x_1, x_2, \ldots, x_m$ from the general formula eq. (10.1.9). Doing this we obtain

$$\boldsymbol{z} = \boldsymbol{c}^{\mathrm{T}} \boldsymbol{x} = z_0 + (c_{m+1} - z_{m+1}) x_{m+1} + (c_{m+2} - z_{m+2}) x_{m+2} + \ldots + (c_n - z_n) x_n \tag{10.1.12}$$

where

$$z_j = y_{1j} c_1 + y_{2j} c_2 + \ldots + y_{mj} c_m, \quad m+1 \leqslant j \leqslant n \tag{10.1.13}$$

which is the fundamental relation required to determine the pivot column. The important point is that this equation gives the values of the objective function z for any solution of $\boldsymbol{Ax} = \boldsymbol{b}$ in terms of the variables $x_{m+1}, \ldots, x_n$. From it we can determine if there is any advantage in changing the basic solution by introducing one of the non-basic variables. For example, if $c_j - z_j$ is negative for some j, $m + 1 \leqslant j \leqslant n$, then increasing x_j from zero to some positive value would decrease the total cost, and therefore would yield a better solution. The formulae eq. (10.1.12) and eq. (10.1.13) automatically take into account the changes that would be required in the values of the basic variables $x_1, x_2, \ldots, x_m$ to accommodate the change in x_j.

Let us derive these relations from a different viewpoint. Let y_j be the i-th column of the tableau. Then any solution satisfies

$$x_1 e_1 + x_2 e_2 + \ldots + x_n e_n = y_0 - x_{m+1} y_{m+1} - x_{m+2} y_{m+2} - \ldots - x_n y_n$$

Taking the inner product of this vector equation with $\boldsymbol{c}_B^{\mathrm{T}}$, we have

$$\sum_{i=1}^{m} c_i x_i = \boldsymbol{c}_B^{\mathrm{T}} y_0 - \sum_{j=m+1}^{n} z_j x_j ,$$

where $z_j = \boldsymbol{c}_B^{\mathrm{T}} y_j$. Thus, adding $\sum_{j=m+1}^{n} c_j x_j$ to both sides,

$$\boldsymbol{c}^{\mathrm{T}} \boldsymbol{x} = z_0 + \sum_{j=m+1}^{n} (c_j - z_j) x_j \tag{10.1.14}$$

Theorem 10.1.2: (Improvement of basic feasible solution). Given a non-degenerate basic feasible solution with corresponding objective value z_0, suppose that for some j there holds $c_j - z_j < 0$. Then there is a feasible solution with objective value $z < z_0$. If the column a_j can be substituted for some vector in the original basis to yield a new basic feasible solution, this new solution will have $z < z_0$. If a_j cannot be substituted to yield a basic feasible solution, then the solution set K is unbounded and the objective function can be made arbitrarily small (toward minus infinity).

Optimality condition theorem: If for some basic feasible solution $c_j - z_j \geqslant 0$ for all j, then that solution is optimal.

Since the constants $c_j - z_j$ play such a central role in the development of the simplex method, it is convenient to introduce the somewhat abbreviated notation $r_j = c_j - z_j$ and refer to r_j as the relative cost coefficients or, alternatively, the reduced cost coefficients (both terms occur in common usage). These coefficients measure the cost of a variable relative to a given basis.

10.2 Many Computational Procedures to Simplex Method

In this section we assume a basic feasible solution and the tableau corresponding to $\boldsymbol{Ax} = \boldsymbol{b}$ is in the canonical form for this solution. We append a row at the bottom consisting of the relative cost coefficients and the negative of the current cost. The result is a simplex tableau.

Thus, if we assume the basic variables are (in order) $x_1, x_2, \ldots, x_m$, the simplex tableau takes the initial form shown in Fig. 10.2.1.

a_1	a_2	...	a_m	a_{m+1}	a_{m+2}	...	a_j	...	a_n	b
1	0	...	0	$y_{1,m+1}$	$y_{1,m+2}$	...	y_{1j}	...	y_{1n}	y_{10}
0	1		.	.	.		.		.	.
.	.		.	.	.		.		.	.
.	.		.	.	.		.		.	.
0	0		.	$y_{i,m+1}$	$y_{i,m+2}$	...	y_{ij}	...	y_{in}	y_{i0}
.	.		.	.	.		.		.	.
.	.		.	.	.		.		.	.
0	0	...	1	$y_{m,m+1}$	$y_{m,m+2}$	...	y_{mj}	...	y_{mn}	y_{m0}
0	0	...	0	r_{m+1}	r_{m+2}	...	r_j	...	r_n	$-z_0$

Fig. 10.2.1 Canonical simplex tableau

The basic solution corresponding to this tableau is

$$x_i = \begin{cases} y_{i0} & 0 \leqslant i \leqslant m \\ 0 & m+1 \leqslant i \leqslant n \end{cases}$$

which we have assumed is feasible, that is, $y_{i0} \geqslant 0, i = 1, 2, \ldots, m$. The corresponding value of the objective function is z_0.

The relative cost coefficients r_j indicate whether the value of the objective will increase or decrease if x_j is pivoted into the solution. If these coefficients are all nonnegative, then the indicated solution is optimal. If some of them are negative, an improvement can be made (assuming non-degeneracy) by bringing the corresponding component into the solution. When more than one of the relative cost coefficients is negative, any one of them may be selected to determine in which column to pivot. Common practice is to select the most negative value.

Some more discussions of the relative cost coefficients and the last row of the tableau are warranted. We may regard z as an additional variable and

$$c_1x_1 + c_2x_2 + \ldots + c_nx_n - z = 0$$

as another equation. A basic solution to the augmented system will have $m+1$ basic variables, but we can require that z be one of them. For this reason, it is not necessary to add a column corresponding to z, since it would always be $(0, 0, \ldots, 0, 1)$. Thus, initially, a last row consisting of c_i and a right-hand side of zero can be appended to the standard array to represent this additional equation. Using standard pivot operations, the elements in this row corresponding to basic variables can be reduced to zero. This is equivalent to transforming the additional equation to the form

$$r_{m+1}x_{m+1} + r_{m+2}x_{m+2} + \dots + r_n x_n - z = -z_0 \qquad (10.2.1)$$

This must be equivalent to eq. (10. 1. 14), and hence r_j obtained is the relative cost coefficients. Thus, the last row can be treated operationally like any other row: just start with c_j and reduce the terms corresponding to basic variables to zero by row operations.

After a column q is selected in which to pivot, the final selection of the pivot element is made by computing the ratio y_{i0}/y_{iq} for the positive elements $y_{iq}, i = 1, 2, \dots, m$, of the q-th column and selecting the element p yielding the minimum ratio. Pivoting on this element will maintain feasibility as well as (assuming non-degeneracy) decrease the value of the objective function. If there are ties, any element yielding the minimum can be used. If there are no non-negative elements in the column, the problem is unbounded. After updating the entire tableau with y_{pq} as pivot and transforming the last row in the same manner as all other rows (except row q), we obtain a new tableau in canonical form. The new value of the objective function again appears in the lower right-hand corner of the tableau.

The simplex algorithm can be summarized by the following steps:

Step 0. Form a tableau as in Fig. 10. 2. 1 corresponding to a basic feasible solution. The relative cost coefficients can be found by row reduction.

Step 1. If each $r_j \geqslant 0$, stop; the current basic feasible solution is optimal.

Step 2. Select q such that $r_q < 0$ to determine which nonbasic variable is to become basic.

Step 3. Calculate the ratios y_{i0}/y_{iq} for $y_{iq} > 0, i = 1, 2, \dots, m$. If no $y_{iq} > 0$, stop; the problem is unbounded. Otherwise, select p as the index i corresponding to the minimum ratio.

Step 4. Pivot on the pq-th element, updating all rows including the last. Return to Step 1.

Matrix Form of the Simplex Method

A preliminary observation in the development is that the tableau at any point in the simplex procedure can be determined solely by a knowledge of which variables are basic. As before, we denote by $\boldsymbol{B}$ the submatrix of the original $\boldsymbol{A}$ matrix consisting of the m columns of $\boldsymbol{A}$ corresponding to the basic variables. These columns are linearly independent and hence the columns of $\boldsymbol{B}$ form a basis for E^m.

We refer to $\boldsymbol{B}$ as the basis matrix.

As usual, let us assume that $\boldsymbol{B}$ consists of the first m columns of $\boldsymbol{A}$. Then by partitioning $\boldsymbol{A}, \boldsymbol{x}$, and $\boldsymbol{c}^{\mathrm{T}}$ as

$$\boldsymbol{A} = [\boldsymbol{B}, \boldsymbol{D}]$$

$$\boldsymbol{x} = (\boldsymbol{x}_B, \boldsymbol{x}_D), \boldsymbol{c}^{\mathrm{T}} = [\boldsymbol{c}_B^{\mathrm{T}}, \boldsymbol{c}_D^{\mathrm{T}}]$$

the standard linear program becomes

$$\begin{aligned} &\text{minimize} \quad \boldsymbol{c}_B^{\mathrm{T}}\boldsymbol{x}_B + \boldsymbol{c}_D^{\mathrm{T}}\boldsymbol{x}_D \\ &\text{subject to} \quad \boldsymbol{B}\boldsymbol{x}_B + \boldsymbol{D}\boldsymbol{x}_D = \boldsymbol{b} \\ &\qquad \boldsymbol{x}_B \geqslant \boldsymbol{0}, \boldsymbol{x}_D \geqslant 0 \end{aligned} \tag{10.2.2}$$

We assume the basic solution is also feasible, and if setting $\boldsymbol{x}_D = \boldsymbol{0}$ and corres-ponding to the basis $\boldsymbol{B}$, it is $\boldsymbol{x} = (\boldsymbol{x}_B, \boldsymbol{0})$ where $\boldsymbol{x}_B = \boldsymbol{B}^{-1}\boldsymbol{b}$. However, for any value of $\boldsymbol{x}_D$, the necessary value of $\boldsymbol{x}_B$ can be computed from eq. (10.2.2) as

$$\boldsymbol{x}_B = \boldsymbol{B}^{-1}\boldsymbol{b} - \boldsymbol{B}^{-1}\boldsymbol{D}\boldsymbol{x}_D \tag{10.2.3}$$

and when substituting in the cost function, this general expression yields

$$\begin{aligned} z &= \boldsymbol{c}_B^{\mathrm{T}}(\boldsymbol{B}^{-1}\boldsymbol{b} - \boldsymbol{B}^{-1}\boldsymbol{D}\boldsymbol{x}_D) + \boldsymbol{c}_D^{\mathrm{T}}\boldsymbol{x}_D \\ &= \boldsymbol{c}_B^{\mathrm{T}}\boldsymbol{B}^{-1}\boldsymbol{b} - (\boldsymbol{c}_D^{\mathrm{T}} - \boldsymbol{c}_B^{\mathrm{T}}\boldsymbol{B}^{-1}\boldsymbol{D})\boldsymbol{x}_D \end{aligned} \tag{10.2.4}$$

which expresses the cost of any solution to eq. (10.2.2) in terms of $\boldsymbol{x}_D$. Thus

$$\boldsymbol{r}_D^{\mathrm{T}} = \boldsymbol{c}_D^{\mathrm{T}} - \boldsymbol{c}_B^{\mathrm{T}}\boldsymbol{B}^{-1}\boldsymbol{D} \tag{10.2.5}$$

is the relative cost vector (for non-basic variables). It is the components of this vector that are used to determine which vector to bring into the basis.

Having derived the vector expression for the relative cost, it is now possible to write the simplex tableau in matrix form.

The initial tableau takes the form

$$\begin{bmatrix} \boldsymbol{A} & \boldsymbol{b} \\ \boldsymbol{c}^{\mathrm{T}} & \boldsymbol{0} \end{bmatrix} = \begin{bmatrix} \boldsymbol{B} & \boldsymbol{D} & \boldsymbol{b} \\ \boldsymbol{c}_B^{\mathrm{T}} & \boldsymbol{c}_D^{\mathrm{T}} & \boldsymbol{0} \end{bmatrix} \tag{10.2.6}$$

which is not in general in canonical form and does not correspond to a point in the simplex procedure. If the matrix $\boldsymbol{B}$ is used as a basis, then the corresponding tableau becomes

$$\boldsymbol{T} = \begin{bmatrix} \boldsymbol{I} & \boldsymbol{B}^{-1}\boldsymbol{D} & \boldsymbol{B}^{-1}\boldsymbol{b} \\ \boldsymbol{0} & \boldsymbol{c}_D^{\mathrm{T}} - \boldsymbol{c}_B^{\mathrm{T}}\boldsymbol{B}^{-1}\boldsymbol{b} & -\boldsymbol{c}_B^{\mathrm{T}}\boldsymbol{B}^{-1}\boldsymbol{b} \end{bmatrix} \tag{10.2.7}$$

which is the matrix form of the simplex method.

The Revised Simplex Method

Extensive experience with the simplex procedure applied to problems from various fields, and having various values of n and m, has indicated that the method can be expected to converge to an optimum solution in about m, or perhaps $3m/2$, pivot operations. Thus, particularly if m is much smaller than n, that is, if the matrix $\boldsymbol{A}$ has far fewer rows than columns, pivots will occur in only a small fraction of the columns during the course of optimization. The revised simplex method is a scheme for ordering the computations required of the simplex method so that unnecessary calculations are avoided.

The revised form of the simplex method is as follows: Given the inverse $\boldsymbol{B}^{-1}$ of a current basis, and the current solution $\boldsymbol{x}_B = \boldsymbol{y}_0 = \boldsymbol{B}^{-1}\boldsymbol{b}$,

Step 1. Calculate the current relative cost coefficients $\boldsymbol{r}_D^{\mathrm{T}} = \boldsymbol{c}_D^{\mathrm{T}} - \boldsymbol{c}_B^{\mathrm{T}}\boldsymbol{B}^{-1}\boldsymbol{D}$. This can best be done by first calculating $\boldsymbol{\lambda}^{\mathrm{T}} = \boldsymbol{c}_B^{\mathrm{T}}\boldsymbol{B}^{-1}$ and then the relative cost vector $\boldsymbol{r}_D^{\mathrm{T}} = \boldsymbol{c}_D^{\mathrm{T}} - \boldsymbol{\lambda}^{\mathrm{T}}\boldsymbol{D}$. If $\boldsymbol{r}_D \geqslant \boldsymbol{0}$, stop; the current solution is optimal.

Step 2. Determine which vector $\boldsymbol{a}_q$ is to enter the basis by selecting the most negative cost coefficient; and calculate $\boldsymbol{y}_q = \boldsymbol{B}^{-1}\boldsymbol{a}_q$ which gives the vector $\boldsymbol{a}_q$ expressed in terms of the current basis.

Step 3. If no $y_{iq} > 0$, stop; the problem is unbounded. Otherwise, calculate the ratios y_{i0}/y_{iq} for $y_{iq} > 0$ to determine which vector is to leave the basis.

Step 4. Update $\boldsymbol{B}^{-1}$ and the current solution $\boldsymbol{B}^{-1}\boldsymbol{b}$. Return to Step 1.

To begin the procedure one requires, as always, an initial basic feasible solution (and hence also its inverse) be an identity matrix, resulting either from slack or surplus variables or from artificial variables.

The Dantzig-Wolfe Decomposition Method

Large linear programming problems usually have some special structural form that can be exploited to develop efficient computational procedures. One common structure is where there are a number of separate activity areas that are linked through common resource constraints. The Dantzig-Wolfe decomposition method is an iterative process where at each step a number of separate subproblems are solved. The subproblems are themselves linear programs within the separate areas. The objective functions of these subproblems are varied from iteration to iteration and are determined by a separate calculation based on the results of the previous iteration. This action coordinates the individual subproblems so that, ultimately, the solution to the overall problem is solved. The method can be derived as a special version of the revised simplex method, where the subproblems correspond to evaluation of reduced cost coefficients for the main problem.

Suppose the $\boldsymbol{A}$ matrix has the special "block-angular" structure as

$$\boldsymbol{A}=\begin{bmatrix} \boldsymbol{L}_1 & \boldsymbol{L}_2 & \cdots & \boldsymbol{L}_N \\ \boldsymbol{A}_1 & & & \\ & \boldsymbol{A}_2 & & \\ & & \ddots & \\ & & & \boldsymbol{A}_N \end{bmatrix} \tag{10.2.8}$$

By partitioning the vectors $\boldsymbol{x}$, $\boldsymbol{c}^{\mathrm{T}}$, and $\boldsymbol{b}$ consistent with this partition of $\boldsymbol{A}$, the problem can be rewritten as

$$\begin{aligned} &\text{minimize } \sum_{i=1}^{N} \boldsymbol{c}_i^{\mathrm{T}}\boldsymbol{x}_i \\ &\text{subject to } \sum_{i=1}^{N} \boldsymbol{L}_i\boldsymbol{x}_i = \boldsymbol{b}_0 \\ &\boldsymbol{A}_i\boldsymbol{x}_i = \boldsymbol{b}_i \\ &\boldsymbol{x}_i \geqslant \boldsymbol{0}, \quad i=1,\ldots,N \end{aligned} \tag{10.2.9}$$

This may be viewed as a problem of minimizing the total cost of N different linear programs that are independent except for the first constraint, which is a linking constraint of, say, dimension m.

Each of the subproblems is of the form

$$\begin{aligned} &\text{minimize } \boldsymbol{c}_i^{\mathrm{T}}\boldsymbol{x}_i \\ &\text{subject to } \boldsymbol{A}_i\boldsymbol{x}_i = \boldsymbol{b}_i \\ &\boldsymbol{x}_i \geqslant \boldsymbol{0} \end{aligned} \tag{10.2.10}$$

The constraint set $\boldsymbol{S}_i = \{\boldsymbol{x}_i : \boldsymbol{A}_i\boldsymbol{x}_i = \boldsymbol{b}_i, \boldsymbol{x}_i \geqslant \boldsymbol{0}\}$ for the i-th subproblem is a polytope and the intersection of a finite number of closed half-space. There is no guarantee that each $\boldsymbol{S}_i$ is bounded, even if the original linear program has a bounded constraint set. We shall assume for simplicity, however, that each of the polytopes $\boldsymbol{S}_i, i=1,\ldots,N$, is indeed bounded and hence is a polyhedron. One may guarantee that this assumption is satisfied by placing artificial (large) upper bounds on each $\boldsymbol{x}_i$.

Under the boundedness assumption, each polyhedron $\boldsymbol{S}_i$ consists entirely of points that are convex combinations of its extreme points. Thus, if the extreme points of $\boldsymbol{S}_i$ are $\{x_{i1}, x_{i2}, \ldots, x_{ik_i}\}$, then any point $\boldsymbol{x}_i \in \boldsymbol{S}_i$ can be expressed in the form

$$x_i = \sum_{j=1}^{k_i} \alpha_{ij} x_{ij} \tag{10.2.11}$$

where weighting coefficients of the extreme points $\sum_{j=1}^{k_i} \alpha_{ij} = 1$ and $\alpha_{ij} \geqslant 0, j = 1, \ldots, K_i$.

We now convert the original linear program to an equivalent master problem, of which the objective is to find the optimal weighting coefficients for each polyhedron, $\boldsymbol{S}_i$. Corresponding to each extreme point x_{ij} in $\boldsymbol{S}_i$, define $p_{ij} = \boldsymbol{c}_i^{\mathrm{T}} x_{ij}$ and $\boldsymbol{q}_{ij} = \boldsymbol{L}_i x_{ij}$. Clearly p_{ij} is the equivalent cost of the extreme point x_{ij}, and $\boldsymbol{q}_{ij}$ is its equivalent activity vector in the linking constraints.

Then the original linear program eq. (10.2.7) is equivalent, using eq. (10.2.11), to the master problem

$$\begin{aligned} &\text{minimize } \sum_{i=1}^{N} \sum_{j=1}^{k_i} p_{ij} \alpha_{ij} \\ &\text{subject to } \sum_{i=1}^{N} \sum_{j=1}^{k_i} \boldsymbol{q}_{ij} \alpha_{ij} = \boldsymbol{b}_0 \\ &\left. \begin{aligned} &\sum_{j=1}^{k_i} \alpha_{ij} = 1 \\ &\alpha_{ij} \geqslant 0, j = 1, \ldots, K_i \end{aligned} \right\} i = 1, \ldots, N \end{aligned} \tag{10.2.12}$$

This master problem has variables

$$\boldsymbol{\alpha} = (\alpha_{11}, \ldots, \alpha_{1k_1}, \alpha_{21}, \ldots, \alpha_{1k_2}, \ldots, \alpha_{N1}, \ldots, \alpha_{Nk_N})$$

and can be expressed more compactly as

$$\begin{aligned} &\text{minimize } \boldsymbol{p}^{\mathrm{T}} \boldsymbol{\alpha} \\ &\text{subject to } \boldsymbol{Q}\boldsymbol{\alpha} = \boldsymbol{g} \\ &\boldsymbol{\alpha} \geqslant \boldsymbol{0} \end{aligned} \tag{10.2.13}$$

where $\boldsymbol{g}^{\mathrm{T}} = [\boldsymbol{b}_0^{\mathrm{T}}, 1, 1, \ldots, 1]$; the element of $\boldsymbol{p}$ associated with α_{ij} is p_{ij}; and the column of $\boldsymbol{Q}$ associated with α_{ij} is $\begin{bmatrix} \boldsymbol{q}_{ij} \\ \boldsymbol{e}_i \end{bmatrix}$, with $\boldsymbol{e}_i$ denoting the i-th unit vector in E^N.

Suppose that at some stage of the revised simplex method for the master problem we know the basis $\boldsymbol{B}$ and corresponding simplex multipliers $\boldsymbol{\lambda}^{\mathrm{T}} = \boldsymbol{p}_B^{\mathrm{T}} \boldsymbol{B}^{-1}$. The corresponding relative cost vector is $\boldsymbol{r}_D^{\mathrm{T}} = \boldsymbol{c}_D^{\mathrm{T}} - \boldsymbol{\lambda}^{\mathrm{T}} \boldsymbol{D}$, having components

$$\boldsymbol{r}_{ij} = p_{ij} - \boldsymbol{\lambda}^{\mathrm{T}} \begin{bmatrix} \boldsymbol{q}_{ij} \\ \boldsymbol{e}_i \end{bmatrix} \tag{10.2.14}$$

It is not necessary to calculate all the $\boldsymbol{r}_{ij}$; it is only necessary to determine the minimal $\boldsymbol{r}_{ij}$. If the minimal value is nonnegative, the current solution is optimal and the process terminates. If, on the other hand, the minimal element is negative, the corresponding column should enter the basis.

The search for the minimal element in eq. (10.2.14) is normally made with respect to nonbasic columns only. The search can be formally extended to include basic columns as well, however, since for basic elements

$$p_{ij} - \boldsymbol{\lambda}^{\mathrm{T}} \begin{bmatrix} \boldsymbol{q}_{ij} \\ \boldsymbol{e}_i \end{bmatrix} = 0$$

The extra zero values do not influence the subsequent procedure, since a new column will enter only if the minimal value is less than zero.

We therefore define $\boldsymbol{r}^*$ as the minimum relative cost coefficient for all possible basis vectors. That is,

$$\boldsymbol{r}^* = \underset{i \in \{1,\ldots,N\}}{\mathrm{minimum}} = \left\{ \boldsymbol{r}_i^* = \underset{j \in \{1,\ldots,k_i\}}{\mathrm{minimum}} \left\{ p_{ij} - \boldsymbol{\lambda}^{\mathrm{T}} \begin{bmatrix} \boldsymbol{q}_{ij} \\ \boldsymbol{e}_i \end{bmatrix} \right\} \right\}$$

Using the definitions of p_{ij} and $\boldsymbol{q}_{ij}$, this becomes

$$\boldsymbol{r}_i^* = \underset{j \in \{1,\ldots,k_i\}}{\mathrm{minimum}} \{ \boldsymbol{c}_i^{\mathrm{T}} x_{ij} - \boldsymbol{\lambda}_0^{\mathrm{T}} \boldsymbol{L}_i x_{ij} - \boldsymbol{\lambda}_{m+i} \} \tag{10.2.15}$$

where $\boldsymbol{\lambda}_0$ is the vector made up of the first m elements of $\boldsymbol{\lambda}$, m being the number of rows of $\boldsymbol{L}_i$ (the number of linking constraints in eq. (10.2.9)).

The minimization problem in eq. (10.2.15) is actually solved by the i-th subproblem:

$$\begin{aligned} &\mathrm{minimize}(\boldsymbol{c}_i^{\mathrm{T}} - \boldsymbol{\lambda}_0^{\mathrm{T}} \boldsymbol{L}_i) \boldsymbol{x}_i \\ &\text{subject to } \boldsymbol{A}_i \boldsymbol{x}_i = \boldsymbol{b}_i \\ &\boldsymbol{x}_i \geqslant \boldsymbol{0} \end{aligned} \tag{10.2.16}$$

This follows the fact that $\boldsymbol{\lambda}_{m+i}$ is independent of the extreme point index j, and that the solution of eq. (10.2.16) must be that extreme point of $\boldsymbol{S}_i$, say $\boldsymbol{x}_{ik}$, of minimum cost, using the adjusted cost coefficients $\boldsymbol{c}_i^{\mathrm{T}} - \boldsymbol{\lambda}_0^{\mathrm{T}} \boldsymbol{L}_i$.

Thus, an algorithm for this special version of the revised simplex method applied to the master problem is the following: Given a basis $\boldsymbol{B}$

Step 1. Calculate the current basic solution $\boldsymbol{x}_B$, and solve $\boldsymbol{\lambda}^{\mathrm{T}} \boldsymbol{B} = \boldsymbol{c}_B^{\mathrm{T}}$ for $\boldsymbol{\lambda}$.

Step 2. For each $i = 1, 2, \ldots, N$, determine the optimal solution $\boldsymbol{x}_i^*$ of the i-th sub-problem eq. (10.2.16) and calculate

$$\boldsymbol{r}_i^* = (\boldsymbol{c}_i^{\mathrm{T}} - \boldsymbol{\lambda}_0^{\mathrm{T}} \boldsymbol{L}_i)\boldsymbol{x}_i^* - \boldsymbol{\lambda}_{m+i} \tag{10.2.17}$$

If all $\boldsymbol{r}_i^* > \mathbf{0}$, stop; the current solution is optimal.

Step 3. Determine which column is to enter the basis by selecting the minimal $\boldsymbol{r}_i^*$.

Step 4. Update the basis of the master problem as usual.

10.3 Duality

10.3.1 Dual Linear Programs

In this section, we define the dual program that is associated with a given linear program. Initially, we depart from our usual strategy of considering programs in standard form, since the duality relationship is most symmetric for programs expressed solely in terms of inequalities. Specifically then, we define duality through the pair of programs displayed below.

$$\begin{array}{llll} & \text{Primal} & & \text{Dual} \\ \text{minimize} & \boldsymbol{c}^{\mathrm{T}}\boldsymbol{x} & \text{maximize} & \boldsymbol{\lambda}^{\mathrm{T}}\boldsymbol{b} \\ \text{subject to} & \boldsymbol{A}\boldsymbol{x} \geqslant \boldsymbol{b} & \text{subject to} & \boldsymbol{\lambda}^{\mathrm{T}}\boldsymbol{A} \leqslant \boldsymbol{c}^{\mathrm{T}} \\ & \boldsymbol{x} \geqslant \mathbf{0} & & \boldsymbol{\lambda} \geqslant \mathbf{0} \end{array} \tag{10.3.1}$$

If $\boldsymbol{A}$ is an $m \times n$ matrix, then $\boldsymbol{x}$ is an n-dimensional column vector, $\boldsymbol{b}$ is an n-dimensional column vector, $\boldsymbol{c}^{\mathrm{T}}$ is an n-dimensional row vector, and $\boldsymbol{\lambda}^{\mathrm{T}}$ is an m-dimensional row vector. The vector $\boldsymbol{x}$ is the variable of the primal program, and $\boldsymbol{\lambda}$ is the variable of the dual program.

The pair of programs eq. (10.3.1) is called the symmetric form of duality and, as explained below, can be used to define the dual of any linear program. It is important to note that the role of primal and dual can be reversed. Thus, studying in detail the process by which the dual is obtained from the primal: interchange of cost and constraint vectors, transposition of coefficient matrix, reversal of constraint inequalities, and change of minimization to maximization; we see that this same process applied to the dual yields the primal. Put another way, if the dual is transformed, by multiplying the objective and the constraints by minus unity, so that it has the structure of the primal (but is still expressed in terms of $\boldsymbol{\lambda}$), its corresponding dual will be equivalent to the original primal.

The dual of any linear program can be found by converting the program to the form of the primal shown above. For example, given a linear program in standard form

$$\begin{array}{ll} \text{minimize} & \boldsymbol{c}^{\mathrm{T}}\boldsymbol{x} \\ \text{subject to} & \boldsymbol{A}\boldsymbol{x}=\boldsymbol{b} \\ & \boldsymbol{x}\geqslant\boldsymbol{0} \end{array}$$

we write it in the equivalent form

$$\begin{array}{ll} \text{minimize} & \boldsymbol{c}^{\mathrm{T}}\boldsymbol{x} \\ \text{subject to} & \boldsymbol{A}\boldsymbol{x}\geqslant\boldsymbol{b} \\ & -\boldsymbol{A}\boldsymbol{x}\geqslant-\boldsymbol{b} \\ & \boldsymbol{x}\geqslant\boldsymbol{0} \end{array}$$

which is in the form of the primal of eq. (10.3.1) but with coefficient matrix $\begin{bmatrix} \boldsymbol{A} \\ \cdots \\ -\boldsymbol{A} \end{bmatrix}$. Using a dual vector partitioned as $(\boldsymbol{u},\boldsymbol{v})$, the corresponding dual is

$$\begin{array}{ll} \text{minimize} & \boldsymbol{u}^{\mathrm{T}}\boldsymbol{b}-\boldsymbol{v}^{\mathrm{T}}\boldsymbol{b} \\ \text{subject to} & \boldsymbol{u}^{\mathrm{T}}\boldsymbol{A}-\boldsymbol{v}^{\mathrm{T}}\boldsymbol{A}\leqslant\boldsymbol{c}^{\mathrm{T}} \\ & \boldsymbol{u}\geqslant\boldsymbol{0} \\ & \boldsymbol{v}\geqslant\boldsymbol{0} \end{array}$$

Letting $\boldsymbol{\lambda}=\boldsymbol{u}-\boldsymbol{v}$ we may simplify the representation of the dual program so that we obtain the pair of problems displayed below:

$$\begin{array}{llll} \multicolumn{2}{c}{\text{Primal}} & \multicolumn{2}{c}{\text{Dual}} \\ \text{minimize} & \boldsymbol{c}^{\mathrm{T}}\boldsymbol{x} & \text{maximize} & \boldsymbol{\lambda}^{\mathrm{T}}\boldsymbol{b} \\ \text{subject to} & \boldsymbol{A}\boldsymbol{x}\geqslant\boldsymbol{b} & \text{subject to} & \boldsymbol{\lambda}^{\mathrm{T}}\boldsymbol{A}\leqslant\boldsymbol{c}^{\mathrm{T}} \\ & \boldsymbol{x}\geqslant\boldsymbol{0} & & \boldsymbol{\lambda}\geqslant\boldsymbol{0} \end{array} \qquad (10.3.2)$$

This is the asymmetric form of the duality relation. In this form the dual vector $\boldsymbol{\lambda}$ (which is really a composite of $\boldsymbol{u}$ and $\boldsymbol{v}$) is not restricted to be nonnegative.

Similar transformations can be worked out for any linear program to first get the primal in the form eq. (10.3.1), calculate the dual, and then simplify the dual to account for special structure.

In general, if some of the linear inequalities in the primal eq. (10.3.1) are changed to equality, the corresponding components of $\boldsymbol{\lambda}$ in the dual become free variables. If some of the components of $\boldsymbol{x}$ in the primal are free variables, then the corresponding inequalities in $\boldsymbol{\lambda}^{\mathrm{T}}\boldsymbol{A} \leqslant \boldsymbol{c}^{\mathrm{T}}$ are changed to equality in the dual. We mention again that these are not arbitrary rules but are direct consequences of the original definition and the equivalence of various forms of linear programs.

10.3.2 The Duality Theorem

Lemma: (Weak Duality Lemma). If $\boldsymbol{x}$ and $\boldsymbol{\lambda}$ are feasible for eq. (10.3.2), respectively, then $\boldsymbol{c}^{\mathrm{T}}\boldsymbol{x} \geqslant \boldsymbol{\lambda}^{\mathrm{T}}\boldsymbol{b}$.

Corollary: If $\boldsymbol{x}_0$ and $\boldsymbol{\lambda}_0$ are feasible for eq. (10.3.2), respectively, and if $\boldsymbol{c}^{\mathrm{T}}\boldsymbol{x}_0 \geqslant \boldsymbol{\lambda}_0^{\mathrm{T}}\boldsymbol{b}$, then $\boldsymbol{x}_0$ and $\boldsymbol{\lambda}_0$ are optimal for their respective problems.

Duality theorem of linear programming: If either of the problems eq. (10.3.2) has a finite optimal solution, so does the other, and the corresponding values of the objective functions are equal. If either problem has an unbounded objective, the other problem has no feasible solution.

Suppose eq. (10.3.2) has a finite optimal solution with value z_0. In the space E^{m+1} define the convex set

$$C = \{(\boldsymbol{r},\boldsymbol{w}) : \boldsymbol{r} = tz_0 - \boldsymbol{c}^{\mathrm{T}}\boldsymbol{x}, \boldsymbol{w} = t\boldsymbol{b} - \boldsymbol{A}\boldsymbol{x}, \boldsymbol{x} \geqslant \boldsymbol{0}, t \geqslant 0\}.$$

It is easily verified that $\boldsymbol{C}$ is in fact a closed convex cone. We show that the point $(1,0)$ is not in $\boldsymbol{C}$. If $\boldsymbol{w} = t_0\boldsymbol{b} - \boldsymbol{A}\boldsymbol{x}_0 = \boldsymbol{0}$ with $t_0 > \boldsymbol{0}, \boldsymbol{x}_0 \geqslant \boldsymbol{0}$, then $\boldsymbol{x} = \boldsymbol{x}_0/t_0$ is feasible for eq. (10.3.2) and hence $\boldsymbol{r}/t_0 = z_0 - \boldsymbol{c}^{\mathrm{T}}\boldsymbol{x} \leqslant \boldsymbol{0}$, which means $\boldsymbol{r} \leqslant \boldsymbol{0}$. If $\boldsymbol{w} = -\boldsymbol{A}\boldsymbol{x}_0 = \boldsymbol{0}$ with $\boldsymbol{x}_0 \geqslant \boldsymbol{0}$ and $\boldsymbol{c}^{\mathrm{T}}\boldsymbol{x}_0 \geqslant \boldsymbol{\lambda}_0^{\mathrm{T}}\boldsymbol{b}$, and if $\boldsymbol{x}$ is any feasible solution to eq. (10.3.2), then $\boldsymbol{x} + \alpha\boldsymbol{x}_0$ is feasible for any $\alpha \geqslant 0$ and gives arbitrarily small objective values as is increased. This contradicts our assumption on the existence of a finite optimum and thus we conclude that no such $\boldsymbol{x}_0$ exists. Hence $(1,0) \notin \boldsymbol{C}$.

Now since $\boldsymbol{C}$ is a closed convex set, a hyperplane separating $(1,0)$ and $\boldsymbol{C}$. Thus there is a nonzero vector $[s,\boldsymbol{\lambda}] \in E^{m+1}$ and a constant c such that

$$s < c = \inf\{s\boldsymbol{r} + \boldsymbol{\lambda}^{\mathrm{T}}\boldsymbol{w} : (\boldsymbol{r},\boldsymbol{w}) \in \boldsymbol{C}\}$$

Now since $\boldsymbol{C}$ is a cone, it follows that $c \geqslant 0$. For if there were $(\boldsymbol{r},\boldsymbol{w}) \in \boldsymbol{C}$ such that $s\boldsymbol{r} + \boldsymbol{\lambda}^{\mathrm{T}}\boldsymbol{w} < 0$, then $\alpha(\boldsymbol{r},\boldsymbol{w})$ for large α would violate the hyperplane inequality. On the other hand, since $(0,0) \in \boldsymbol{C}$ we must have $c \leqslant 0$. Thus $c = 0$. As a consequence $s < 0$, and without loss of generality we may assume $s = -1$.

We have to this point established the existence of $\boldsymbol{\lambda} \in E^m$ such that $-\boldsymbol{r} + \boldsymbol{\lambda}^{\mathrm{T}}\boldsymbol{w} \geqslant 0$ for all $(\boldsymbol{r}, \boldsymbol{w}) \in \boldsymbol{C}$. Equivalently, using the definition of $\boldsymbol{C}$,

$$(\boldsymbol{c} - \boldsymbol{\lambda}^{\mathrm{T}}\boldsymbol{A})\boldsymbol{x} - tz_0 + t\boldsymbol{\lambda}^{\mathrm{T}}\boldsymbol{b} \geqslant \boldsymbol{0}$$

for all $\boldsymbol{x} \geqslant \boldsymbol{0}, t \geqslant 0$. Setting $t = 0$ yields $\boldsymbol{\lambda}^{\mathrm{T}}\boldsymbol{A} \leqslant \boldsymbol{c}^{\mathrm{T}}$, which says $\boldsymbol{\lambda}$ is feasible for the dual. Setting $\boldsymbol{x} = \boldsymbol{0}$ and $t = 1$ yields $\boldsymbol{\lambda}^{\mathrm{T}}\boldsymbol{b} \geqslant z_0$, which in view of Lemma 1 and its corollary shows that $\boldsymbol{\lambda}$ is optimal for the dual.

10.3.3 Relations to the Simplex Procedure

Suppose that for the linear program

$$\begin{aligned} \text{minimize} \quad & \boldsymbol{c}^{\mathrm{T}}\boldsymbol{x} \\ \text{subject to} \quad & \boldsymbol{A}\boldsymbol{x} = \boldsymbol{b} \\ & \boldsymbol{x} \geqslant \boldsymbol{0} \end{aligned} \tag{10.3.3}$$

we have the optimal basic feasible solution $\boldsymbol{x} = (\boldsymbol{x}_B, \boldsymbol{0})$ with corresponding basis $\boldsymbol{B}$. We shall determine a solution of the dual program

$$\begin{aligned} \text{maximize} \quad & \boldsymbol{\lambda}^{\mathrm{T}}\boldsymbol{b} \\ \text{subject to} \quad & \boldsymbol{\lambda}^{\mathrm{T}}\boldsymbol{A} \leqslant \boldsymbol{c}^{\mathrm{T}} \end{aligned} \tag{10.3.4}$$

in terms of $\boldsymbol{B}$.

We partition $\boldsymbol{A}$ as $\boldsymbol{A} = [\boldsymbol{B}, \boldsymbol{D}]$. Since the basic feasible solution $\boldsymbol{x}_B = \boldsymbol{B}^{-1}\boldsymbol{b}$ is optimal, the relative cost vector $\boldsymbol{r}$ must be nonnegative in each component. We have $\boldsymbol{r}_D^{\mathrm{T}} = \boldsymbol{c}_D^{\mathrm{T}} - \boldsymbol{c}_B^{\mathrm{T}}\boldsymbol{B}^{-1}\boldsymbol{D}$ and since $\boldsymbol{r}_D$ is nonnegative in each component, we have $\boldsymbol{c}_B^{\mathrm{T}}\boldsymbol{B}^{-1}\boldsymbol{D} \leqslant \boldsymbol{c}_D^{\mathrm{T}}$.

Now define $\boldsymbol{\lambda}^{\mathrm{T}} = \boldsymbol{c}_B^{\mathrm{T}}\boldsymbol{B}^{-1}$, we show that this choice of $\boldsymbol{\lambda}$ solves the dual problem. We have

$$\boldsymbol{\lambda}^{\mathrm{T}}\boldsymbol{A} = [\boldsymbol{\lambda}^{\mathrm{T}}\boldsymbol{B}, \boldsymbol{\lambda}^{\mathrm{T}}\boldsymbol{D}] = [\boldsymbol{c}_B^{\mathrm{T}}, \boldsymbol{c}_B^{\mathrm{T}}\boldsymbol{B}^{-1}\boldsymbol{D}] \leqslant [\boldsymbol{c}_B^{\mathrm{T}}, \boldsymbol{c}_D^{\mathrm{T}}] = \boldsymbol{c}^{\mathrm{T}}$$

Thus since $\boldsymbol{\lambda}^{\mathrm{T}}\boldsymbol{A} \leqslant \boldsymbol{c}^{\mathrm{T}}$, $\boldsymbol{\lambda}$ is feasible for the dual. On the other hand, $\boldsymbol{\lambda}^{\mathrm{T}}\boldsymbol{b} = \boldsymbol{c}_B^{\mathrm{T}}\boldsymbol{B}^{-1}\boldsymbol{b} = \boldsymbol{c}_B^{\mathrm{T}}\boldsymbol{x}_B$, and thus the value of the dual objective function for this $\boldsymbol{\lambda}$ is equal to the value of the primal problem. This may establish the optimality of $\boldsymbol{\lambda}$ for the dual. The above discussion yields an alternative derivation of the main portion of the Duality Theorem.

Theorem 10.3.1: Let the linear program eq. (10.3.3) have an optimal basic feasible solution corresponding to the basis $\boldsymbol{B}$. Then the vector $\boldsymbol{\lambda}$ satisfying $\boldsymbol{\lambda}^{\mathrm{T}} = \boldsymbol{c}_B^{\mathrm{T}}\boldsymbol{B}^{-1}$ is an optimal solution to the dual program eq. (10.3.4). The optimal values of both problems are equal.

We turn now to a discussion of how the dual solution can be obtained directly from the final simplex tableau of the primal. Suppose that embedded in the original matrix $\boldsymbol{A}$ is an $m \times m$ identity matrix. This will be the case if, for example, m slack variables are employed to convert inequalities to equalities. Then in the final tableau the matrix $\boldsymbol{B}^{-1}$ appears where the identity appeared in the beginning. Furthermore, in the last row the components corresponding to this identity matrix will be $\boldsymbol{c}_i^{\mathrm{T}} - \boldsymbol{c}_B^{\mathrm{T}}\boldsymbol{B}^{-1}$, where $\boldsymbol{c}_i$ is the m-vector representing the cost coefficients of the variables corresponding to the columns of the original identity matrix. Thus by subtracting these cost coefficients from the corresponding elements in the last row, the negative of the solution $\boldsymbol{\lambda}^{\mathrm{T}} = \boldsymbol{c}_B^{\mathrm{T}}\boldsymbol{B}^{-1}$ to the dual is obtained. In particular, if, as is the case with slack variables, $\boldsymbol{c}_1 = \boldsymbol{0}$, then the elements in the last row under $\boldsymbol{B}^{-1}$ are equal to the negative of components of the solution to the dual.

Simplex Multipliers

We conclude this section by giving an economic interpretation of the relation between the simplex basis and the vector $\boldsymbol{\lambda}$. At any point in the simplex procedure, we may form the vector $\boldsymbol{\lambda}$ satisfying $\boldsymbol{\lambda}^{\mathrm{T}} = \boldsymbol{c}_B^{\mathrm{T}}\boldsymbol{B}^{-1}$. This vector is not a solution to the dual unless $\boldsymbol{B}$ is an optimal basis for the primal. Furthermore, as we have seen in the development of the revised simplex method, this $\boldsymbol{\lambda}$ vector can be used at every step to calculate the relative cost coefficients. For this reason, $\boldsymbol{\lambda}^{\mathrm{T}} = \boldsymbol{c}_B^{\mathrm{T}}\boldsymbol{B}^{-1}$ is often called the vector of simplex multipliers.

Let us pursue the economic interpretation of these simplex multipliers. As usual, denote the columns of $\boldsymbol{A}$ by $\boldsymbol{a}_1, \boldsymbol{a}_2, \ldots, \boldsymbol{a}_n$ and denote E^m by the m unit vectors $\boldsymbol{e}_1, \boldsymbol{e}_2, \ldots, \boldsymbol{e}_m$. The components of $\boldsymbol{a}_i$ and $\boldsymbol{b}$ tell how to construct these vectors from $\boldsymbol{e}_i$.

Given any basis $\boldsymbol{B}$, however, consisting of m columns of $\boldsymbol{A}$, any other vector can be constructed (synthetically) as a linear combination of these basis vectors. If there is a unit cost $\boldsymbol{c}_i$ associated with each basis vector $\boldsymbol{a}_i$, then the cost of a (synthetic) vector constructed from the basis can be calculated as the corresponding linear combination of $\boldsymbol{c}_i$ associated with the basis. In particular, the cost of the j-th unit vector $\boldsymbol{e}_j$, when constructed from the basis $\boldsymbol{B}$, is $\boldsymbol{\lambda}_j$, the j-th component of $\boldsymbol{\lambda}^{\mathrm{T}} = \boldsymbol{c}_B^{\mathrm{T}}\boldsymbol{B}^{-1}$. Thus $\boldsymbol{\lambda}_j$ can be interpreted as synthetic prices of the unit vectors.

Now, any vector can be expressed in terms of the basis $\boldsymbol{B}$ in two steps: (i) express the unit vectors in terms of the basis, and then (ii) express the desired vector as a linear combination of unit vectors. The corresponding synthetic cost of a vector constructed from the basis $\boldsymbol{B}$ can correspondingly be computed directly by: (i) finding the synthetic price of the unit vectors, and then (ii) using these prices to evaluate the cost of the linear combination of unit vectors. Thus, the simplex multipliers can be used to quickly evaluate the synthetic cost of any vector that is expressed in terms of the unit vectors. The difference between the true cost of this vector and the synthetic cost is the relative cost. The process of calculating the synthetic cost of a vector, with respect to a given basis, by using the simplex multipliers is sometimes referred to as pricing out the vector.

Optimality of the primal corresponds to the situation where every vector $\boldsymbol{a}_1$, $\boldsymbol{a}_2, \ldots, \boldsymbol{a}_n$ is cheaper when constructed from the basis than when purchased directly at its own price. Thus we have $\boldsymbol{\lambda}^{\mathrm{T}}\boldsymbol{a}_i \leqslant \boldsymbol{c}_i$ for $i = 1, 2, \ldots, n$ or equivalently $\boldsymbol{\lambda}^{\mathrm{T}}\boldsymbol{A} \leqslant \boldsymbol{c}^{\mathrm{T}}$.

Complementary Slackness

Theorem 10.3.2: (Complementary slackness asymmetric form). Let $\boldsymbol{x}$ and $\boldsymbol{\lambda}$ be feasible solutions for the primal and dual programs, respectively, in the pair (10.3.2). A necessary and sufficient condition that they are both optimal solutions is that for all i,

i) $\boldsymbol{x}_i > \boldsymbol{0} \Rightarrow \boldsymbol{\lambda}^{\mathrm{T}}\boldsymbol{a}_i = \boldsymbol{c}_i$,

ii) $\boldsymbol{x}_i = \boldsymbol{0} \Leftarrow \boldsymbol{\lambda}^{\mathrm{T}}\boldsymbol{a}_i < \boldsymbol{c}_i$.

Theorem 10.3.3: (Complementary slackness symmetric form). Let $\boldsymbol{x}$ and $\boldsymbol{\lambda}$ be feasible solutions for the primal and dual programs, respectively, in the pair (10.3.1). A necessary and sufficient condition that they are both optimal solutions is that for all i and j,

i) $\boldsymbol{x}_i > \boldsymbol{0} \Rightarrow \boldsymbol{\lambda}^{\mathrm{T}}\boldsymbol{a}_i = \boldsymbol{c}_i$,

ii) $\boldsymbol{x}_i = \boldsymbol{0} \Leftarrow \boldsymbol{\lambda}^{\mathrm{T}}\boldsymbol{a}_i < \boldsymbol{c}_i$,

iii) $\boldsymbol{\lambda}_j > \boldsymbol{0} \Rightarrow \boldsymbol{a}^j\boldsymbol{x} = \boldsymbol{b}_j$,

iv) $\boldsymbol{\lambda}_j = \boldsymbol{0} \Leftarrow \boldsymbol{a}^j\boldsymbol{x} > \boldsymbol{b}_j$,

where $\boldsymbol{a}^j$ is the j-th row of $\boldsymbol{A}$.

10.4 Interior-point Methods

10.4.1 Elements of Complexity Theory

In this section, we focus on a particular resource, namely, computing time. In complexity theory, however, one is not interested in the execution time of a program implemented in a particular programming language, running on a particular computer over a particular input. This involves too many contingent factors. Instead, one wishes to associate to an algorithm more intrinsic measures of its time requirements.

Roughly speaking, to do so one needs to define:

- a notion of input size,
- a set of basic operations, and
- a cost for each basic operation.

The last two allow one to associate a cost of a computation. For a complex event, it is important to save costs and computing time, so the related content are as follows: If $\boldsymbol{x}$ is any input, the cost $\boldsymbol{C}(\boldsymbol{x})$ of the computation with input $\boldsymbol{x}$ is the sum of the costs of all the basic operations performed during this computation.

Let $\boldsymbol{A}$ be an algorithm and $\boldsymbol{J}_n$ be the set of all its inputs having size n. The worst-case cost function of $\boldsymbol{A}$ is the function T_A^W defined by

$$T_A^W(n) = \sup_{\boldsymbol{x} \in \boldsymbol{J}_n} \boldsymbol{C}(\boldsymbol{x})$$

If there is a probability structure on $\boldsymbol{J}_n$ it is possible to define the average-case cost function T_A^a given by

$$T_A^a(n) = \boldsymbol{E}_n(\boldsymbol{C}(x))$$

where $\boldsymbol{E}_n$ is the expectation over $\boldsymbol{J}_n$. However, the average is usually more difficult to find, and there is of course the issue of what probabilities to assign.

10.4.2 The Analytic Center

Consider a set S in a subset of χ of E^n defined by a group of inequalities as

$$S = \{x \in \chi : g_j(\boldsymbol{x}) \geqslant 0, j = 1, 2, \ldots, m\}$$

and assume that the functions g_j are continuous. S has a nonempty interior $\mathring{S} =$

$\{x \in \chi : g_j(x) > 0, \text{for all } j\}$. Associated with this definition of the set is the potential function $\psi(x) = -\sum_{j=1}^{m} \log g_j(x)$ defined on $\mathring{S}$.

The analytic center of S is the vector (or set of vectors) that minimizes the potential function; that is, the vector (or vectors) that solve

$$\min \psi(x) = \min\left\{-\sum_{j=1}^{m} \log g_j(x) : x \in \chi, g_j(x) > 0, \text{for each } j\right\}$$

10.4.3 The Central Path

Consider a primal linear program in standard form

$$(LP) \quad \begin{aligned} &\text{minimize} && c^{\mathrm{T}}x \\ &\text{subject to} && Ax = b \\ & && x \geqslant 0 \end{aligned} \tag{10.4.1}$$

We denote the feasible region of this program by F_p. We assume that $\mathring{F}_p = \{x : Ax = b, x > 0\}$ is nonempty and the optimal solution set of the problem is bounded.

Associated with this problem and for $\mu \geqslant 0$, we define the barrier problem

$$(BP) \quad \begin{aligned} &\text{minimize} && c^{\mathrm{T}}x - \mu \sum_{j=1}^{n} \log x_j \\ &\text{subject to} && Ax = b \\ & && x > 0 \end{aligned} \tag{10.4.2}$$

It is clear that $\mu = 0$ corresponds to the original problem eq. (10.4.1). As $\mu \to \infty$, the solution approaches the analytic center of the feasible region (when it is bounded), since the barrier term swamps out $c^{\mathrm{T}}x$ in the objective. As μ is varied continuously toward 0, there is a path $x(\mu)$ defined by the solution to (BP). This path $x(\mu)$ is termed as the primal central path. As $\mu \to 0$ this path converges to the analytic center of the optimal face $\{x : c^{\mathrm{T}}x = z^*, Ax = b, x \geqslant 0\}$ where z^* is the optimal value of (LP).

A strategy for solving (LP) is to solve (BP) for smaller and smaller values of μ and thereby approach a solution to (LP). This is indeed the basic idea of interior-point methods.

At any $\mu > 0$, the necessary and sufficient conditions for a unique and bounded solution are obtained by introducing a Lagrange multiplier vector y for the linear equality constraints to form the Lagrangian $c^{\mathrm{T}}x - \mu \sum_{j=1}^{n} \log x_j - y^{\mathrm{T}}(Ax - b)$.

The derivatives with respect to $\boldsymbol{x}_j$ are set to zero, leading to the conditions $c_j - \mu/\boldsymbol{x}_j - \boldsymbol{y}^{\mathrm{T}}\boldsymbol{a}_j = 0$, for each j, or equivalently

$$\mu \boldsymbol{X}^{-1}\mathbf{1} + \boldsymbol{A}^{\mathrm{T}}\boldsymbol{y} = \boldsymbol{c} \tag{10.4.3}$$

where as before $\boldsymbol{a}_j$ is the j-th column of $\boldsymbol{A}$, $\mathbf{1}$ is the vector of $\mathbf{1}$, and $\boldsymbol{X}$ is the diagonal matrix whose diagonal entries are the components of $\boldsymbol{x} > \mathbf{0}$. Setting $\boldsymbol{s}_j = \mu/\boldsymbol{x}_j$ the complete set of conditions can be rewritten

$$\begin{aligned} \boldsymbol{x} \circ \boldsymbol{s} &= \mu \mathbf{1} \\ \boldsymbol{A}\boldsymbol{x} &= \boldsymbol{b} \\ \boldsymbol{A}^{\mathrm{T}}\boldsymbol{y} + \boldsymbol{s} &= \boldsymbol{c} \end{aligned} \tag{10.4.4}$$

Note that $\boldsymbol{y}$ is a dual feasible solution and $\boldsymbol{c} - \boldsymbol{A}^{\mathrm{T}}\boldsymbol{y} > \mathbf{0}$.

Dual Central Path

Now consider the dual problem

$$\begin{aligned} (LD)\ \text{maximize} \quad & \boldsymbol{y}^{\mathrm{T}}\boldsymbol{b} \\ \text{subject to} \quad & \boldsymbol{y}^{\mathrm{T}}\boldsymbol{A} + \boldsymbol{s}^{\mathrm{T}} = \boldsymbol{c}^{\mathrm{T}} \\ & \boldsymbol{s} \geqslant \mathbf{0} \end{aligned}$$

We may apply the barrier approach to this problem by formulating the problem

$$\begin{aligned} (BD)\ \text{maximize} \quad & \boldsymbol{y}^{\mathrm{T}}\boldsymbol{b} + \mu \sum_{j=1}^{n} \log s_j \\ \text{subject to} \quad & \boldsymbol{y}^{\mathrm{T}}\boldsymbol{A} + \boldsymbol{s}^{\mathrm{T}} = \boldsymbol{c}^{\mathrm{T}} \\ & \boldsymbol{s} > \mathbf{0} \end{aligned}$$

We assume that the dual feasible set F_d has an interior $\mathring{F}_d = \{(\boldsymbol{y}, \boldsymbol{s}) : \boldsymbol{y}^{\mathrm{T}}\boldsymbol{A} + \boldsymbol{s}^{\mathrm{T}} = \boldsymbol{c}^{\mathrm{T}}, \boldsymbol{s} > \mathbf{0}\}$ is nonempty and the optimal solution set of (LD) is bounded. Then, as μ is varied continuously toward 0, there is a path $(\boldsymbol{y}(\mu), \boldsymbol{s}(\mu))$ defined by the solution to (BD). This path is termed as the dual central path.

To work out the necessary and sufficient conditions we introduce $\boldsymbol{x}$ as a Lagrange multiplier and form the Lagrangian $\boldsymbol{y}^{\mathrm{T}}\boldsymbol{b} + \mu \sum_{j=1}^{n} \log s_j - (\boldsymbol{y}^{\mathrm{T}}\boldsymbol{A} + \boldsymbol{s}^{\mathrm{T}} - \boldsymbol{c}^{\mathrm{T}})\boldsymbol{x}$.

Setting to zero the derivative with respect to $\boldsymbol{y}_i$ leads to

$$\boldsymbol{b}_i - \boldsymbol{a}^i\boldsymbol{x} = \mathbf{0}, \text{for all } i$$

where $\boldsymbol{a}^i$ is the i-th row of $\boldsymbol{A}$. Setting to zero the derivative with respect to $\boldsymbol{s}_j$ leads to

$$\mu/\boldsymbol{s}_j - \boldsymbol{x}_j = \mathbf{0}, \text{for all } j$$

Combining these equations and including the original constraint yield the complete set of conditions

$$\boldsymbol{x} \circ \boldsymbol{s} = \mu \mathbf{1}$$
$$\boldsymbol{A}\boldsymbol{x} = \boldsymbol{b}$$
$$\boldsymbol{A}^{\mathrm{T}}\boldsymbol{y} + \boldsymbol{s} = \boldsymbol{c}$$

These are identical to the optimality conditions for the primal central path eq. (10.4.4). Note that $\boldsymbol{x}$ is a primal feasible solution and $\boldsymbol{x} > \mathbf{0}$.

To see the geometric representation of the dual central path, consider the dual level set

$$\Omega(z) = \{\boldsymbol{y} : \boldsymbol{c}^{\mathrm{T}} - \boldsymbol{A}^{\mathrm{T}}\boldsymbol{y} \geqslant \mathbf{0}, \boldsymbol{y}^{\mathrm{T}}\boldsymbol{b} \geqslant z\}$$

for any $z < z^*$ where z^* is the optimal value of (LD). Then, the analytic center $(\boldsymbol{y}(z), \boldsymbol{s}(z))$ of $\Omega(z)$ coincides with the dual central path as z tends to the optimal value z^* from below.

Primal-Dual Central Path

Suppose the feasible region of the primal (LP) has interior points and its optimal solution set is bounded. Then, the dual also has interior points. The primal-dual path is defined to be the set of vectors $(\boldsymbol{x}(\mu), \boldsymbol{y}(\mu), \boldsymbol{s}(\mu))$ that satisfy the conditions

$$\begin{aligned} &\boldsymbol{x} \circ \boldsymbol{s} = \mu \mathbf{1} \\ &\boldsymbol{A}\boldsymbol{x} = \boldsymbol{b} \\ &\boldsymbol{A}^{\mathrm{T}}\boldsymbol{y} + \boldsymbol{s} = \boldsymbol{c} \\ &\boldsymbol{x} \geqslant \mathbf{0}, \boldsymbol{s} \geqslant \mathbf{0} \end{aligned} \tag{10.4.5}$$

for $0 \leqslant \mu \leqslant \infty$. Hence the central path is defined without explicit reference to an optimization problem. It is simply defined in terms of the set of equality and inequality conditions.

Since conditions eq. (10.4.4) and eq. (10.4.5) are identical, the primal-dual central path can be split into two components by projecting onto the relevant space, as described in the following proposition.

Proposition: Suppose the feasible sets of the primal and dual programs contain interior points. Then the primal-dual central path $(\boldsymbol{x}(\mu), \boldsymbol{y}(\mu), \boldsymbol{s}(\mu))$ exists for all $0 \leqslant \mu < \infty$. Furthermore, $\boldsymbol{x}(\mu)$ is the primal central path, and $(\boldsymbol{y}(\mu), \boldsymbol{s}(\mu))$ is the dual central path. Moreover, $\boldsymbol{x}(\mu)$ and $(\boldsymbol{y}(\mu), \boldsymbol{s}(\mu))$ converge to the analytic centers of the optimal primal solution and dual solution faces, respectively, as $\mu \rightarrow 0$.

Duality Gap

Let$(\boldsymbol{x}(\mu), \boldsymbol{y}(\mu), \boldsymbol{s}(\mu))$ be on the primal-dual central path. Then from eq. (10.4.5) it follows that $\boldsymbol{c}^{\mathrm{T}}\boldsymbol{x} - \boldsymbol{y}^{\mathrm{T}}\boldsymbol{b} = \boldsymbol{y}^{\mathrm{T}}\boldsymbol{A}\boldsymbol{x} + \boldsymbol{s}^{\mathrm{T}}\boldsymbol{x} - \boldsymbol{y}^{\mathrm{T}}\boldsymbol{b} = \boldsymbol{s}^{\mathrm{T}}\boldsymbol{x} = n\mu$.

The value $\boldsymbol{c}^{\mathrm{T}}\boldsymbol{x} - \boldsymbol{y}^{\mathrm{T}}\boldsymbol{b} = \boldsymbol{s}^{\mathrm{T}}\boldsymbol{x}$ is the difference between the primal objective value and the dual objective value. This value is always nonnegative and is termed as the duality gap.

The duality gap provides a measure of closeness to optimality. For any primal feasible $\boldsymbol{x}$, the value $\boldsymbol{c}^{\mathrm{T}}\boldsymbol{x}$ gives an upper bound as $\boldsymbol{c}^{\mathrm{T}}\boldsymbol{x} \geqslant z^*$ where z^* is the optimal value of the primal. Likewise, for any dual feasible pair $(\boldsymbol{y}, \boldsymbol{s})$, the value $\boldsymbol{y}^{\mathrm{T}}\boldsymbol{b}$ gives a lower bound as $\boldsymbol{y}^{\mathrm{T}}\boldsymbol{b} \leqslant z^*$. The difference, the duality gap $\boldsymbol{g} = \boldsymbol{c}^{\mathrm{T}}\boldsymbol{x} - \boldsymbol{y}^{\mathrm{T}}\boldsymbol{b}$, provides a bound on z^* as $z^* \geqslant \boldsymbol{c}^{\mathrm{T}}\boldsymbol{x} - \boldsymbol{g}$. Hence if at a feasible point $\boldsymbol{x}$, a dual feasible $(\boldsymbol{y}, \boldsymbol{s})$ is available, the quality of $\boldsymbol{x}$ can be measured as $\boldsymbol{c}^{\mathrm{T}}\boldsymbol{x} - z^* \leqslant \boldsymbol{g}$.

At any point on the primal-dual central path, the duality gap is equal to $n\mu$. It is clear that as $\mu \to 0$ the duality gap goes to zero, and hence both $\boldsymbol{x}(\mu)$ and $(\boldsymbol{y}(\mu), \boldsymbol{s}(\mu))$ approach optimality for the primal and dual, respectively.

10.4.4 Solution Strategies

The various definitions of the central path directly suggest corresponding strategies for solution of a linear program. We outline three general approaches here: the primal barrier or path-following method, the primal-dual path-following method and the primal-dual potential-reduction method, although the details of their implementation and analysis must be deferred to later chapters after study of general nonlinear methods.

Primal Barrier Method

A direct approach is to use the barrier construction and solve the problem

$$
\begin{aligned}
\text{minimize} \quad & \boldsymbol{c}^{\mathrm{T}}\boldsymbol{x} - \mu \sum_{j=1}^{n} \log x_j \\
\text{subject to} \quad & \boldsymbol{A}\boldsymbol{x} = \boldsymbol{b} \\
& \boldsymbol{x} > \boldsymbol{0}
\end{aligned}
\qquad (10.4.6)
$$

for a very small value of μ. In fact, if we desire to reduce the duality gap to ε, it is only necessary to solve the problem for $\mu = \varepsilon/n$. Unfortunately, when μ is small, the problem eq. (10.4.6) could be highly ill-conditioned in the sense that the necessary conditions are nearly singular. This makes it difficult to directly solve the problem for small μ.

An overall strategy, therefore, is to start with a moderately large μ (say μ = 100) and solve that problem approximately. The corresponding solution is a point approximately on the primal central path, but it is likely to be quite distant from the point corresponding to the limit of $\mu \to 0$. However this solution point at μ = 100 can be used as the starting point for the problem with a slightly smaller μ, for this point is likely to be close to the solution of the new problem. The value of μ might be reduced at each stage by a specific factor, giving $\mu_{k+1} = \gamma\mu_k$, where γ is a fixed positive parameter less than one and k is the stage count.

If the strategy begins with a value μ_0, then at the k-th stage we have $\mu_k = \gamma^k\mu_0$. Hence to reduce μ_k/μ_0 to below ε requires $k = \dfrac{\log\varepsilon}{\log\gamma}$ stages.

Often a version of Newton's method for minimization is used to solve each of the problems. For the current strategy, Newton's method works on problem eq. (10.4.6) with fixed μ by considering the central path eq. (10.4.4)

$$\begin{aligned} \boldsymbol{x} \circ \boldsymbol{s} &= \mu \mathbf{1} \\ \boldsymbol{A}\boldsymbol{x} &= \boldsymbol{b} \\ \boldsymbol{A}^{\mathrm{T}}\boldsymbol{y} + \boldsymbol{s} &= \boldsymbol{c} \end{aligned} \tag{10.4.7}$$

From a given point $\boldsymbol{x} \in \mathring{F}_p$, Newton's method moves to a closer point $\boldsymbol{x}^+ \in \mathring{F}_p$ by moving in the directions $\boldsymbol{d}_x$, $\boldsymbol{d}_y$ and $\boldsymbol{d}_s$ determined from the linearized version of eq. (10.4.7)

$$\begin{aligned} \mu \boldsymbol{X}^{-2}\boldsymbol{d}_x + \boldsymbol{d}_s &= \mu \boldsymbol{X}^{-1}\mathbf{1} - \boldsymbol{c} \\ \boldsymbol{A}\boldsymbol{d}_x &= \mathbf{0} \\ -\boldsymbol{A}^{\mathrm{T}}\boldsymbol{d}_y - \boldsymbol{d}_s &= \mathbf{0} \end{aligned} \tag{10.4.8}$$

(Recall that $\boldsymbol{X}$ is the diagonal matrix whose diagonal entries are components of $\boldsymbol{x} > \mathbf{0}$.) The new point is then updated by taking a step in the direction of $\boldsymbol{d}_x$, as $\boldsymbol{x}^+ = \boldsymbol{x} + \boldsymbol{d}_x$.

Notice that if $\boldsymbol{x} \circ \boldsymbol{s} = \mu\mathbf{1}$ for some $\boldsymbol{s} = \boldsymbol{c} - \boldsymbol{A}^{\mathrm{T}}\boldsymbol{y}$, then $\boldsymbol{d} \equiv (\boldsymbol{d}_x, \boldsymbol{d}_y, \boldsymbol{d}_s) = 0$ because the current point satisfies $\boldsymbol{A}\boldsymbol{x} = \boldsymbol{b}$ and hence is already the central path solution for μ. If some component of $\boldsymbol{x} \circ \boldsymbol{s}$ is less than μ, then $\boldsymbol{d}$ will tend to increment the solution so as to increase that component. The converse will occur for components of $\boldsymbol{x} \circ \boldsymbol{s}$ greater than μ.

This process may be repeated several times until a point close enough to the proper solution to the barrier problem for the given value of μ is obtained. That is, until the necessary and sufficient conditions eq. (10.4.3) are (approximately) satisfied.

To solve eq. (10.4.8), pre-multiplying both sides by $\boldsymbol{X}^2$, we have

$$\mu \boldsymbol{d}_x + \boldsymbol{X}^2 \boldsymbol{d}_s = \mu \boldsymbol{X}\boldsymbol{1} - \boldsymbol{X}^2 \boldsymbol{c}$$

Then, premultiplying by $\boldsymbol{A}$ and using $\boldsymbol{A}\boldsymbol{d}_x = \boldsymbol{0}$, we have

$$\boldsymbol{A}\boldsymbol{X}^2 \boldsymbol{d}_s = \mu \boldsymbol{A}\boldsymbol{X}\boldsymbol{1} - \boldsymbol{A}\boldsymbol{X}^2 \boldsymbol{c}$$

Using $\boldsymbol{d}_s = -\boldsymbol{A}^{\mathrm{T}} \boldsymbol{d}_y$ we have

$$(\boldsymbol{A}\boldsymbol{X}^2 \boldsymbol{A}^{\mathrm{T}}) \boldsymbol{d}_y = -\mu \boldsymbol{A}\boldsymbol{X}\boldsymbol{1} + \boldsymbol{A}\boldsymbol{X}^2 \boldsymbol{c}$$

Thus, $\boldsymbol{d}_y$ can be computed by solving the above linear system of equations. Then $\boldsymbol{d}_s$ can be found from the third equation in eq. (10.4.8) and finally $\boldsymbol{d}_x$ can be found from the first equation in eq. (10.4.8), together this amounts to $O(nm^2 + m^3)$ arithmetic operations for each Newton step.

Primal-Dual Path-Following

Another strategy for solving a linear program is to follow the central path from a given initial primal-dual solution pair. Consider a linear program in standard form

$$\begin{aligned} (LP)\ \text{minimize} \quad & \boldsymbol{c}^{\mathrm{T}} \boldsymbol{x} \\ \text{subject to} \quad & \boldsymbol{A}\boldsymbol{x} = \boldsymbol{b} \\ & \boldsymbol{x} \geqslant \boldsymbol{0} \\ (LD)\ \text{maximize} \quad & \boldsymbol{y}^{\mathrm{T}} \boldsymbol{b} \\ \text{subject to} \quad & \boldsymbol{y}^{\mathrm{T}} \boldsymbol{A} + \boldsymbol{s}^{\mathrm{T}} = \boldsymbol{c}^{\mathrm{T}} \\ & \boldsymbol{s} \geqslant \boldsymbol{0} \end{aligned}$$

Assume that $\mathring{F}_p \neq \phi$; that is, both

$$\mathring{F}_p = \{\boldsymbol{x} : \boldsymbol{A}\boldsymbol{x} = \boldsymbol{b}, \boldsymbol{x} > \boldsymbol{0}\} \neq \phi$$

and

$$\mathring{F}_d = \{(\boldsymbol{y}, \boldsymbol{s}) : \boldsymbol{s} = \boldsymbol{c} - \boldsymbol{A}^{\mathrm{T}} \boldsymbol{y} > \boldsymbol{0}\} \neq \phi$$

and denote by z^* the optimal objective value.

The central path can be expressed as

$$\ell = \left\{ (\boldsymbol{x}, \boldsymbol{y}, \boldsymbol{s}) \in \mathring{F} : \boldsymbol{x} \circ \boldsymbol{s} = \frac{\boldsymbol{x}^{\mathrm{T}} \boldsymbol{s}}{n} \boldsymbol{1} \right\}$$

in the primal-dual form. On the path we have $\boldsymbol{x} \circ \boldsymbol{s} = \mu \boldsymbol{1}$ and hence $\boldsymbol{s}^{\mathrm{T}} \boldsymbol{x} = n\mu$. A neighborhood of the central path ℓ is of the form

$$N(\eta)=\{(\boldsymbol{x},\boldsymbol{y},\boldsymbol{s})\in\mathring{F}:|\boldsymbol{s}\circ\boldsymbol{x}-\mu\mathbf{1}|\leqslant\eta\mu,\text{where }\mu=\boldsymbol{s}^{\mathrm{T}}\boldsymbol{x}/n\}\quad(10.4.9)$$

for some $\eta\in(0,1)$, say $\eta=1/4$. This can be thought of as a tube whose center is the central path.

The idea of the path-following method is to move within a tubular neighborhood of the central path toward the solution point. A suitable initial point $(\boldsymbol{x}^0,\boldsymbol{y}^0,\boldsymbol{s}^0)\in N(\eta)$ can be found by solving the barrier problem for some fixed μ_0 or from an initialization phase proposed later. After that, step by step moves are made, alternating between a predictor step and a corrector step. After each pair of steps, the point achieved is again in the fixed given neighborhood of the central path, but closer to the linear program's solution set.

The predictor step is designed to move essentially parallel to the true central path. The step $\boldsymbol{d}\equiv(\boldsymbol{d}_x,\boldsymbol{d}_y,\boldsymbol{d}_s)$ is determined from the linearized version of the primal-dual central path equations of eq. (10.4.5), as

$$\begin{aligned}\boldsymbol{s}\circ\boldsymbol{d}_x+\boldsymbol{x}\circ\boldsymbol{d}_s&=\gamma\mu\mathbf{1}-\boldsymbol{x}\circ\boldsymbol{s}\\ \boldsymbol{A}\boldsymbol{d}_x&=\mathbf{0}\\ -\boldsymbol{A}^{\mathrm{T}}\boldsymbol{d}_y-\boldsymbol{d}_s&=\mathbf{0}\end{aligned}\quad(10.4.10)$$

where one selects $\gamma=0$. (To show the dependence of $\boldsymbol{d}$ on the current pair $(\boldsymbol{x},\boldsymbol{s})$ and the parameter γ, we write $\boldsymbol{d}=\boldsymbol{d}(\boldsymbol{x},\boldsymbol{s},\gamma)$.)

The new point is then found by taking a step in the direction of $\boldsymbol{d}$, as $(\boldsymbol{x}^+,\boldsymbol{y}^+,\boldsymbol{s}^+)=(\boldsymbol{x},\boldsymbol{y},\boldsymbol{s})+\alpha(\boldsymbol{d}_x,\boldsymbol{d}_y,\boldsymbol{d}_s)$, where α is the step-size. Note that $\boldsymbol{d}_x^{\mathrm{T}}\boldsymbol{d}_s=\boldsymbol{d}_x^{\mathrm{T}}\boldsymbol{A}^{\mathrm{T}}\boldsymbol{d}_y=\mathbf{0}$, then

$$(\boldsymbol{x}^+)^{\mathrm{T}}\boldsymbol{s}^+=(\boldsymbol{x}+\alpha\boldsymbol{d}_x)^{\mathrm{T}}(\boldsymbol{s}+\alpha\boldsymbol{d}_s)=\boldsymbol{x}^{\mathrm{T}}\boldsymbol{s}+\alpha(\boldsymbol{d}_x^{\mathrm{T}}\boldsymbol{s}+\boldsymbol{x}^{\mathrm{T}}\boldsymbol{d}_s)=(1-\alpha)\boldsymbol{x}^{\mathrm{T}}\boldsymbol{s}$$

where the last step follows by multiplying the first equation in eq. (10.4.10) by $\mathbf{1}^{\mathrm{T}}$. Thus, the predictor step reduces the duality gap by a factor $1-\alpha$. The maximum possible step-size α in that direction is made in that parallel direction without going outside of the neighborhood $N(2\eta)$.

The corrector step essentially moves perpendicular to the central path in order to get closer to it. This step moves the solution back to within the neighborhood $N(\eta)$ and the step is determined by selecting $\gamma=1$ in eq. (10.4.10) with $\mu=\boldsymbol{x}^{\mathrm{T}}\boldsymbol{s}/n$. Notice that if $\boldsymbol{x}\circ\boldsymbol{s}=\mu\mathbf{1}$, then $\boldsymbol{d}=\mathbf{0}$ because the current point is already a central path solution.

This corrector step is identical to one step of the barrier method. Note, however, that the predictor-corrector method requires only one sequence of steps, each consisting of a single predictor and corrector. This contrasts with the barrier method which requires a complete sequence for each μ to get back to the central path, and then an outer sequence to reduce μ.

One can prove that for any $(\boldsymbol{x},\boldsymbol{y},\boldsymbol{s}) \in N(\eta)$ with $\mu = \boldsymbol{x}^{\mathrm{T}}\boldsymbol{s}/n$, the step-size in the predictor stop satisfies

$$\alpha \geqslant \frac{1}{2\sqrt{n}}$$

Thus, the iteration complexity of the method is $O(\sqrt{n}\log(1/\varepsilon))$ to achieve $\mu/\mu_0 \leqslant \varepsilon$ where $n\mu_0$ is the initial duality gap. Moreover, one can prove that the step-size $\alpha \to 1$ as $\boldsymbol{x}^{\mathrm{T}}\boldsymbol{s} \to 0$, that is, the duality reduction speed is accelerated as the gap becomes smaller.

Primal-Dual Potential Function

For $\boldsymbol{x} \in \mathring{F}_p$ and $(\boldsymbol{y},\boldsymbol{s}) \in \mathring{F}_d$ the primal-dual potential function is defined by

$$\psi_{n+\rho}(\boldsymbol{x},\boldsymbol{s}) \equiv (n+p)\log(\boldsymbol{x}^{\mathrm{T}}\boldsymbol{s}) - \sum_{j=1}^{n} \log(x_j s_j) \qquad (10.4.11)$$

where $\rho \geqslant 0$.

From the arithmetic and geometric mean inequality we can derive that

$$n\log(\boldsymbol{x}^{\mathrm{T}}\boldsymbol{s}) - \sum_{j=1}^{n} \log(x_j s_j) \geqslant n\log n$$

Then

$$\psi_{n+\rho}(\boldsymbol{x},\boldsymbol{s}) = \rho\log(\boldsymbol{x}^{\mathrm{T}}\boldsymbol{s}) + n\log(\boldsymbol{x}^{\mathrm{T}}\boldsymbol{s}) - \sum_{j=1}^{n} \log(x_j s_j) \geqslant \rho\log(\boldsymbol{x}^{\mathrm{T}}\boldsymbol{s}) + n\log n \qquad (10.4.12)$$

Thus, for $\rho > 0$, $\psi_{n+\rho}(\boldsymbol{x},\boldsymbol{s}) \to -\infty$ implies that $\boldsymbol{x}^{\mathrm{T}}\boldsymbol{s} \to 0$. More precisely, we have from eq. (10.4.12)

$$\boldsymbol{x}^{\mathrm{T}}\boldsymbol{s} \leqslant \exp\left(\frac{\psi_{n+\rho}(\boldsymbol{x},\boldsymbol{s}) - n\log n}{\rho}\right)$$

Hence the primal-dual potential function gives an explicit bound on the magnitude of the duality gap.

The objective of this method is to drive the potential function down toward minus infinity. In this case we select $\gamma = n/(n+\rho)$ in eq. (10.4.10). Notice that it is a combination of a predictor and corrector choice. The predictor uses $\gamma = 0$ and the corrector uses $\gamma = 1$. The primal-dual potential method uses something in between.

For $\rho \geqslant \sqrt{n}$, there is in fact a guaranteed decrease in the potential function by a fixed amount δ. Specifically,

$$\psi_{n+\rho}(\boldsymbol{x}^+, \boldsymbol{s}^+) - \psi_{n+\rho}(\boldsymbol{x}, \boldsymbol{s}) \leqslant -\delta \tag{10.4.13}$$

for a constant $\delta \geqslant 0.2$. This result provides a theoretical bound on the number of required iterations and the bound is competitive with other methods. However, a faster algorithm may be achieved by conducting a line search along direction $\boldsymbol{d}$ to achieve the greatest reduction in the primal-dual potential function at each iteration.

We outline the algorithm here:

Step1. Start at a point $(\boldsymbol{x}_0, \boldsymbol{y}_0, \boldsymbol{s}_0) \in \mathring{F}$ with

$$\psi_{n+\rho}(\boldsymbol{x}_0, \boldsymbol{s}_0) \leqslant -\rho\log((\boldsymbol{s}_0)^{\mathrm{T}}\boldsymbol{x}_0) + n\log n + O(\sqrt{n}\log n)$$

which is determined by an initiation procedure. Set $\rho \geqslant \sqrt{n}$. Set $k = 0$ and $\gamma = n/(n + \rho)$. Select an accuracy parameter $\varepsilon > 0$.

Step 2. Set $(\boldsymbol{x}, \boldsymbol{s}) = (\boldsymbol{x}_k, \boldsymbol{s}_k)$ and compute $(\boldsymbol{d}_x, \boldsymbol{d}_y, \boldsymbol{d}_s)$ from eq. (10.4.10).

Step 3. Let $\boldsymbol{x}_{k+1} = \boldsymbol{x}_k + \bar{\alpha}\boldsymbol{d}_x$, $\boldsymbol{y}_{k+1} = \boldsymbol{y}_k + \bar{\alpha}\boldsymbol{d}_y$, and $\boldsymbol{s}_{k+1} = \boldsymbol{s}_k + \bar{\alpha}\boldsymbol{d}_s$ where

$$\bar{\alpha} = \arg\min_{\alpha \geqslant 0} \psi_{n+\rho}(\boldsymbol{x}_k + \alpha\boldsymbol{d}_x, \boldsymbol{s}_k + \alpha\boldsymbol{d}_s)$$

Step 4. Let $k = k + 1$. If $\dfrac{\boldsymbol{s}_k^{\mathrm{T}}\boldsymbol{x}_k}{\boldsymbol{s}_0^{\mathrm{T}}\boldsymbol{x}_0} \leqslant \varepsilon$, Stop. Otherwise return to Step 2.

Theorem: The algorithm above terminates in at most $O(\rho\log(n/\varepsilon))$ iterations with $\dfrac{(\boldsymbol{s}_k)^{\mathrm{T}}\boldsymbol{x}_k}{(\boldsymbol{s}_0)^{\mathrm{T}}\boldsymbol{x}_0} \leqslant \varepsilon$.

Iteration Complexity

The computation of each iteration basically requires solving eq. (10.4.10) for $\boldsymbol{d}$. Note that the first equation of eq. (10.4.10) can be written as

$$\boldsymbol{S}\boldsymbol{d}_x + \boldsymbol{X}\boldsymbol{d}_s = \gamma\mu\mathbf{1} - \boldsymbol{X}\boldsymbol{S}\mathbf{1}$$

where $\boldsymbol{X}$ and $\boldsymbol{S}$ are two diagonal matrices whose diagonal entries are components of $\boldsymbol{x} > 0$ and $\boldsymbol{s} > 0$, respectively.

Pre-multiplying both sides by $\boldsymbol{S}^{-1}$, we have

$$\boldsymbol{d}_x + \boldsymbol{S}^{-1}\boldsymbol{X}\boldsymbol{d}_s = \gamma\mu\boldsymbol{S}^{-1}\mathbf{1} - \boldsymbol{x}$$

Then, pre-multiplying by $\boldsymbol{A}$ and using $\boldsymbol{A}\boldsymbol{d}_x = \mathbf{0}$, we have

$$\boldsymbol{A}\boldsymbol{S}^{-1}\boldsymbol{X}\boldsymbol{d}_s = \gamma\mu\boldsymbol{S}^{-1}\mathbf{1} - \boldsymbol{A}\boldsymbol{x} = \gamma\mu\boldsymbol{S}^{-1}\mathbf{1} - \boldsymbol{b}$$

Using $\boldsymbol{d}_s = -\boldsymbol{A}^{\mathrm{T}}\boldsymbol{d}_y$, we have

$$(\boldsymbol{A}\boldsymbol{S}^{-1}\boldsymbol{X}\boldsymbol{A}^{\mathrm{T}})\boldsymbol{d}_y = \boldsymbol{b} - \gamma\mu\boldsymbol{A}\boldsymbol{S}^{-1}\mathbf{1}$$

Thus, the primary computational cost of each iteration of the interior-point algorithm discussed in this section is to form and invert the normal matrix $\boldsymbol{A}\boldsymbol{X}\boldsymbol{S}^{-1}\boldsymbol{A}^{\mathrm{T}}$, which typically requires $O(nm^2 + m^3)$ arithmetic operations. However, an approximation of this matrix can be updated and inverted using far fewer arithmetic operations. In fact, using a rank-one technique to update the approximate inverse of the normal matrix during the iterative progress, one can reduce the average number of arithmetic operations per iteration to $O(\sqrt{n}m^2)$. Thus, if the relative tolerance ε is viewed as a variable, we have the following total arithmetic operation complexity bound to solve a linear program.

Corollary: Let $\rho = \sqrt{n}$, then the algorithm of above Theorem terminates in at most $O(nm^2\log(n/\varepsilon))$ arithmetic operations.

Chapter 11

Unconstrained Problems

11.1 Transportation and Network Flow Problems

The problem is to find the shipping pattern between origins and destinations that satisfies all the requirements and minimizes the total shipping cost. The chapter is roughly divided into two parts. In the first part the transportation problem is examined from the viewpoint of the revised simplex method, which takes an extremely simple form for this problem. The second part of the chapter introduces graphs and network flows. The transportation algorithm is generalized and given new interpretations. Next, a special, highly efficient algorithm, the tree algorithm, is developed for solution of the maximal flow problem.

11.1.1 The Transportation Problem

There are m origins that contain various amounts of a commodity that must be shipped to n destinations to meet requirements. Specifically, origin i contains an amount a_i, and destination j has a requirement of amount b_j. It is assumed that the system is balanced in the sense that the total supply equals the total demand. That is,

$$\sum_{i=1}^{m} a_i = \sum_{j=1}^{n} b_j \tag{11.1.1}$$

The numbers a_i and $b_j (i = 1, 2, ..., m, j = 1, 2, ..., n)$ are assumed to be nonnegative, and in many applications they are in fact nonnegative integers. There is a unit cost c_{ij} associated with the shipping of the commodity from origin i to destination j.

In mathematical terms the above problem can be expressed as finding a set of $x_{ij} (i = 1, 2, ..., m, j = 1, 2, ..., n)$ satisfying

$$\begin{aligned} \text{minimize} \quad & \sum_{i=1}^{m} \sum_{j=1}^{n} c_{ij} x_{ij} \\ \text{subject to} \quad & \sum_{j=1}^{n} x_{ij} = a_i \text{ for } i = 1, 2, ..., m \\ & \sum_{i=1}^{m} x_{ij} = b_j \quad \text{for } j = 1, 2, ..., n \\ & x_{ij} \geqslant 0 \quad \text{for all } i \text{ and all } j \end{aligned} \tag{11.1.2}$$

This mathematical problem, together with the assumption eq. (11.1.1), is the general transportation problem. In the shipping context, the variables x_{ij} represent the amounts of the commodity shipped from origin i to destination j. The structure of the problem can be seen more clearly by writing the constraint equations in standard form:

$$\begin{array}{cccccccccc} x_{11} + x_{12} + ... + x_{1n} & & & & & & & & & = a_1 \\ & & & x_{21} + x_{22} + ... + x_{2n} & & & & & & = a_2 \\ & & & & & & & & & \vdots \\ & & & & & & & x_{m1} + x_{m2} + ... + x_{mn} & & = a_m \\ \hline x_{11} & & & + x_{21} & & & & x_{m1} & & = b_1 \\ & x_{12} & & & + x_{22} & & & + x_{m2} & & = b_2 \\ & & & & & & & & & \vdots \\ & & x_{1n} & & & + x_{2n} & & & + x_{mn} & = b_n \end{array} \tag{11.1.3}$$

The structure is perhaps even more evident when the coefficient matrix $\boldsymbol{A}$ of the above system equations is expressed in vector-matrix notation as

$$A = \begin{bmatrix} \mathbf{1}^{\mathrm{T}} & & & & \\ & \mathbf{1}^{\mathrm{T}} & & & \\ & & \cdot & & \\ & & & \cdot & \\ & & & & \cdot \\ & & & & \mathbf{1}^{\mathrm{T}} \\ \boldsymbol{I} & \boldsymbol{I} & \cdot & \cdot & \cdot & \boldsymbol{I} \end{bmatrix} \qquad (11.1.4)$$

where $\mathbf{1} = (1,1,\ldots,1)$ is n-dimensional, and where each $\boldsymbol{I}$ is an $n \times n$ identity matrix.

In practice it is usually unnecessary to write out the constraint equations of the transportation problem in the explicit form eq. (11.1.3). A specific transportation problem is generally defined by simply presenting the data in compact form, such as:

$$\boldsymbol{a} = (a_1, a_2, \ldots, a_m)$$

$$\boldsymbol{b} = (b_1, b_2, \ldots, b_n)$$

$$\boldsymbol{C} = \begin{bmatrix} c_{11} & c_{12} & \cdots & c_{1n} \\ c_{21} & c_{22} & \cdots & c_{2n} \\ \cdots & \cdots & \cdots & \cdots \\ c_{m1} & c_{m2} & \cdots & c_{mn} \end{bmatrix}$$

The solution can also be represented by an $m \times n$ array, and as we shall see, all computations can be made on arrays of a similar dimension.

Feasibility and Redundancy

There are several important steps in the study of the structure of the transportation problem below.

Step 1. Show that there is always a feasible solution. Let S be equal to the total supply or the total demand, we can easily verify that $x_{ij} = a_i b_j / S$ is a bounded feasible solution. Furthermore, we also note that a bounded program with a feasible solution has an optimal solution, so a transportation problem always has an optimal solution.

Step 2. Examine the total $n + m$ constraint equations, including m origin constraints equations and n destination constraints equations. For example, the sum of the origin equations is

$$\sum_{i=1}^{m}\sum_{j=1}^{n} x_{ij} = \sum_{i=1}^{m} a_i \tag{11.1.5}$$

and the sum of the destination equations is

$$\sum_{j=1}^{n}\sum_{i=1}^{m} x_{ij} = \sum_{j=1}^{n} b_j \tag{11.1.6}$$

Their left-hand sides are equal, but the original system equations are not independent. The right-hand sides of eq. (11.1.5) and eq. (11.1.6) are equal from the system's balance, therefore, the two equations are consistent. So a transportation problem always has a feasible solution, but there is exactly one redundant equality constraint.

11.1.2 The Northwest Corner Rule

This procedure is conducted on the solution array shown below:

$$\begin{array}{|ccccc|c}
\hline
x_{11} & x_{12} & x_{13} & \cdots & x_{1n} & a_1 \\
x_{12} & x_{22} & x_{23} & \cdots & x_{2n} & a_2 \\
\vdots & & & & \vdots & \vdots \\
x_{m1} & x_{m2} & x_{m3} & \cdots & x_{mn} & a_m \\
\hline
b_1 & b_2 & b_3 & \cdots & b_n &
\end{array} \tag{11.1.7}$$

The individual elements of the array appear in cells and represent a solution. An empty cell denotes a value of zero.

Beginning with all empty cells, the procedure of the Northwest Corner Rule is given by the following steps:

Step 1. Start with the cell in the upper left-hand corner.

Step 2. Allocate the maximum feasible amount consistent with row and column sum requirements involving that cell. (At least one of these requirements will then be met.)

Step 3. Move one cell to the right if there is any remaining row requirement (supply). Otherwise move one cell down. If all requirements are met, stop; otherwise go to Step 2.

11.1.3 Basic Network Concepts

Flows in Networks

A flow in a given directed arc(i,j) is a number $x_{ij} \geqslant 0$. Flows in the arcs of the network must jointly satisfy a conservation criterion at each node. Specifically, unless the node is a source or sink as discussed below, flow cannot be created or lost at a node; the total flow into a node must equal the total flow out of the node. Thus at each such node i

$$\sum_{j=1}^{n} x_{ij} - \sum_{k=1}^{n} x_{ki} = 0$$

The first sum is the total flow from i, and the second sum is the total flow to i. (Of course x_{ij} does not exist if there is no arc from i to j.) It should be clear that for nonzero flows to exist in a network without sources or sinks, the network must contain a cycle.

Minimum Cost Flow

Consider a network having n nodes. Corresponding to each node i, there is a number b_i representing the available supply at the node. (If $b_i < 0$, then there is a required demand.) We assume that the network is balanced in the sense that

$$\sum_{i=1}^{n} b_i = 0$$

Associated with each arc (i,j) is a number c_{ij}, representing the unit cost for flow along this arc. The minimal cost flow problem is that of determining flows $x_{ij} \geqslant 0$ in each arc of the network so that the net flow into each node i is b_i while minimizing the total cost. In mathematical terms the problem is

$$\begin{aligned} \text{minimize} \quad & \sum c_{ij} x_{ij} \\ \text{subject to} \quad & \sum_{j=1}^{n} x_{ij} - \sum_{k=1}^{n} x_{ki} = b_i, \quad i = 1,2,\ldots,n \\ & x_{ij} \geqslant 0, \quad i,j = 1,2,\ldots,n \end{aligned} \tag{11.1.8}$$

11.1.4 Maximal Flow

Tree Procedure is to determine whether a path from node 1 to node m exists. At each step of the algorithm, each node is either unlabeled, labeled but unscanned, or labeled and scanned. The procedure consists of these steps:

Step 1. Label node 1 with any mark. All other nodes are unlabeled.

Step 2. For any labeled but unscanned node i, scan the node by finding all unlabeled nodes reachable from i by a single arc. Label these nodes with an i.

Step 3. If node m is labeled, stop; a breakthrough has been achieved—a path exists. If no unlabeled nodes can be labeled, stop; no connecting path exists.

Otherwise, go to Step 2.

The process is illustrated in Fig. 11.1.1, where a path between nodes 1 and 10 is sought. The nodes have been labeled and scanned in the order 1,2,3,5,6, 8,4,7,9,10. The labels are indicated close to the nodes. The arcs that were used in the scanning processes are indicated by heavy lines. Note that the collection of nodes and arcs selected by the process, regarded as an undirected graph, form a tree—a graph without cycles. This, of course, accounts for the name of the process, the tree procedure. If one is interested only in determining whether a connecting path exists and does not need to find the path itself, then the labels need only be simple check marks rather than node indices. However, if node indices are used as labels, then after successful completion of the algorithm, the actual connecting path can be found by tracing backward from node m by following the labels. In the example, one begins at 10 and moves to node 7 as indicated; then to 6,3, and 1. The path follows the reverse of this sequence.

It is easy to prove that the algorithm does indeed resolve the issue of the existence of a connecting path. At each stage of the process, either a new node is labeled, it is impossible to continue, or node m is labeled and the process is successfully terminated. Clearly, the process can continue for at most $n-1$ stages, where n is the number of nodes in the graph. Suppose at some stage it is impossible to continue. Let S be the set of labeled nodes at that stage and let $\overline{S}$ be the set of unlabeled nodes. Clearly, node 1 is contained in S, and node m is contained in $\overline{S}$. If there were a path connecting node 1 with node m, then there must be an arc in that path from a node k in S to a node in $\overline{S}$. However, this would imply that node k was not scanned, which is a contradiction. Conversely, if the algorithm does

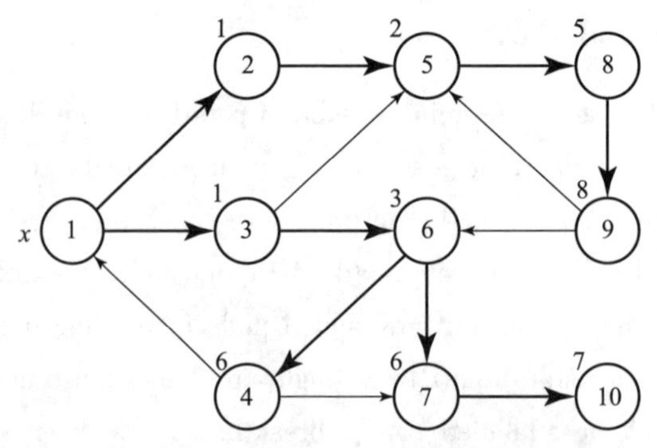

Fig. 11.1.1 The scanning procedure

continue until reaching node m, then it is clear that a connecting path can be constructed backward as outlined above.

Capacitated Networks

Definition. A capacitated network is a network in which some arcs are assigned nonnegative capacities, which define the maximum allowable flow in those arcs. The capacity of an arc (i,j) is denoted k_{ij}, and this capacity is indicated on the graph by placing the number k_{ij} adjacent to the arc.

The Maximal Flow Problem

Consider a capacitated network in which two special nodes, called the source and the sink, are distinguished. Say they are nodes 1 and m, respectively. All other nodes must satisfy the strict conservation requirement; that is, the net flow into these nodes must be zero. However, the source may have a net outflow and the sink a net inflow. The outflow f of the source will equal the inflow of the sink as a consequence of the conservation at all other nodes. A set of arc flows satisfying these conditions is said to be a flow in the network of value f. The maximal flow problem is that of determining the maximal flow that can be established in such a network. When written out, it takes the form

$$\begin{aligned} \text{minimize} \quad & f \\ \text{subject to} \quad & \sum_{j=1}^{n} x_{1j} - \sum_{j=1}^{n} x_{j1} - f = 0 \\ & \sum_{j=1}^{n} x_{ij} - \sum_{j=1}^{n} x_{ji} = 0, \quad i \neq 1, m \\ & \sum_{j=1}^{n} x_{mj} - \sum_{j=1}^{n} x_{jm} + f = 0 \\ & 0 \leqslant x_{ij} \leqslant k_{ij}, \quad \text{all } i,j \end{aligned} \tag{11.1.9}$$

where only those i, j pairs corresponding to arcs are allowed.

The maximal flow problem can be expressed more compactly in terms of the node-arc incidence matrix as follows

$$\begin{aligned} \text{maximize} \quad & f \\ \text{subject to} \quad & \boldsymbol{Ax} - f\boldsymbol{e} = \boldsymbol{0} \\ & \boldsymbol{x} \leqslant \boldsymbol{k} \end{aligned} \tag{11.1.10}$$

where $\boldsymbol{x}$ is the vector of arc flows x_{ij} (ordered in any way), $\boldsymbol{A}$ is the corresponding node-arc incidence matrix and $\boldsymbol{e}$ is a vector with dimension equal to the number of nodes and having $a+1$ component on node 1, $a-1$ on node m, and all other components zero.

It is clear that the coefficient matrix of this problem is equal to the node-arc incidence matrix with an additional column for the flow variable f. Any basis of this matrix is triangular, and hence as indicated by the theory in the earlier part of this chapter, the simplex method can be effectively employed to solve this problem. However, instead of the simplex method, a more efficient algorithm based on the tree algorithm can be used.

The basic strategy of the algorithm is quite simple. First we recognize that it is possible to send nonzero flow from node 1 to node m only if node m is reachable from node 1. The tree procedure of the previous section can be used to determine if m is in fact reachable; and if it is reachable, the algorithm will produce a path from 1 to m. By examining the arcs along this path, we can determine the one with minimum capacity. We may then construct a flow equal to this capacity from 1 to m by using this path. This gives us a strictly positive (and integer-valued) initial flow.

Next consider the nature of the network at this point in terms of additional flows that might be assigned. If there is already flow x_{ij} in the arc (i,j), then the effective capacity of that arc is reduced by x_{ij} (to $k_{ij} - x_{ij}$), since that is the maximal amount of additional flow that can be assigned to that arc. On the other hand, the effective reverse capacity, on the arc (j,i), is increased by x_{ij} (to $k_{ji} + x_{ij}$), since a small incremental backward flow is actually realized as a reduction in the forward flow through that arc. Once these changes in capacities have been made, the tree procedure can again be used to find a path from node 1 to node m on which to assign additional flow. (Such a path is termed as an augmenting path.) Finally, if m is not reachable from node 1, no additional flow can be assigned, and the procedure is complete.

It is seen that the method outlined above is based on repeated application of the tree procedure, which is implemented by labeling and scanning. By including slightly more information in the labels than in the basic tree algorithm, the minimum arc capacity of the augmenting path can be determined during the initial scanning, instead of by reexamining the arcs after the path is found. A typical label at a node i has the form (k, c_i), where k denotes a precursor node and c_i is the maximal flow that can be sent from the source to node i through the path created by the previous labeling and scanning.

The complete procedure about the maximal flow problem is as follows:

Step 0. Set all $x_{ij}=0$ and $f=0$.

Step 1. Label node 1 $(-,\infty)$. All other nodes are unlabeled.

Step 2. Select any labeled node i for scanning. Say it has label (k,c_i). For all unlabeled nodes j such that (i,j) is an arc with $x_{ij}<k_{ij}$, assign the label (i,c_j), where $c_j=\min\{c_i,k_{ij}-x_{ij}\}$. For all unlabeled nodes j such that (j,i) is an arc with $x_{ji}>0$, assign the label (i,c_j), where $c_j=\min\{c_i,x_{ji}\}$.

Step 3. Repeat Step 2 until either node m is labeled or until no more labels can be assigned. In this latter case, the current solution is optimal.

Step 4. (Augmentation) If the node m is labeled (i,c_m), then increase f and the flow on arc (i,m) by c_m. Continue to work backward along the augmenting path determined by the nodes, increasing the flow on each arc of the path by c_m. Return to Step 1.

Proposition: The maximal flow algorithm converges in at most a finite number of iterations.

Max Flow-Min Cut Theorem

A great deal of insight and some further results can be obtained through the introduction of the notion of cuts in a network. Given a network with source node 1 and sink node m, divide the nodes arbitrarily into two sets S and $\bar{S}$ such that the source node is in S and the sink is in $\bar{S}$. The set of arcs from S to $\bar{S}$ is a cut and is denoted $(S,\bar{S})$. The capacity of the cut is the sum of the capacities of the arcs in the cut.

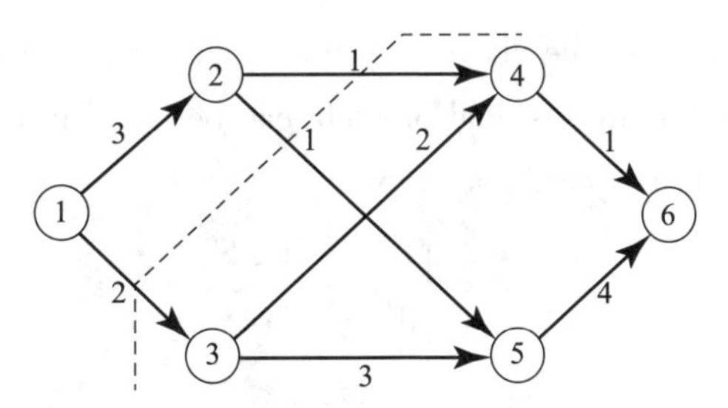

Fig. 11. 1. 2 A cut

An example of a cut is shown in Fig. 11. 1. 2. The set S consists of nodes 1 and 2, while $\bar{S}$ consists of 3, 4, 5, 6. The capacity of this cut is 4.

It should be clear that a path from node 1 to node m must include at least one arc in any cut, for the path must have an arc from the set S to the set $\bar{S}$. Furthermore, it is clear that the maximal amount of flow that can be sent through a cut is equal to its capacity. Thus each cut gives an upper bound on the value of the maximal flow problem. The max flow-min cut theorem states that equality is actually achieved for some cut. That is, the maximal flow is equal to the minimal cut capacity. It should be noted that the proof of the theorem also establishes the maximality of the flow obtained by the maximal flow algorithm.

Max flow-min cut theorem: In a network the maximal flow between a source and a sink is equal to the minimal cut capacity of all cuts separating the source and sink.

Duality

The maximal flow problem is a linear program, which is expressed formally by eq. (11.1.10). The dual problem is found to be

$$\begin{aligned} \text{minimize} \quad & \boldsymbol{w}^{\mathrm{T}}\boldsymbol{k} \\ \text{subject to} \quad & \boldsymbol{u}^{\mathrm{T}}\boldsymbol{A} = \boldsymbol{w}^{\mathrm{T}} \\ & \boldsymbol{u}^{\mathrm{T}}\boldsymbol{e} = \mathbf{1} \\ & \boldsymbol{w} \geqslant \mathbf{0} \end{aligned} \tag{11.1.11}$$

When written out in detail, the dual is

$$\begin{aligned} \text{minimize} \quad & \sum_{ij} w_{ij}k_{ij} \\ \text{subject to} \quad & u_i - u_j = w_{ij} \\ & u_1 - u_m = 1 \\ & w_{ij} \geqslant 0 \end{aligned} \tag{11.1.12}$$

A pair (i,j) is included in the above only if (i,j) is an arc of the network.

A feasible solution to this dual problem can be found in terms of any cut set $(S,\bar{S})$. In particular, it is easily seen that

$$u_i = \begin{cases} 1 & \text{if } i \in S \\ 0 & \text{if } i \in \bar{S} \end{cases} \tag{11.1.13}$$

$$w_{ij} = \begin{cases} 1 & \text{if } (i,j) \in (S,\bar{S}) \\ 0 & \text{otherwise} \end{cases}$$

is a feasible solution. The value of the dual problem corresponding to this solution is the cut capacity. If we take the cut set to be the one determined by the labeling procedure of the maximal flow algorithm as described in the proof of the theorem above, it can be seen to be optimal by verifying the complementary slackness conditions. The minimum value of the dual is therefore equal to the minimum cut capacity.

11.2 Basic Properties of Solutions and Algorithms

In this section, we consider optimization problems of the form

$$\begin{aligned} &\text{minimize} \quad f(\boldsymbol{x}) \\ &\text{subject to} \quad \boldsymbol{x} \in \Omega, \end{aligned} \tag{11.2.1}$$

where f is a real-valued function and the feasible set Ω is a subset of E^n, most cases $\Omega = E^n$, but sometimes Ω is some particularly simple subset of E^n.

11.2.1 First-order Necessary Conditions

In an investigation of the general problem eq. (11.2.1) we distinguish two kinds of solution points: *local minimum points*, and *global minimum points*.

Definition: A point $\boldsymbol{x}^* \in \Omega$ is said to be a relative minimum point or a local minimum point of f over Ω if there is an $\varepsilon > 0$ such that $f(\boldsymbol{x}) \geqslant f(\boldsymbol{x}^*)$ for all $\boldsymbol{x} \in \Omega$ within a distance ε of $\boldsymbol{x}^*$ (that is, $\boldsymbol{x} \in \Omega$ and $|\boldsymbol{x} - \boldsymbol{x}^*| < \varepsilon$). If $f(\boldsymbol{x}) > f(\boldsymbol{x}^*)$ for all $\boldsymbol{x} \in \Omega, \boldsymbol{x} \neq \boldsymbol{x}^*$, within a distance ε of $\boldsymbol{x}^*$, then $\boldsymbol{x}^*$ is said to be a strict relative minimum point of f over Ω.

Definition: A point $\boldsymbol{x}^* \in \Omega$ is said to be a global minimum point of f over Ω if $f(\boldsymbol{x}) \geqslant f(\boldsymbol{x}^*)$ for all $\boldsymbol{x} \in \Omega$. If $f(\boldsymbol{x}) > f(\boldsymbol{x}^*)$ for all $\boldsymbol{x} \in \Omega, \boldsymbol{x} \neq \boldsymbol{x}^*$, then $\boldsymbol{x}^*$ is said to be a strict relative minimum point of f over Ω.

Feasible Directions

To derive necessary conditions satisfied by a relative minimum point $\boldsymbol{x}^*$, the basic idea is to consider movement away from the point in some given direction. Along any given direction the objective function can be regarded as a function of a single variable, the parameter defining movement in this direction, and hence the ordinary calculus of a single variable is applicable. Thus given $\boldsymbol{x} \in \Omega$ we are motivated to say that a vector $\boldsymbol{d}$ is a feasible direction at $\boldsymbol{x}$ if there is an $\bar{\alpha} > 0$ such that $\boldsymbol{x} + \alpha \boldsymbol{d} \in \Omega$ for all $\alpha, 0 \leqslant \alpha \leqslant \bar{\alpha}$. With this simple concept we can state some simple conditions satisfied by relative minimum points.

Proposition (First-order necessary conditions): Let Ω be a subset of E^n and let $f \in C^1$ be a function on Ω. If $\boldsymbol{x}^*$ is a relative minimum point of f over Ω, then for any $\boldsymbol{d} \in E^n$ that is a feasible direction at $\boldsymbol{x}^*$, we have $\nabla f(\boldsymbol{x}^*)\boldsymbol{d} \geqslant 0$.

Corollary: (Unconstrained case). Let Ω be a subset of E^n, and let $f \in C^1$ be a function on Ω. If $\boldsymbol{x}^*$ is a relative minimum point of f over Ω and if $\boldsymbol{x}^*$ is an interior point of Ω, then $\nabla f(\boldsymbol{x}^*) = 0$.

11.2.2 Second-order Conditions

Proposition 1 (second-order necessary conditions): Let Ω be a subset of E^n and let $f \in C^2$ be a function on Ω. If $\boldsymbol{x}^*$ is a relative minimum point of f over Ω, then for any $\boldsymbol{d} \in E^n$ that is a feasible direction at $\boldsymbol{x}^*$ we have

1) $\nabla f(\boldsymbol{x}^*)\boldsymbol{d} \geqslant 0$,

2) if $\nabla f(\boldsymbol{x}^*)\boldsymbol{d} = 0$, then $\boldsymbol{d}^T \nabla^2 f(\boldsymbol{x}^*)\boldsymbol{d} \geqslant 0$.

Proposition 2 (second-order necessary conditions—unconstrained case): Let $\boldsymbol{x}^*$ be an interior point of the set Ω, and suppose $\boldsymbol{x}^*$ is a relative minimum point over Ω of the function $f \in C^2$, then

1) $\nabla f(\boldsymbol{x}^*) = 0$,

2) for all $\boldsymbol{d}$, $\boldsymbol{d}^T \nabla^2 f(\boldsymbol{x}^*)\boldsymbol{d} \geqslant 0$.

Proposition 3 (second-order sufficient conditions—unconstrained case): Let $f \in C^2$ be a function defined on a region in which the point $\boldsymbol{x}^*$ is an interior point. Suppose in addition that

1) $\nabla f(\boldsymbol{x}^*) = 0$,

2) $F(\boldsymbol{x}^*)$ is positive definite,

then $\boldsymbol{x}^*$ is a strict relative minimum point of f.

11.2.3 Minimization and Maximization of Convex Functions

Theorem11.2.1: Let f be a convex function defined on the convex set Ω, then the set Γ where f achieves its minimum is convex, and any relative minimum of f is a global minimum.

Theorem11.2.2: Let $f \in C^1$ be convex on the convex set Ω. If there is a point $\boldsymbol{x}^* \in \Omega$ such that, for all $\boldsymbol{y} \in \Omega$, $\nabla f(\boldsymbol{x}^*)(\boldsymbol{y} - \boldsymbol{x}^*) \geqslant 0$, then $\boldsymbol{x}^*$ is a global minimum point of f over Ω.

Theorem11.2.3: Let f be a convex function defined on the bounded, closed convex set Ω. If f has a maximum over Ω it is achieved at an extreme point of Ω.

11.2.4 Zero-order Conditions

We have considered the problem

$$\begin{aligned} &\text{minimize} \quad f(\boldsymbol{x}) \\ &\text{subject to} \quad \boldsymbol{x} \in \Omega \end{aligned} \tag{11.2.2}$$

to be unconstrained because there are no functional constraints of the form $g(\boldsymbol{x}) \leqslant b$ or $h(\boldsymbol{x}) = c$. However, the problem is of course constrained by the set Ω. This constraint influences the first-order and second-order necessary and sufficient conditions through the relation between feasible directions and derivatives of the function f.

Proposition(zero – order necessary conditions): If $\boldsymbol{x}^*$ solves eq. (11.2.2) under the stated convexity conditions, then there is a nonzero vector $\boldsymbol{\lambda} \in E^n$ such that $\boldsymbol{x}^*$ is a solution to the two problems:

$$\begin{aligned} &\text{minimize} \quad f(\boldsymbol{x}) + \boldsymbol{\lambda}^{\mathrm{T}}\boldsymbol{x} \\ &\text{subject to} \quad \boldsymbol{x} \in E^n \end{aligned} \tag{11.2.3}$$

and

$$\begin{aligned} &\text{maximize} \quad \boldsymbol{\lambda}^{\mathrm{T}}\boldsymbol{x} \\ &\text{subject to} \quad \boldsymbol{x} \in \Omega \end{aligned} \tag{11.2.4}$$

Notice that problem eq. (11.2.3) is completely unconstrained, since $\boldsymbol{x}$ may range over all of E^n. The second problem eq. (11.2.4) is constrained by Ω but has a linear objective function.

If the optimal solution $\boldsymbol{x}^*$ is in the interior of Ω, then the second problem eq. (11.2.4) implies that $\boldsymbol{\lambda} = \boldsymbol{0}$, for otherwise there would be a direction of movement from $\boldsymbol{x}^*$ that increases the product $\boldsymbol{\lambda}^{\mathrm{T}}\boldsymbol{x}$ above $\boldsymbol{\lambda}^{\mathrm{T}}\boldsymbol{x}^*$. The hyperplane is horizontal in that case. The zeroth-order conditions provide no new information in this situation. However, when the solution is on a boundary point of Ω the conditions give very useful information.

Sufficient conditions: The conditions of Proposition 1 are sufficient for $\boldsymbol{x}^*$ to be a minimum even without the convexity assumptions.

Proposition(zero-order sufficiency conditions): If there is a $\boldsymbol{\lambda}$ such that $\boldsymbol{x}^* \in \Omega$ solves the problems eq. (11.2.3) and eq. (11.2.4), then $\boldsymbol{x}^*$ solves eq. (11.2.2).

11.2.5 Global Convergence of Descent Algorithms

Algorithms

We think of an algorithm as a mapping. Given a point $\boldsymbol{x}$ in some space X, the output of an algorithm applied to $\boldsymbol{x}$ is a new point. Operated iteratively, an algorithm is repeatedly reapplied to the new points it generates so as to produce a whole sequence of points. Thus, as a preliminary definition, we might formally define an algorithm A as a mapping taking points in a space X into (other) points in X. Operated iteratively, the algorithm A initiated at $\boldsymbol{x}_0 \in X$ would generate the sequence $\{\boldsymbol{x}_k\}$ defined by

$$\boldsymbol{x}_{k+1} \in A(\boldsymbol{x}_k)$$

In practice, the mapping A might be defined explicitly by a simple mathematical expression or it might be defined implicitly by, say, a lengthy complex computer program. Given an input vector, both define a corresponding output.

With this intuitive idea of an algorithm in mind, we now generalize the concept somewhat so as to provide greater flexibility in our analyses.

Definition: An algorithm A is a mapping defined on a space X that assigns to every point $\boldsymbol{x} \in X$ a subset of X.

In this definition the term "space" can be interpreted loosely. Usually X is the vector space E^n but it may be only a subset of E^n or even a more general metric space. The most important aspect of the definition, however, is that the mapping A, rather than being a point-to-point mapping of X, is a point-to-set mapping of X.

An algorithm A generates a sequence of points in the following way. Given $\boldsymbol{x}_k \in X$ the algorithm yields $A(\boldsymbol{x}_k)$ which is a subset of X. From this subset an arbitrary element $\boldsymbol{x}_{k+1}$ is selected. In this way, given an initial point $\boldsymbol{x}_0$, the algorithm generates sequences through the iteration

$$\boldsymbol{x}_{k+1} \in A(\boldsymbol{x}_k)$$

It is clear that, unlike the case where A is a point-to-point mapping, the sequence generated by the algorithm A cannot, in general, be predicted solely from knowledge of the initial point $\boldsymbol{x}_0$. This degree of uncertainty is designed to reflect uncertainty that we may have in practice as to specific details of an algorithm.

Descent

Definition: Let $\Gamma \subset X$ be a given solution set and let A be an algorithm on X. A continuous real-valued function Z on X is said to be a descent function for Γ and A if it satisfies

1) if $\boldsymbol{x} \notin \Gamma$ and $\boldsymbol{y} \in A(\boldsymbol{x})$, then $Z(\boldsymbol{y}) < Z(\boldsymbol{x})$,

2) if $\boldsymbol{x} \in \Gamma$ and $\boldsymbol{y} \in A(\boldsymbol{x})$, then $Z(\boldsymbol{y}) \leqslant Z(\boldsymbol{x})$.

There are a number of ways a solution set, algorithm, and descent function can be defined. A natural set-up for the problem

$$\begin{aligned} &\text{minimize} \quad f(\boldsymbol{x}) \\ &\text{subject to} \quad \boldsymbol{x} \in \Omega \end{aligned} \tag{11.2.5}$$

is to let Γ be the set of minimizing points, and define an algorithm A on Ω in such a way that f decreases at each step and thereby serves as a descent function. Indeed, this is the procedure followed in a majority of cases. Another possibility for unconstrained problems is to let Γ be the set of points $\boldsymbol{x}$ satisfying $\nabla f(\boldsymbol{x}) = 0$. In this case, we might design an algorithm for which $|\nabla f(\boldsymbol{x})|$ or $f(\boldsymbol{x})$ serves as a descent function.

Closed Mappings

Definition: A point-to-set mapping A from X to Y is said to be closed at $\boldsymbol{x} \in X$ if the assumptions

1) $\boldsymbol{x}_k \to \boldsymbol{x}, \boldsymbol{x}_k \in X$,

2) $\boldsymbol{y}_k \to \boldsymbol{y}, \boldsymbol{y}_k \in A(\boldsymbol{x}_k)$,

imply

3) $\boldsymbol{y} \in A(\boldsymbol{x})$.

The point-to-set map A is said to be closed on X if it is closed at each point of X.

Definition: Let $A: X \to Y$ and $B: Y \to Z$ be point-to-set mappings. The composite mapping $C = BA$ is defined as the point-to-set mapping $C: X \to Z$ with $C(\boldsymbol{x}) = \bigcup_{\boldsymbol{y} \in A(\boldsymbol{x})} B(\boldsymbol{y})$.

Proposition: Let $A: X \to Y$ and $B: Y \to Z$ be point-to-set mappings. Suppose A is closed at $\boldsymbol{x}$ and B is closed on $A(\boldsymbol{x})$. Suppose also that if $\boldsymbol{x}_k \to \boldsymbol{x}$ and $\boldsymbol{y}_k \in A(\boldsymbol{x}_k)$, there is a $\boldsymbol{y}$ such that, for some subsequence $\{\boldsymbol{y}_{ki}\}$, $\boldsymbol{y}_{ki} \to \boldsymbol{y}$. Then the composite mapping $C = BA$ is closed at $\boldsymbol{x}$.

Two important corollaries follow immediately.

Corollary 1: Let $A: X \to Y$ and $B: Y \to Z$ be point-to-set mappings. If A is closed at $\boldsymbol{x}$, B is closed on $A(\boldsymbol{x})$ and Y is compact, then the composite map $C = BA$ is closed at $\boldsymbol{x}$.

Corollary 2: Let $A: X \to Y$ be a point-to-point mapping and $B: Y \to Z$ a point-to-set mapping. If A is continuous at $\boldsymbol{x}$ and B is closed at $A(\boldsymbol{x})$, then the composite mapping $C = BA$ is closed at $\boldsymbol{x}$.

Global convergence theorem: Let A be an algorithm on X, and suppose that, given $\boldsymbol{x}_0$ the sequence $\{\boldsymbol{x}_k\}_{k=0}^{\infty}$ is generated satisfying $\boldsymbol{x}_{k+1} \in A(\boldsymbol{x}_k)$. Let a solution set $\Gamma \subset X$ be given, and suppose

1) all points $\boldsymbol{x}_k$ are contained in a compact set $S \subset X$,

2) there is a continuous function Z on X such that

(a) if $\boldsymbol{x} \notin \Gamma$, then $Z(\boldsymbol{y}) < Z(\boldsymbol{x})$ for all $\boldsymbol{y} \in A(\boldsymbol{x})$,

(b) if $\boldsymbol{x} \in \Gamma$, then $Z(\boldsymbol{y}) \leqslant Z(\boldsymbol{x})$ for all $\boldsymbol{y} \in A(\boldsymbol{x})$,

3) the mapping A is closed at points outside Γ,

then the limit of any convergent subsequence of $\{\boldsymbol{x}_k\}$ is a solution.

Corollary: If under the conditions of the Global Convergence Theorem Γ consists of a single point $\bar{\boldsymbol{x}}$, then the sequence $\{\boldsymbol{x}_k\}$ converges to $\bar{\boldsymbol{x}}$.

11.2.6 Speed of Convergence

Order of Convergence

Consider a sequence of real numbers $\{r_k\}_{k=0}^{\infty}$ converging to the limit r^*. We define several notions related to the speed of convergence of such a sequence.

Definition: Let the sequence $\{r_k\}$ converge to r^*. The order of convergence of $\{r_k\}$ is defined as the supremum of the nonnegative numbers p satisfying $0 \leqslant \overline{\lim_{k\to\infty}} \frac{|r_{k+1} - r^*|}{|r_k - r^*|^p} < \infty$.

To ensure that the definition is applicable to any sequence, it is stated in terms of limit superior rather than just limit and 0/0 (which occurs if $r_k = r^*$ for all k) is regarded as finite. But these technicalities are rarely necessary in actual analysis, since the sequences generated by algorithms are generally quite well behaved.

It should be noted that the order of convergence, as with all other notions related to speed of convergence that are introduced, is determined only by the properties of the sequence that hold as $k \to \infty$. Somewhat loosely but picturesquely, we are therefore led to refer to the tail of a sequence—that part of the sequence that is arbitrarily far out. In this language we might say that the order of convergence is a measure of how good the worst part of the tail is. Larger values of the order p imply, in a sense, faster convergence, since the distance from the limit r^* is reduced, at least in the tail, by the p-th power in a single step. Indeed, if the sequence has order p and (as is the usual case) the limit

$$\beta = \lim_{k\to\infty} \frac{|r_{k+1} - r^*|}{|r_k - r^*|^p}$$

exists, then asymptotically we have

$$|r_{k+1} - r^*| = \beta |r_k - r^*|^p$$

Linear Convergence

Definition: If the sequence $\{r_k\}$ converges to r^* in such a way that

$$\lim_{k\to\infty} \frac{|r_{k+1} - r^*|}{|r_k - r^*|} = \beta < 1$$

the sequence is said to converge linearly to r^* with convergence ratio (or rate) β.

Linear convergence is, for our purposes, without doubt the most important type of convergence behavior. A linearly convergent sequence, with convergence ratio β, can be said to have a tail that converges at least as fast as the geometric sequence $c\beta^k$ for some constant c. Thus linear convergence is sometimes referred to as geometric convergence, although in this book we reserve that phrase for the case when a sequence is exactly geometric.

As a rule, when comparing the relative effectiveness of two competing algorithms both of which produce linearly convergent sequences, the comparison is based on their corresponding convergence ratios—the smaller the ratio is, the faster the rate is. The ultimate case where $\beta = 0$ is referred to as superlinear convergence. We note immediately that convergence of any order greater than unity is superlinear, but it is also possible for superlinear convergence to correspond to unity order.

Convergence of Vectors

Suppose $\{\boldsymbol{x}_k\}_{k=0}^{\infty}$ is a sequence of vectors in E^n converging to a vector $\boldsymbol{x}^*$. The convergence properties of such a sequence are defined with respect to some particular function that converts the sequence of vectors into a sequence of numbers. Thus, if f is a given continuous function on E^n, the convergence properties of $\{\boldsymbol{x}_k\}$ can be defined with respect to f by analyzing the convergence of $f(\boldsymbol{x}_k)$ to $f(\boldsymbol{x}^*)$. The function f used in this way to measure convergence is called the error function.

In optimization theory it is common to choose the error function by which to measure convergence as the same function that defines the objective function of the original optimization problem. This means we measure convergence by how fast the objective converges to its minimum. Alternatively, we sometimes use the function $|\boldsymbol{x} - \boldsymbol{x}^*|^2$ and thereby measure convergence by how fast the (squared) distance from the solution point decreases to zero.

Proposition: Let f and g be two error functions satisfying $f(\boldsymbol{x}^*) = g(\boldsymbol{x}^*) = 0$ and, for all $\boldsymbol{x}$, a relation of the form

$$0 \leqslant a_1 g(\boldsymbol{x}) \leqslant f(\boldsymbol{x}) \leqslant a_2 g(\boldsymbol{x})$$

for some fixed $a_1 > 0, a_2 > 0$. If the sequence $\{\boldsymbol{x}_k\}_{k=0}^{\infty}$ converges to $\boldsymbol{x}^*$ linearly with average ratio β with respect to one of these functions, it also does so with respect to the other.

11.3 Basic Descent Methods

11.3.1 Fibonacci and Golden Section Search

The method determines the minimum value of a function f over a closed interval $[c_1, c_2]$. In applications, f may in fact be defined over a broader domain, but for this method a fixed interval of search must be specified. The only property that is assumed of f is that it is unimodal, that is, it has a single relative minimum. The minimum point of f is to be determined, at least approximately, by measuring the value of f at a certain number of points. It should be imagined, as is indeed the case in the setting of nonlinear programming, that each measurement of f is somewhat costly—of time if nothing more.

To develop an appropriate search strategy, that is, a strategy for selecting measurement points based on the previously obtained values, we pose the following problem: Find how to successively select N measurement points so that, without explicit knowledge of f, we can determine the smallest possible region of uncertainty in which the minimum must lie. In this problem the region of uncertainty is determined in any particular case by the relative values of the measured points in conjunction with our assumption that f is unimodal. Thus, after values are known at N points $x_1, x_2, \ldots, x_N$ with

$$c_1 \leqslant x_1 < x_2 < \ldots < x_{N-1} < x_N \leqslant c_2$$

the region of uncertainty is the interval $[x_{k-1}, x_{k+1}]$ where x_k is the minimum point among the N, and we define $x_0 = c_1, x_{N+1} = c_2$ for consistency. The minimum of f must lie somewhere in this interval.

Search by Golden Section

If the number N of allowed measurement points in a Fibonacci search is made to approach infinity, we obtain the golden section method. It can be argued, based on the optimal property of the finite Fibonacci method, that the corresponding infinite version yields a sequence of intervals of uncertainty whose widths tend to zero faster than that which would be obtained by other methods.

The solution to the Fibonacci difference equation

$$F_N = F_{N-1} + F_{N-2} \tag{11.3.1}$$

is of the form

$$F_N = A\ \tau_1^N + B\ \tau_2^N \tag{11.3.2}$$

where τ_1 and τ_2 are roots of the characteristic equation $\tau^2 = \tau + 1$. Explicitly,

$$\tau_1 = \frac{1+\sqrt{5}}{2}, \qquad \tau_2 = \frac{1-\sqrt{5}}{2}$$

(The number $\tau_1 > 1.618$ is known as the golden section ratio and was considered by early Greeks to be the most aesthetic value for the ratio of two adjacent sides of a rectangle.) For large N the first term on the right side of eq. (11.3.2) dominates the second, and hence

$$\lim_{N\to\infty} \frac{F_{N-1}}{F_N} = \frac{1}{\tau_1} = 0.618$$

It follows that the interval of uncertainty at any point in the process has width

$$d_k = \left(\frac{1}{\tau_1}\right)^{k-1} d_1 \tag{11.3.3}$$

and from this it follows that

$$\frac{d_{k-1}}{d_k} = \frac{1}{\tau_1} = 0.618 \tag{11.3.4}$$

Therefore, we conclude that, with respect to the width of the uncertainty interval, the search by golden section converges linearly to the overall minimum of the function f with convergence ratio $1/\tau_1 = 0.618$.

11.3.2 Closedness of Line Search Algorithms

Since searching along a line for a minimum point is a component part of most nonlinear programming algorithms, it is desirable to establish at once that this procedure is closed; that is, the end product of the iterative procedures outlined above, when viewed as a single algorithmic step finding a minimum along a line, defines closed algorithms. That is the objective of this section.

To initiate a line search with respect to a function f, two vectors must be specified: the initial point $\boldsymbol{x}$ and the direction $\boldsymbol{d}$ in which the search is to be made. The result of the search is a new point. Thus we define the search algorithm S as a mapping from E^{2n} to E^n. We assume that the search is to be made over the semi-infinite line emanating from $\boldsymbol{x}$ in the direction $\boldsymbol{d}$. We also assume, for simplicity, that the search is not made in vain; that is, we assume that there is a minimum point along the line. This will be the case, for instance, if f is continuous and increases without bound as $\boldsymbol{x}$ tends toward infinity.

Definition: The mapping $S: E^{2n} \to E^n$ is defined by

$$S(\boldsymbol{x},\boldsymbol{d}) = \{\boldsymbol{y}: \boldsymbol{y} = \boldsymbol{x} + \alpha\boldsymbol{d} \text{ for some } \alpha \geqslant 0, f(\boldsymbol{y}) = \min_{0 \leqslant \alpha \leqslant \infty} f(\boldsymbol{x} + \alpha\boldsymbol{d})\} \tag{11.3.5}$$

In some cases there may be many vectors $\boldsymbol{y}$ yielding the minimum, so S is a set-valued mapping. We must verify that S is closed.

Theorem 11.3.1: Let f be continuous on E^n, then the mapping defined by eq. (11.3.5) is closed at $(\boldsymbol{x},\boldsymbol{d})$ if $\boldsymbol{d} \neq \boldsymbol{0}$.

Proof: Suppose $\{\boldsymbol{x}_k\}$ and $\{\boldsymbol{d}_k\}$ are sequences with $\boldsymbol{x}_k \to \boldsymbol{x}, \boldsymbol{d}_k \to \boldsymbol{d} \neq \boldsymbol{0}$. Suppose also that $\boldsymbol{y}_k \in S(\boldsymbol{x}_k, \boldsymbol{d}_k)$ and that $\boldsymbol{y}_k \to \boldsymbol{y}$. We must show that $\boldsymbol{y} \in S(\boldsymbol{x},\boldsymbol{d})$.

For each k we have $\boldsymbol{y}_k = \boldsymbol{x}_k + \alpha_k \boldsymbol{d}_k$ for some α. Furthermore, we have

$$\alpha_k = \frac{|\boldsymbol{y}_k - \boldsymbol{x}_k|}{|\boldsymbol{d}_k|}$$

Taking the limit of the right-hand side of the above, we see that

$$\alpha_k \to \bar{\alpha} \equiv \frac{|\boldsymbol{y} - \boldsymbol{x}|}{|\boldsymbol{d}|}$$

Then, it then follows that $\boldsymbol{y} = \boldsymbol{x} + \bar{\alpha}\boldsymbol{d}$. It still remains to be shown that $\boldsymbol{y} \in S(\boldsymbol{x}, \boldsymbol{d})$.

For each k and each $\alpha, 0 \leqslant \alpha < \infty$,

$$f(\boldsymbol{y}_k) \leqslant f(\boldsymbol{x}_k + \alpha\boldsymbol{d}_k)$$

Letting $k \to \infty$, we obtain

$$f(\boldsymbol{y}) \leqslant f(\boldsymbol{x} + \alpha\boldsymbol{d})$$

Thus

$$f(\boldsymbol{y}) \leqslant \min_{0\leqslant\alpha\leqslant\infty} f(\boldsymbol{x}+\alpha\boldsymbol{d})$$

and hence $\boldsymbol{y} \in S(\boldsymbol{x},\boldsymbol{d})$. The requirement that $\boldsymbol{d} \neq \boldsymbol{0}$ is natural both theoretically and practically. From a practical point of view this condition implies that, when constructing algorithms, the choice $\boldsymbol{d}=\boldsymbol{0}$ had better occur only in the solution set; but it is clear that if $\boldsymbol{d}=\boldsymbol{0}$, no search will be made. Theoretically, the map S can fail to be closed at $\boldsymbol{d}=\boldsymbol{0}$, as illustrated below.

Example: On E^1 define $f(x)=(x-1)^2$. Then $S(x,d)$ is not closed at $x=0, d=0$. To see this we note that for any $d>0$

$$\min_{0\leqslant\alpha\leqslant\infty} f(\alpha d) = f(1)$$

and hence

$$S(0,d)=1$$

But

$$\min_{0\leqslant\alpha\leqslant\infty} f(\alpha\cdot 0) = f(0)$$

so that

$$S(0,0)=0$$

Thus as $d\to 0$, $\lim_{d\to 0} S(0,d) \neq S(0,0)$.

11.3.3 Line Search

Percentage test: One important inaccurate line search algorithm is the one that determines the search parameter α to within a fixed percentage of its true value. Specifically, a constant c, $0<c<1$, is selected ($c=0.10$ is reasonable) and the parameter α in the line search is found so as to satisfy $|\alpha-\bar{\alpha}| \leqslant c\,\bar{\alpha}$ where $\bar{\alpha}$ is the true minimizing value of the parameter.

Armijo's rule. A practical and popular criterion for terminating a line search is Armijo's rule. The essential idea is that the rule should first guarantee that the selected α is not too large, and next it should not be too small. Let us define the function $\phi(\alpha)=f(\boldsymbol{x}_k+\alpha\boldsymbol{d}_k)$.

Armijo's rule is implemented by consideration of the function $\phi(0) = \varepsilon\phi'(0)\alpha$ for fixed ε, $0<\varepsilon<1$. A value of α is considered to be not too large if the corresponding function value lies below the dashed line; that is, if

$$\phi(\alpha) \leqslant \phi(0)+\varepsilon\phi'(0)\alpha \tag{11.3.6}$$

To insure that α is not too small, a value $\eta>1$ is selected, and α is then considered to be not too small if

$$\phi(\eta\alpha) > \phi(0) + \varepsilon\phi'(0)\eta\alpha$$

This means that if α is increased by the factor η, it will fail to meet the test eq. (11.3.6). The acceptable region is defined by the Armijo rule when $\eta = 2$.

Sometimes in practice, the Armijo test is used to define a simplified line search technique that does not employ curve fitting methods. One begins with an arbitrary α. If it satisfies eq. (11.3.6), it is repeatedly increased by η($\eta = 2$ or $\eta = 10$ and $\varepsilon = 0.2$ are often used) until eq. (11.3.6) is not satisfied, and then the penultimate α is selected. If, on the other hand, the original α does not satisfy eq. (11.3.6), it is repeatedly divided by η until the resulting α does satisfy eq. (11.3.6).

Goldstein test: Another line search accuracy test that is frequently used is the Goldstein test. As in the Armijo rule, a value of α is considered not too large if it satisfies eq. (11.3.6), with a given ε, $0 < \varepsilon < (1/2)$. A value of α is considered not too small in the Goldstein test if

$$\phi(\alpha) > \phi(0) + (1 - \varepsilon)\phi'(0)\alpha \tag{11.3.7}$$

In terms of the original notation, the Goldstein criterion for an acceptable value of α, with corresponding $\boldsymbol{x}_{k+1} = \boldsymbol{x}_k + \alpha\boldsymbol{d}_k$, is

$$\varepsilon \leqslant \frac{f(\boldsymbol{x}_{k+1}) - f(\boldsymbol{x}_k)}{\alpha \nabla f(\boldsymbol{x}_k)\boldsymbol{d}_k} \leqslant 1 - \varepsilon$$

We now show that the Goldstein test leads to a closed line search algorithm.

Theorem 11.3.2: Let $f \in C^2$ on E^n, Fixed $0 < \varepsilon < 1/2$, then the mapping S: $E^{2n} \to E^n$ is defined by

$$S(\boldsymbol{x}, \boldsymbol{d}) = \left\{ \boldsymbol{y} : \boldsymbol{y} = \boldsymbol{x} + \alpha\boldsymbol{d} \text{ for some } \alpha > 0, \varepsilon \leqslant \frac{f(\boldsymbol{y}) - f(\boldsymbol{x})}{\alpha \nabla f(\boldsymbol{x})\boldsymbol{d}} \leqslant 1 - \varepsilon \right\}$$

is closed at $(\boldsymbol{x}, \boldsymbol{d})$ if $\boldsymbol{d} \neq \boldsymbol{0}$.

Wolfe test: If derivatives of the objective function, as well as its values, can be evaluated relatively easily, then the Wolfe test, which is a variation of the above, is sometimes preferred. In this case ε is selected with $0 < \varepsilon < 1/2$, and α is required to satisfy eq. (11.3.6) and $\phi'(\alpha) \geqslant (1 - \varepsilon)\phi'(0)$.

An advantage of this test is that this last criterion is invariant to scale-factor changes, whereas eq. (11.3.7) in the Goldstein test is not.

Backtracking: Here a good initial choice is $\alpha = 1$. Backtracking is defined by the initial guess α and two positive parameters $\eta > 1$ and $\varepsilon < 1$ (usually $\varepsilon <$

0.5). The stopping criterion used is the same as the first part of Amijo's rule or the Goldstein test. That is, defining $\phi(\alpha) \equiv f(\boldsymbol{x}_k + \alpha \boldsymbol{d}_k)$, the procedure is terminated at the current α if $\phi(\alpha) \leqslant \varphi(0) + \varepsilon\phi'(0)\alpha$. If this criterion is not satisfied, then α is reduced by the factor $1/\eta$. That is, $\alpha_{\text{new}} = \alpha_{\text{old}}/\eta$. (here $\eta = 1.1$ or $\eta = 1.2$ is often used.)

If the intial α (such as $\alpha = 1$) satisfies the test, then it is taken as the step size. Otherwise, α is reduced by $1/\eta$. Repeating this successively, the first α that satisfies the test is declared the final value. By definition it is known that the previous value $\alpha_{\text{new}} = \alpha_{\text{old}}\eta$ does not pass the first test, and this means that it passes the second condition of Amijo's rule.

11.3.4 The Steepest Descent Method

The Method

Let f have continuous first partial derivatives on E^n. We will frequently have need for the gradient vector of f and therefore we introduce some simplifying notation. The gradient $\nabla f(\boldsymbol{x})$ is, according to our conventions, defined as an n-dimensional row vector. For convenience, we define the n-dimensional column vector $\boldsymbol{g}(\boldsymbol{x}) = \nabla f(\boldsymbol{x})^{\mathrm{T}}$. When there is no chance for ambiguity, we sometimes suppress the argument $\boldsymbol{x}$ and, for example, $\boldsymbol{g}(\boldsymbol{x}_k) = \nabla f(\boldsymbol{x}_k)^{\mathrm{T}}$.

The method of steepest descent is defined by the iterative algorithm

$$\boldsymbol{x}_{k+1} = \boldsymbol{x}_k - \alpha_k \boldsymbol{g}_k$$

where α_k is a nonnegative scalar minimizing $f(\boldsymbol{x}_k - \alpha \boldsymbol{g}_k)$. In words, from the point $\boldsymbol{x}_k$ we search along the direction of the negative gradient $-\boldsymbol{g}_k$ to a minimum point on this line; this minimum point is taken to be $\boldsymbol{x}_{k+1}$.

In formal terms, the overall algorithm $\boldsymbol{A}: E^n \to E^n$, which gives $\boldsymbol{x}_{k+1} \in \boldsymbol{A}(\boldsymbol{x}_k)$ can be decomposed in the form $\boldsymbol{A} = \boldsymbol{SG}$. Here $\boldsymbol{G}: E^n \to E^{2n}$ is defined by $\boldsymbol{G}(\boldsymbol{x}) = (\boldsymbol{x}, -\boldsymbol{g}(\boldsymbol{x}))$, giving the initial point and direction of a line search. This is followed by the line search $\boldsymbol{S}: E^{2n} \to E^n$ defined in eq. (11.3.2).

Global Convergence

We define the solution set to be the points $\boldsymbol{x}$ where $\nabla f(\boldsymbol{x}) = \boldsymbol{0}$, then $\boldsymbol{Z}(\boldsymbol{x}) = f(\boldsymbol{x})$ is a descent function for $\boldsymbol{A}$, since for $\nabla f(\boldsymbol{x}) \neq \boldsymbol{0}$,

$$\lim_{0 \leqslant \alpha < \infty} f(\boldsymbol{x} - \alpha \boldsymbol{g}(\boldsymbol{x})) < f(\boldsymbol{x})$$

Thus by the Global Convergence Theorem, if the sequence $\{\boldsymbol{x}_k\}$ is bounded, it will have limit points and each of these is a solution.

The Quadratic Case

Consider

$$f(\boldsymbol{x}) = \frac{1}{2}\boldsymbol{x}^{\mathrm{T}}\boldsymbol{Q}\boldsymbol{x} - \boldsymbol{x}^{\mathrm{T}}\boldsymbol{b} \tag{11.3.8}$$

where $\boldsymbol{Q}$ is a positive definite symmetric $n \times n$ matrix and its eigenvalues are ordered: $0 < m = \lambda_1 \leqslant \lambda_2 \leqslant \cdots \leqslant \lambda_n = M$. Further f is strictly convex.

The unique minimum point of f can be found directly, by setting the gradient to zero, as the vector $\boldsymbol{x}^*$ satisfying

$$\boldsymbol{Q}\boldsymbol{x}^* = \boldsymbol{b} \tag{11.3.9}$$

Moreover, introducing the function

$$\boldsymbol{E}(\boldsymbol{x}) = \frac{1}{2}(\boldsymbol{x} - \boldsymbol{x}^*)^{\mathrm{T}}\boldsymbol{Q}(\boldsymbol{x} - \boldsymbol{x}^*) \tag{11.3.10}$$

we have $\boldsymbol{E}(\boldsymbol{x}) = f(\boldsymbol{x}) + (1/2)\boldsymbol{x}^{*\mathrm{T}}\boldsymbol{Q}\boldsymbol{x}^* \sqrt{b^2 - 4ac}$, which shows that the function $\boldsymbol{E}$ differs from f only by a constant. For many purposes then, it will be convenient to consider that we are minimizing $\boldsymbol{E}$ rather than f.

The gradient (of both f and $\boldsymbol{E}$) is given explicitly by

$$\boldsymbol{g}(\boldsymbol{x}) = \boldsymbol{Q}\boldsymbol{x} - \boldsymbol{b} \tag{11.3.11}$$

Thus the method of steepest descent can be expressed as

$$\boldsymbol{x}_{k+1} = \boldsymbol{x}_k - \alpha_k \boldsymbol{g}_k \tag{11.3.12}$$

where $\boldsymbol{g}_k = \boldsymbol{Q}\boldsymbol{x}_k - \boldsymbol{b}$, α_k minimizes $f(\boldsymbol{x}_k - \alpha\boldsymbol{g}_k)$. We have, by definition eq. (11.3.8)

$$f(\boldsymbol{x}_k - \alpha\boldsymbol{g}_k) = \frac{1}{2}(\boldsymbol{x}_k - \alpha\boldsymbol{g}_k)^{\mathrm{T}}\boldsymbol{Q}(\boldsymbol{x}_k - \alpha\boldsymbol{g}_k) - (\boldsymbol{x}_k - \alpha\boldsymbol{g}_k)^{\mathrm{T}}\boldsymbol{b}$$

which is minimized at

$$\alpha_k = \frac{\boldsymbol{g}_k^{\mathrm{T}}\boldsymbol{g}_k}{\boldsymbol{g}_k^{\mathrm{T}}\boldsymbol{Q}\boldsymbol{g}_k} \tag{11.3.13}$$

Hence the method of steepest descent eq. (11.2.12) takes the explicit form

$$\boldsymbol{x}_{k+1} = \boldsymbol{x}_k - \left(\frac{\boldsymbol{g}_k^{\mathrm{T}}\boldsymbol{g}_k}{\boldsymbol{g}_k^{\mathrm{T}}\boldsymbol{Q}\boldsymbol{g}_k}\right)\boldsymbol{g}_k \tag{11.3.14}$$

where $\boldsymbol{g}_k = \boldsymbol{Q}\boldsymbol{x}_k - \boldsymbol{b}$.

Lemma 1: The iterative process eq. (11.3.14) satisfies

$$\boldsymbol{E}(\boldsymbol{x}_{k+1}) = \left\{1 - \frac{(\boldsymbol{g}_k^{\mathrm{T}}\boldsymbol{g}_k)^2}{(\boldsymbol{g}_k^{\mathrm{T}}\boldsymbol{Q}\boldsymbol{g}_k)(\boldsymbol{g}_k^{\mathrm{T}}\boldsymbol{Q}^{-1}\boldsymbol{g}_k)}\right\}\boldsymbol{E}(\boldsymbol{x}_k) \tag{11.3.15}$$

Kantorovich inequality: For any $n \times n$ positive definite symmetric matrix

Q, and any vector x, there holds

$$\frac{(x^{T}x)^{2}}{(x^{T}Qx)(x^{T}Q^{-1}x)} \geqslant \frac{4mM}{(m+M)^{2}} \qquad (11.3.16)$$

where m and M are defined as the above.

Theorem 11.3.3: (Steepest descent—quadratic case). For any $x_0 \in E^n$ the method of steepest descent eq. (11.3.14) converges to the unique minimum point x^* of f. Furthermore, with $E(x) = \frac{1}{2}(x - x^*)^T Q(x - x^*)$, there holds at every step k

$$E(x_{k+1}) \leqslant \left(\frac{M-m}{M+m}\right)^2 E(x_k) \qquad (11.3.17)$$

11.3.5 Coordinate Descent Methods

Let f be a function on E^n having continuous first partial derivatives. Given a point $x = (x_1, x_2, \ldots, x_n)$, descent with respect to the coordinate x_i (i being fixed) means that one solves $\underset{x_i}{\text{minimize}} f(x_1, x_2, \ldots, x_n)$.

Thus only changes in the single component x_i are allowed in seeking a new and better vector x. In our general terminology, each such descent can be regarded as a descent in the direction e_i (or $-e_i$) where e_i is the i-th unit vector. By sequentially minimizing with respect to different components, a relative minimum f might ultimately be determined.

Global convergence: It is simple to prove global convergence for cyclic coordinate descent. The algorithmic map A is the composition of $2n$ maps

$$A = SC^n SC^{n-1} \ldots SC^1$$

where $C^i(x) = (x, e_i)$ with e_i equal to the i-th unit vector, and S is the usual line search algorithm but over the doubly infinite line rather than the semi-infinite line. The map C^i is obviously continuous and S is closed. If we assume that points are restricted to a compact set, then we know A is closed. We define the solution set $\Gamma = \{x : \nabla f(x) = 0\}$. If we impose the mild assumption on f that a search along any coordinate direction yields a unique minimum point, then the function $Z(x) \equiv f(x)$ serves as a continuous descent function for A with respect to Γ. This is because a search along any coordinate direction either must yield a decrease or, by the uniqueness assumption, it cannot change position. Therefore, if at a point x we have $\nabla f(x) \neq 0$, then at least one component of $\nabla f(x)$ does not vanish and a search along the corresponding coordinate direction must yield a decrease.

Local convergence rate: There holds for the Gauss-Southwell method

$$E(\boldsymbol{x}_{k+1}) \leqslant \left(1 - \frac{m}{M(n-1)}\right) E(\boldsymbol{x}_k) \tag{11.3.18}$$

where m and M are defined as above and n is the dimension of the problem. Since

$$\left(\frac{M-m}{M+m}\right)^2 \leqslant \left(1 - \frac{m}{M}\right) \leqslant \left(1 - \frac{m}{M(n-1)}\right)^{n-1} \tag{11.3.19}$$

we see that the bound we have for steepest descent is better than the bound we have for $n-1$ applications of the Gauss-Southwell scheme. Hence we might argue that it takes essentially $n-1$ coordinate searches to be as effective as a single gradient search. This is admittedly a crude guess, but the overall conclusion is consistent with the results of many experiments. Indeed, unless the variables of a problem are essentially uncoupled from each other (corresponding to a nearly diagonal Hessian matrix) coordinate descent methods seem to require about n line searches to equal the effect of one step of steepest descent.

11.4 Conjugate Direction Methods

11.4.1 Conjugate Directions

Definition: Given a symmetric matrix $\boldsymbol{Q}$, two vectors $\boldsymbol{d}_1$ and $\boldsymbol{d}_2$ are said to be $\boldsymbol{Q}$-orthogonal, or conjugate with respect to $\boldsymbol{Q}$, if $\boldsymbol{d}_1^{\mathrm{T}}\boldsymbol{Q}\boldsymbol{d}_2 = \boldsymbol{0}$.

In the applications that we consider, the matrix $\boldsymbol{Q}$ will be positive definite but this is not inherent in the basic definition. Thus if $\boldsymbol{Q} = \boldsymbol{0}$, any two vectors are conjugate, while if $\boldsymbol{Q} = \boldsymbol{I}$, conjugacy is equivalent to the usual notion of orthogonality. A finite set of vectors $\boldsymbol{d}_0, \boldsymbol{d}_1, \ldots, \boldsymbol{d}_k$ is said to be a $\boldsymbol{Q}$-orthogonal set if $\boldsymbol{d}_i^{\mathrm{T}}\boldsymbol{Q}\boldsymbol{d}_j = \boldsymbol{0}$ for all $i \neq j$.

Proposition: If $\boldsymbol{Q}$ is positive definite and the set of nonzero vectors $\boldsymbol{d}_0, \boldsymbol{d}_1, \boldsymbol{d}_2, \ldots, \boldsymbol{d}_k$ are $\boldsymbol{Q}$-orthogonal, then these vectors are linearly independent.

Before discussing the general conjugate direction algorithm, let us investigate just why the notion of $\boldsymbol{Q}$-orthogonality is useful in the solution of the quadratic problem

$$\text{minimize} \quad \frac{1}{2}\boldsymbol{x}^{\mathrm{T}}\boldsymbol{Q}\boldsymbol{x} - \boldsymbol{b}^{\mathrm{T}}\boldsymbol{x} \tag{11.4.1}$$

when $\boldsymbol{Q}$ is positive definite. Recall that the unique solution to this problem is also the unique solution to the linear equation

$$\boldsymbol{Q}\boldsymbol{x} = \boldsymbol{b} \tag{11.4.2}$$

and hence the quadratic minimization problem is equivalent to a linear equation problem.

Corresponding to the $n \times n$ positive definite matrix $\boldsymbol{Q}$, let $\boldsymbol{d}_0, \boldsymbol{d}_1, \ldots, \boldsymbol{d}_{n-1}$ be n nonzero $\boldsymbol{Q}$-orthogonal vectors. By the above proposition they are linearly independent, which implies that the solution $\boldsymbol{x}^*$ of eq. (11.4.1) or eq. (11.4.2) can be expanded in terms of them as

$$\boldsymbol{x}^* = \alpha_0 \boldsymbol{d}_0 + \ldots + \alpha_{n-1}\boldsymbol{d}_{n-1} \tag{11.4.3}$$

for some set of α_i. In fact, multiplying by $\boldsymbol{Q}$ and then taking the scalar product with $\boldsymbol{d}_i$ yields directly

$$\alpha_i = \frac{\boldsymbol{d}_i^{\mathrm{T}}\boldsymbol{Q}\boldsymbol{x}^*}{\boldsymbol{d}_i^{\mathrm{T}}\boldsymbol{Q}\boldsymbol{d}_i} = \frac{\boldsymbol{d}_i^{\mathrm{T}}\boldsymbol{b}}{\boldsymbol{d}_i^{\mathrm{T}}\boldsymbol{Q}\boldsymbol{d}_i} \tag{11.4.4}$$

This shows that α_i and consequently the solution $\boldsymbol{x}^*$ can be found by evaluation of simple scalar products. The end result is

$$\boldsymbol{x}^* = \sum_{i=0}^{n-1} \frac{\boldsymbol{d}_i^{\mathrm{T}}\boldsymbol{b}}{\boldsymbol{d}_i^{\mathrm{T}}\boldsymbol{Q}\boldsymbol{d}_i}\boldsymbol{d}_i \tag{11.4.5}$$

There are two basic ideas imbedded in eq. (11.4.5). The first is the idea of selecting an orthogonal set of $\boldsymbol{d}_i$ so that by taking an appropriate scalar product, all terms on the right side of eq. (11.4.3), except the i-th, vanish. This could, of course, have been accomplished by making $\boldsymbol{d}_i$ orthogonal in the ordinary sense instead of making them $\boldsymbol{Q}$-orthogonal. The second basic observation, however, is that by using $\boldsymbol{Q}$-orthogonality the resulting equation for α_i can be expressed in terms of the known vector $\boldsymbol{b}$ rather than the unknown vector $\boldsymbol{x}^*$; hence the coefficients can be evaluated without knowing $\boldsymbol{x}^*$.

The expansion for $\boldsymbol{x}^*$ can be considered to be the result of an iterative process of n steps where at the i-th step $\alpha_i \boldsymbol{d}_i$ is added. Viewing the procedure this way, and allowing for an arbitrary initial point for the iteration, the basic conjugate direction method is obtained.

Conjugate direction theorem: Let $\{\boldsymbol{d}_i\}_{i=0}^{n-1}$ be a set of nonzero $\boldsymbol{Q}$-orthogonal vectors. For any $\boldsymbol{x}_0 \in E^n$ the sequence $\{\boldsymbol{x}_k\}$ generated according to

$$\boldsymbol{x}_{k+1} = \boldsymbol{x}_k + \alpha_k \boldsymbol{d}_k, \qquad k \geqslant 0 \tag{11.4.6}$$

with

$$\alpha_k = -\frac{\boldsymbol{g}_k^{\mathrm{T}} \boldsymbol{d}_k}{\boldsymbol{d}_k^{\mathrm{T}} \boldsymbol{Q} \boldsymbol{d}_k} \tag{11.4.7}$$

and

$$\boldsymbol{g}_k = \boldsymbol{Q}\boldsymbol{x}_k - \boldsymbol{b}$$

converges to the unique solution, $\boldsymbol{x}^*$, of $\boldsymbol{Q}\boldsymbol{x} = \boldsymbol{b}$ after n steps, that is, $\boldsymbol{x}_n = \boldsymbol{x}^*$.

11.4.2 Descent Properties of the Conjugate Direction Method

We define $\boldsymbol{B}_k$ as the subspace of E^n spanned by $\{\boldsymbol{d}_0, \boldsymbol{d}_1, \dots, \boldsymbol{d}_{k-1}\}$. We shall show that as the method of conjugate directions progresses each $\boldsymbol{x}_k$ minimizes the objective over the k-dimensional linear variety $\boldsymbol{x}_0 + \boldsymbol{B}_k$.

Expanding subspace theorem: Let $\{\boldsymbol{d}_i\}_{i=0}^{n-1}$ be a sequence of nonzero $\boldsymbol{Q}$-orthogonal vectors in E^n, then for any $\boldsymbol{x}_0 \in E^n$ the sequence $\{\boldsymbol{x}_k\}$ generated according to

$$\boldsymbol{x}_{k+1} = \boldsymbol{x}_k + \alpha_k \boldsymbol{d}_k \tag{11.4.8}$$

$$\alpha_k = -\frac{\boldsymbol{g}_k^{\mathrm{T}} \boldsymbol{d}_k}{\boldsymbol{d}_k^{\mathrm{T}} \boldsymbol{Q} \boldsymbol{d}_k} \tag{11.4.9}$$

has the property that $\boldsymbol{x}_k$ minimizes $f(\boldsymbol{x}) = \frac{1}{2}\boldsymbol{x}^{\mathrm{T}}\boldsymbol{Q}\boldsymbol{x} - \boldsymbol{b}^{\mathrm{T}}\boldsymbol{x}$ on the line $\boldsymbol{x} = \boldsymbol{x}_{k-1} + \alpha \boldsymbol{d}_{k-1}$, $-\infty < \alpha < \infty$, as well as on the linear variety $\boldsymbol{x}_0 + \boldsymbol{B}_k$.

11.4.3 The Conjugate Gradient Method

There are three primary advantages to this method of direction selection.

First, unless the solution is attained in less than n steps, the gradient is always nonzero and linearly independent of all previous direction vectors. Indeed, the gradient $\boldsymbol{g}_k$ is orthogonal to the subspace $\boldsymbol{B}_k$ generated by $\boldsymbol{d}_0, \boldsymbol{d}_1, \dots, \boldsymbol{d}_{k-1}$.

Second, a more important advantage of the conjugate gradient method is the especially simple formula that is used to determine the new direction vector.

Third, because the directions are based on the gradients, the process makes good uniform progress toward the solution at every step.

Conjugate Gradient Algorithm

Starting at any $\boldsymbol{x}_0 \in E^n$ defines $\boldsymbol{d}_0 = -\boldsymbol{g}_0 = \boldsymbol{b} - \boldsymbol{Q}\boldsymbol{x}_0$ and

$$\boldsymbol{x}_{k+1} = \boldsymbol{x}_k + \alpha_k \boldsymbol{d}_k \tag{11.4.10}$$

$$\alpha_k = -\frac{\boldsymbol{g}_k^{\mathrm{T}}\boldsymbol{d}_k}{\boldsymbol{d}_k^{\mathrm{T}}\boldsymbol{Q}\boldsymbol{d}_k} \tag{11.4.11}$$

$$\boldsymbol{d}_{k+1} = -\boldsymbol{g}_{k+1} + \beta_k \boldsymbol{d}_k \tag{11.4.12}$$

$$\beta_k = \frac{\boldsymbol{g}_{k+1}^{\mathrm{T}}\boldsymbol{Q}\boldsymbol{d}_k}{\boldsymbol{d}_k^{\mathrm{T}}\boldsymbol{Q}\boldsymbol{d}_k} \tag{11.4.13}$$

where $\boldsymbol{g}_k = \boldsymbol{Q}\boldsymbol{x}_k - \boldsymbol{b}$.

Verification of the Algorithm

To verify that the algorithm is a conjugate direction algorithm, it is necessary to verify that the vectors $\{\boldsymbol{d}_k\}$ are $\boldsymbol{Q}$-orthogonal. It is easiest to prove this by simultaneously proving a number of other properties of the algorithm. This is done in the theorem below where the notation $[\boldsymbol{d}_0, \boldsymbol{d}_1, \dots, \boldsymbol{d}_k]$ is used to denote the subspace spanned by the vectors $\boldsymbol{d}_0, \boldsymbol{d}_1, \dots, \boldsymbol{d}_k$.

Conjugate gradient theorem: The conjugate gradient algorithm eq. (11.4.10) – eq. (11.4.13) is a conjugate direction method. If it does not terminate at $\boldsymbol{x}_k$, then

1) $[\boldsymbol{g}_0, \boldsymbol{g}_1, \dots, \boldsymbol{g}_k] = [\boldsymbol{g}_0, \boldsymbol{Q}\boldsymbol{g}_0, \dots, \boldsymbol{Q}^k\boldsymbol{g}_0]$,

2) $[\boldsymbol{d}_0, \boldsymbol{d}_1, \dots, \boldsymbol{d}_k] = [\boldsymbol{g}_0, \boldsymbol{Q}\boldsymbol{g}_0, \dots, \boldsymbol{Q}^k\boldsymbol{g}_0]$,

3) $\boldsymbol{d}_k^{\mathrm{T}}\boldsymbol{Q}\boldsymbol{d}_i = \boldsymbol{0}$ for $i \leqslant k-1$,

4) $\alpha_k = \boldsymbol{g}_k^{\mathrm{T}}\boldsymbol{g}_k / \boldsymbol{d}_k^{\mathrm{T}}\boldsymbol{Q}\boldsymbol{d}_k$,

5) $\beta_k = \boldsymbol{g}_{k+1}^{\mathrm{T}}\boldsymbol{g}_{k+1} / \boldsymbol{g}_k^{\mathrm{T}}\boldsymbol{g}_k$.

11.4.4 The C-G Method as an Optimal Process

Let us consider a new general approach for solving the quadratic minimization problem. Given an arbitrary starting point $\boldsymbol{x}_0$, let

$$\boldsymbol{x}_{k+1} = \boldsymbol{x}_0 + \boldsymbol{P}_k(\boldsymbol{Q})\boldsymbol{g}_0 \tag{11.4.14}$$

where $\boldsymbol{P}_k$ is a polynomial of degree k. Selection of a set of coefficients for each of the polynomials $\boldsymbol{P}_k$ determines a sequence of $\boldsymbol{x}_k$. We have

$$\begin{aligned}\boldsymbol{x}_{k+1} - \boldsymbol{x}^* &= \boldsymbol{x}_0 - \boldsymbol{x}^* + \boldsymbol{P}_k(\boldsymbol{Q})\boldsymbol{Q}(\boldsymbol{x}_0 - \boldsymbol{x}^*) \\ &= [\boldsymbol{I} + \boldsymbol{Q}\boldsymbol{P}_k(\boldsymbol{Q})](\boldsymbol{x}_0 - \boldsymbol{x}^*)\end{aligned} \tag{11.4.15}$$

and hence

$$\begin{aligned}\boldsymbol{E}(\boldsymbol{x}_{k+1}) &= \frac{1}{2}(\boldsymbol{x}_{k+1} - \boldsymbol{x}^*)^{\mathrm{T}}\boldsymbol{Q}(\boldsymbol{x}_{k+1} - \boldsymbol{x}^*) \\ &= \frac{1}{2}(\boldsymbol{x}_0 - \boldsymbol{x}^*)^{\mathrm{T}}\boldsymbol{Q}[\boldsymbol{I} + \boldsymbol{Q}\boldsymbol{P}_k(\boldsymbol{Q})]^2(\boldsymbol{x}_0 - \boldsymbol{x}^*)\end{aligned}$$

(11.4.16)

We may now pose the problem of selecting the polynomial $\boldsymbol{P}_k$ in such a way as to minimize $\boldsymbol{E}(\boldsymbol{x}_{k+1})$ with respect to all possible polynomials of degree k. Expanding eq. (11.4.14), however, we obtain

$$\boldsymbol{x}_{k+1} = \boldsymbol{x}_0 + \gamma_0 \boldsymbol{g}_0 + \gamma_1 \boldsymbol{Q}\boldsymbol{g}_0 + \ldots + \gamma_k \boldsymbol{Q}^k \boldsymbol{g}_0 \tag{11.4.17}$$

where γ_i is the coefficients of $\boldsymbol{P}_k$. In view of

$$\boldsymbol{B}_{k+1} = [\boldsymbol{d}_0, \boldsymbol{d}_1, \ldots, \boldsymbol{d}_k] = [\boldsymbol{g}_0, \boldsymbol{Q}\boldsymbol{g}_0, \ldots, \boldsymbol{Q}^k \boldsymbol{g}_0]$$

the vector $\boldsymbol{x}_{k+1} = \boldsymbol{x}_0 + \alpha_0 \boldsymbol{d}_0 + \alpha_1 \boldsymbol{d}_1 + \ldots + \alpha_k \boldsymbol{d}_k$ generated by the method of conjugate gradients has precisely this form; moreover, according to the Expanding Subspace Theorem, the coefficient γ_i determined by the conjugate gradient process is such as to minimize $\boldsymbol{E}(\boldsymbol{x}_{k+1})$. Therefore, the problem posed of selecting the optimal $\boldsymbol{P}_k$ is solved by the conjugate gradient procedure.

The explicit relation between the optimal coefficient γ_i of $\boldsymbol{P}_k$ and the constants α_i, β_i associated with the conjugate gradient method is, of course, somewhat complicated, as is the relation between the coefficients of $\boldsymbol{P}_k$ and those of $\boldsymbol{P}_{k+1}$. We summarize the above development by the following very useful theorem.

Theorem 1: The point $\boldsymbol{x}_{k+1}$ generated by the conjugate gradient method satisfies

$$\boldsymbol{E}(\boldsymbol{x}_{k+1}) = \min_{P_k} \frac{1}{2} (\boldsymbol{x}_0 - \boldsymbol{x}^*)^{\mathrm{T}} \boldsymbol{Q} [\boldsymbol{I} + \boldsymbol{Q}\boldsymbol{P}_k(\boldsymbol{Q})]^2 (\boldsymbol{x}_0 - \boldsymbol{x}^*) \tag{11.4.18}$$

where the minimum is taken with respect to all polynomials $\boldsymbol{P}_k$ of degree k.

Bounds on Convergence

To use Theorem 1 most effectively it is convenient to recast it in terms of eigenvectors and eigenvalues of the matrix $\boldsymbol{Q}$. Suppose that the vector $\boldsymbol{x}_0 - \boldsymbol{x}^*$ is written in the eigenvector expansion

$$\boldsymbol{x}_0 - \boldsymbol{x}^* = \xi_1 \boldsymbol{e}_1 + \xi_2 \boldsymbol{e}_2 + \ldots + \xi_n \boldsymbol{e}_n$$

where $\boldsymbol{e}_i$ is normalized eigenvectors of $\boldsymbol{Q}$. Then since $\boldsymbol{Q}(\boldsymbol{x}_0 - \boldsymbol{x}^*) = \lambda_1 \xi_1 \boldsymbol{e}_1 + \lambda_2 \xi_2 \boldsymbol{e}_2 + \ldots + \lambda_n \xi_n \boldsymbol{e}_n$ and since the eigenvectors are mutually orthogonal, we have

$$\boldsymbol{E}(\boldsymbol{x}_0) = \frac{1}{2} (\boldsymbol{x}_0 - \boldsymbol{x}^*)^{\mathrm{T}} \boldsymbol{Q} (\boldsymbol{x}_0 - \boldsymbol{x}^*) = \frac{1}{2} \sum_{i=1}^{n} \lambda_i \xi_i^2 \tag{11.4.19}$$

where λ_i is the corresponding eigenvalues of $\boldsymbol{Q}$. Applying the same manipulations to (11.4.18), we find that for any polynomial $\boldsymbol{P}_k$ of degree k, there holds

$$\boldsymbol{E}(\boldsymbol{x}_{k+1}) \leqslant \frac{1}{2} \sum_{i=1}^{n} [1 + \lambda_i \boldsymbol{P}_k(\lambda_i)]^2 \lambda_i \xi_i^2$$

It then follows that

$$E(\boldsymbol{x}_{k+1}) \leqslant \max_{\lambda_i} [1 + \lambda_i \boldsymbol{P}_k(\lambda_i)]^2 \frac{1}{2} \sum_{i=1}^{n} \lambda_i \xi_i^2$$

and finally

$$E(\boldsymbol{x}_{k+1}) \leqslant \max_{\lambda_i} [1 + \lambda_i \boldsymbol{P}_k(\lambda_i)]^2 E(\boldsymbol{x}_0)$$

We summarize this result by the following theorem.

Theorem 2: From the above conjugate gradient method, we have

$$E(\boldsymbol{x}_{k+1}) \leqslant \max_{\lambda_i} [1 + \lambda_i \boldsymbol{P}_k(\lambda_i)]^2 E(\boldsymbol{x}_0) \qquad (11.4.20)$$

for any polynomial $\boldsymbol{P}_k$ of degree k, where the maximum is taken over all eigenvalues λ_i of $\boldsymbol{Q}$.

We note here that every step of the conjugate gradient method is at least as good as a steepest descent step would be from the same point. To see this, suppose $\boldsymbol{x}_k$ has been computed by the conjugate gradient method. From (11.4.17) we know $\boldsymbol{x}_k$ has the form

$$\boldsymbol{x}_k = \boldsymbol{x}_0 + \bar{\gamma}_0 \boldsymbol{g}_0 + \bar{\gamma}_1 \boldsymbol{Q} \boldsymbol{g}_0 + \ldots + \bar{\gamma}_{k-1} \boldsymbol{Q}^{k-1} \boldsymbol{g}_0$$

Now if $\boldsymbol{x}_{k+1}$ is computed from $\boldsymbol{x}_k$ by steepest descent, then $\boldsymbol{x}_{k+1} = \boldsymbol{x}_k - \alpha_k \boldsymbol{g}_k$ for some α_k. In view of part (a) of the Conjugate Gradient Theorem $\boldsymbol{x}_{k+1}$ will have the form eq. (11.4.17). Since the conjugate direction method $E(\boldsymbol{x}_{k+1})$ is lower than any other $\boldsymbol{x}_{k+1}$ of the form eq. (11.4.17), we obtain the desired conclusion.

Typically when some information about the eigenvalue structure of $\boldsymbol{Q}$ is known, that information can be exploited by the construction of a suitable polynomial $\boldsymbol{P}_k$ used in eq. (11.4.20). Suppose, for example, it were known that $\boldsymbol{Q}$ had only $m < n$ distinct eigenvalues. Then it is clear that by suitable choice of $\boldsymbol{P}_{m-1}$ it would be possible to make the m-th degree polynomial $1 - \lambda \boldsymbol{P}_{m-1}(\lambda)$ have its m zeros at the m eigenvalues. Using that particular polynomial in eq. (11.4.20) shows that $E(\boldsymbol{x}_m) = 0$. Thus the optimal solution will be obtained in at most m steps, rather than n steps.

Chapter 12

Constrained Minimization

12.1 Quasi-Newton Methods

12.1.1 Modified Newton Method

A very basic iterative process for solving the problem

$$\text{minimize} \quad f(\boldsymbol{x})$$

which includes as special cases most of our earlier ones is

$$\boldsymbol{x}_{k+1} = \boldsymbol{x}_k - \alpha_k \boldsymbol{S}_k \nabla f(\boldsymbol{x}_k)^{\mathrm{T}} \tag{12.1.1}$$

where $\boldsymbol{S}_k$ is a symmetric $n \times n$ matrix and where, as usual, α_k is chosen to minimize $f(\boldsymbol{x}_{k+1})$. If $\boldsymbol{S}_k$ is the inverse of the Hessian of f, we obtain Newton's method, while if $\boldsymbol{S}_k = \boldsymbol{I}$ we have steepest descent. It would seem to be a good idea, in general, to select $\boldsymbol{S}_k$ as an approximation to the inverse of the Hessian. We examine that philosophy in this section.

First, we note, as in Section 8.8, that in order that the process eq. (12.1.1) is guaranteed to be a descent method for small values of α, it is necessary in general to require that $\boldsymbol{S}_k$ be positive definite. We shall therefore always impose this as a requirement.

Because of the similarity of the algorithm eq. (12.1.1) with steepest descent, it should not be surprising that its convergence properties are similar in character to our earlier results. We derive the actual rate of convergence by considering, as usual, the standard quadratic problem with

$$f(\boldsymbol{x}) = \frac{1}{2}\boldsymbol{x}^{\mathrm{T}}\boldsymbol{Q}\boldsymbol{x} - \boldsymbol{b}^{\mathrm{T}}\boldsymbol{x} \tag{12.1.2}$$

where $\boldsymbol{Q}$ is symmetric and positive definite. For this case we can find an explicit expression for α_k in eq. (12.1.1). The algorithm becomes

$$\boldsymbol{x}_{k+1} = \boldsymbol{x}_k - \alpha_k \boldsymbol{S}_k \boldsymbol{g}_k \tag{12.1.3a}$$

where

$$\boldsymbol{g}_k = \boldsymbol{Q}\boldsymbol{x}_k - \boldsymbol{b} \tag{12.1.3b}$$

$$\alpha_k = \frac{\boldsymbol{g}_k^{\mathrm{T}} \boldsymbol{S}_k \boldsymbol{g}_k}{\boldsymbol{g}_k^{\mathrm{T}} \boldsymbol{S}_k \boldsymbol{Q} \boldsymbol{S}_k \boldsymbol{g}_k} \tag{12.1.3c}$$

Modified newton method theorem (Quadratic case): Let $\boldsymbol{x}^*$ be the unique minimum point of f, $\boldsymbol{E}(\boldsymbol{x}) = \frac{1}{2}(\boldsymbol{x} - \boldsymbol{x}^*)^{\mathrm{T}}\boldsymbol{Q}(\boldsymbol{x} - \boldsymbol{x}^*)$, then for the algorithm eq. (12.1.3) there holds at every step k

$$\boldsymbol{E}(\boldsymbol{x}_{k+1}) \leqslant \left(\frac{B_k - b_k}{B_k + b_k}\right)^2 \boldsymbol{E}(\boldsymbol{x}_k) \tag{12.1.4}$$

where b_k and B_k are respectively the smallest and largest eigenvalues of the matrix $\boldsymbol{S}_k\boldsymbol{Q}$.

A classical method: We conclude this section by mentioning the *classical modified Newton's method*, a standard method for approximating Newton's method without evaluating $\boldsymbol{F}(\boldsymbol{x}_k)^{-1}$ for each k. We set

$$\boldsymbol{x}_{k+1} = \boldsymbol{x}_k - \alpha_k [\boldsymbol{F}(\boldsymbol{x}_0)]^{-1} \nabla f(\boldsymbol{x}_k)^{\mathrm{T}} \tag{12.1.5}$$

In this method the Hessian at the initial point $\boldsymbol{x}_0$ is used throughout the process.

12.1.2 Scaling

Improvement of Eigenvalue Ratio

Interlocking eigenvalues lemma: Let the $n \times n$ symmetric matrix $\boldsymbol{A}$ have eigenvalues $\lambda_1 \leqslant \lambda_2 \leqslant \ldots \leqslant \lambda_n$, $\boldsymbol{a}$ be any vector in E^n and the matrix $\boldsymbol{A} + \boldsymbol{a}\boldsymbol{a}^{\mathrm{T}}$ have eigenvalues by $\mu_1 \leqslant \mu_2 \leqslant \ldots \leqslant \mu_n$, then $\lambda_1 \leqslant \mu_1 \leqslant \lambda_2 \leqslant \mu_2 \leqslant \ldots \leqslant \lambda_n \leqslant \mu_n$.

For convenience we introduce the following definitions:

$$\boldsymbol{R}_k = \boldsymbol{F}_k^{1/2} \boldsymbol{H}_k \boldsymbol{F}_k^{1/2}$$

$$\boldsymbol{r}_k = \boldsymbol{F}_k^{1/2} \boldsymbol{p}_k$$

Then using $q_k = F_k^{1/2} r_k$, it can be readily verified that eq. (12.1.5) is equivalent to

$$R_{k+1}^{\phi} = R_k - \frac{R_k r_k r_k^{\mathrm{T}} R_k}{r_k^{\mathrm{T}} R_k r_k} + \frac{r_k r_k^{\mathrm{T}}}{r_k^{\mathrm{T}} r_k} + \phi z_k z_k^{\mathrm{T}} \tag{12.1.6}$$

where

$$z_k = F^{1/2} v_k = \sqrt{r_k^{\mathrm{T}} R_k r_k} \left(\frac{r_k}{r_k^{\mathrm{T}} r_k} - \frac{R_k r_k}{r_k^{\mathrm{T}} R_k r_k} \right)$$

Since R_k is similar to $H_k F$ (because $H_k F = F^{1/2} R_k F^{1/2}$), both have the same eigenvalues. It is most convenient, however, in view of eq. (12.1.6) to study R_k, obtaining conclusions about $H_k F$ indirectly.

Before proving the general theorem we shall consider the case $\phi = 0$ corresponding to the Davidon-Fletcher-Powell formula. Suppose the eigenvalues of R_k are λ_1, $\lambda_2, \ldots, \lambda_n$ with $0 < \lambda_1 \leqslant \lambda_2 \leqslant \ldots \leqslant \lambda_n$. Suppose also that $1 \in [\lambda_1, \lambda_n]$. We will show that the eigenvalues of R_{k+1} are all contained in the interval $[\lambda_1, \lambda_n]$, which of course implies that R_{k+1} is no worse than R_k in terms of its condition number. Let us first consider the matrix

$$P = R_k - \frac{R_k r_k r_k^{\mathrm{T}} R_k}{r_k^{\mathrm{T}} R_k r_k}$$

We see that $P r_k = 0$, so one eigenvalue of P is zero. If we denote the eigenvalues of P by $\mu_1 \leqslant \mu_2 \leqslant \ldots \leqslant \mu_n$, we have from the above observation and the lemma on interlocking eigenvalues that

$$0 = \mu_1 \leqslant \lambda_1 \leqslant \mu_2 \leqslant \ldots \leqslant \mu_n \leqslant \lambda_n$$

Next we consider

$$R_{k+1} = R_k - \frac{R_k r_k r_k^{\mathrm{T}} R_k}{r_k^{\mathrm{T}} R_k r_k} + \frac{r_k r_k^{\mathrm{T}}}{r_k^{\mathrm{T}} r_k} = P + \frac{r_k r_k^{\mathrm{T}}}{r_k^{\mathrm{T}} r_k}. \tag{12.1.7}$$

Since r_k is an eigenvector of P and since, by symmetry, all other eigenvectors of P are therefore orthogonal to r_k, it follows that the only eigenvalue different in R_{k+1} from in P is the one corresponding to r_k—it now being unity. Thus R_{k+1} has eigenvalues $\mu_2, \mu_3, \ldots, \mu_n$ and unity. These are all contained in the interval $[\lambda_1, \lambda_n]$. Thus updating does not worsen the eigenvalue ratio. It should be noted that this result in no way depends on α_k being selected to minimize f.

We now extend the above to the Broyden class with $0 \leqslant \phi \leqslant 1$.

Theorem: Let the n eigenvalues of $H_k F$ be $\lambda_1, \lambda_2, \ldots, \lambda_n$ with $0 < \lambda_1 \leqslant \lambda_2 \leqslant \ldots \leqslant \lambda_n$. Suppose that $1 \in [\lambda_1, \lambda_n]$. Then for any $\phi, 0 \leqslant \phi \leqslant 1$, the eigenvalues of $H_{k+1}^{\phi}F$, are all contained in $[\lambda_1, \lambda_n]$.

Scale Factors

In view of the result derived above, it is clearly advantageous to scale the matrix H_k so that the eigenvalues of H_kF are spread both below and above unity. Of course in the ideal case of a quadratic problem with perfect line search this is strictly only necessary for H_0, since unity is an eigenvalue of H_kF for $k>0$. But because of the inescapable deviations from the ideal, it is useful to consider the possibility of scaling every H_k.

A scale factor can be incorporated directly into the updating formula. We first multiply H_k by the scale factor γ_k and then apply the usual updating formula. This is equivalent to replacing H_k by $\gamma_k H_k$ in eq. (12.1.6) and leads to

$$H_{k+1} = \left(H_k - \frac{H_k q_k q_k^{\mathrm{T}} H_k}{q_k^{\mathrm{T}} H_k q_k} + \phi_k v_k v_k^{\mathrm{T}}\right)\gamma_k + \frac{p_k p_k^{\mathrm{T}}}{p_k^{\mathrm{T}} q_k} \tag{12.1.8}$$

This defines a two-parameter family of updates that reduces to the Broyden family for $\gamma_k = 1$. Using $\gamma_0, \gamma_1, \cdots$ as arbitrary positive scale factors, we consider the algorithm: Start with any symmetric positive definite matrix H_0 and any point x_0, then starting with $k=0$,

Step 1. Set $d_k = -H_k g_k$;

Step 2. Minimize $f(x_k + \alpha d_k)$ with respect to $\alpha \geqslant 0$ to obtain $x_{k+1}, P_k = \alpha_k d_k$, and g_{k+1};

Step 3. Set $q_k = g_{k+1} - g_k$ and

$$H_{k+1} = \left(H_k - \frac{H_k q_k q_k^{\mathrm{T}} H_k}{q_k^{\mathrm{T}} H_k q_k} + \phi_k v_k v_k^{\mathrm{T}}\right)\gamma_k + \frac{p_k p_k^{\mathrm{T}}}{p_k^{\mathrm{T}} q_k}$$

$$v_k = (q_k^{\mathrm{T}} H q_k)^{1/2}\left(\frac{p_k}{p_k^{\mathrm{T}} q_k} - \frac{H_k q_k}{q_k^{\mathrm{T}} H_k q_k}\right) \tag{12.1.9}$$

The use of scale factors does destroy the property $H_n = F^{-1}$ in the quadratic case, but it does not destroy the conjugate direction property.

1) If H_k is positive definite and $p_k^{\mathrm{T}} q_k > 0$, eq. (12.1.9) yields an H_{k+1} that is positively definite.

2) If f is quadratic with Hessian F, then the vectors $p_0, p_1, \cdots, p_{n-1}$ are mutually F-orthogonal, and, for each k, the vectors $p_0, p_1, \cdots, p_k$ are eigenvectors of $H_{k+1}F$.

A Self-Scaling Quasi-Newton Algorithm

The question that arises next is how to select appropriate scale factors. If $\lambda_1 \leqslant \lambda_2 \leqslant \ldots \leqslant \lambda_n$ are the eigenvalues of $\boldsymbol{H}_k\boldsymbol{F}$, we want to multiply $\boldsymbol{H}_k$ by γ_k where $\lambda_1 \leqslant 1/\gamma_k \leqslant \lambda_n$. This will ensure that the new eigenvalues contain unity in the interval they span.

Note that in terms of our earlier notation

$$\frac{\boldsymbol{q}_k^{\mathrm{T}}\boldsymbol{H}_k\boldsymbol{q}_k}{\boldsymbol{p}_k^{\mathrm{T}}\boldsymbol{q}_k}=\frac{\boldsymbol{r}_k^{\mathrm{T}}\boldsymbol{R}_k\boldsymbol{r}_k}{\boldsymbol{r}_k^{\mathrm{T}}\boldsymbol{r}_k}$$

Recalling that $\boldsymbol{R}_k$ has the same eigenvalues as $\boldsymbol{H}_k\boldsymbol{F}$ and noting that for any $\boldsymbol{r}_k$

$$\lambda_1 \leqslant \frac{\boldsymbol{r}_k^{\mathrm{T}}\boldsymbol{R}_k\boldsymbol{r}_k}{\boldsymbol{r}_k^{\mathrm{T}}\boldsymbol{r}_k} \leqslant \lambda_n$$

we see that

$$\gamma_k=\frac{\boldsymbol{p}_k^{\mathrm{T}}\boldsymbol{q}_k}{\boldsymbol{q}_k^{\mathrm{T}}\boldsymbol{H}_k\boldsymbol{q}_k} \tag{12.1.10}$$

serves as a suitable scale factor.

We now state a complete self-scaling, restarting, quasi-Newton method based on the ideas above. For simplicity we take $\phi=0$ and thus obtain a modification of the DFP method. Start at any point $\boldsymbol{x}_0, k=0$.

Step 1. Set $\boldsymbol{H}_k=\boldsymbol{I}$;

Step 2. Set $\boldsymbol{d}_k=-\boldsymbol{H}_k\boldsymbol{g}_k$;

Step 3. Minimize $f(\boldsymbol{x}_k+\alpha\boldsymbol{d}_k)$ with respect to $\alpha \geqslant 0$ to obtain $\alpha_k, \boldsymbol{x}_{k+1}, \boldsymbol{p}_k=\alpha_k\boldsymbol{d}_k, \boldsymbol{g}_{k+1}$ and $\boldsymbol{q}_k=\boldsymbol{g}_{k+1}-\boldsymbol{g}_k$. (Select α_k accurately enough to ensure $\boldsymbol{p}_k^{\mathrm{T}}\boldsymbol{q}_k>0$.)

Step 4. If k is not an integer multiple of n, set

$$\boldsymbol{H}_{k+1}=\left(\boldsymbol{H}_k-\frac{\boldsymbol{H}_k\boldsymbol{q}_k\boldsymbol{q}_k^{\mathrm{T}}\boldsymbol{H}_k}{\boldsymbol{q}_k^{\mathrm{T}}\boldsymbol{H}_k\boldsymbol{q}_k}\right)\frac{\boldsymbol{p}_k^{\mathrm{T}}\boldsymbol{q}_k}{\boldsymbol{q}_k^{\mathrm{T}}\boldsymbol{H}_k\boldsymbol{q}_k}+\frac{\boldsymbol{p}_k\boldsymbol{p}_k^{\mathrm{T}}}{\boldsymbol{p}_k^{\mathrm{T}}\boldsymbol{q}_k} \tag{12.1.11}$$

12.1.3 Memoryless Quasi-Newton Methods

Consider a simplification of the BFGS Quasi-Newton method where $\boldsymbol{H}_{k+1}$ is defined by a BFGS update applied to $\boldsymbol{H}=\boldsymbol{I}$, rather than to $\boldsymbol{H}_k$. Thus $\boldsymbol{H}_{k+1}$ is determined without reference to the previous $\boldsymbol{H}_k$, and hence the update procedure is memoryless. This update procedure leads to the following algorithm: Start at any point $\boldsymbol{x}_0, k=0$.

Step 1. Set

$$\boldsymbol{H}_k=\boldsymbol{I} \tag{12.1.12}$$

Step 2. Set

$$d_k = -H_k g_k \tag{12.1.13}$$

Step 3. Minimize $f(x_k + \alpha d_k)$ with respect to $\alpha \geqslant 0$ to obtain $\alpha_k, x_{k+1}, p_k = \alpha_k d_k, g_{k+1}$ and $q_k = g_{k+1} - g_k$. (Select α_k accurately enough to ensure $p_k^{\mathrm{T}} q_k > 0$.)

Step 4. If k is not an integer multiple of n, set

$$H_{k+1} = I - \frac{q_k p_k^{\mathrm{T}} + p_k q_k^{\mathrm{T}}}{p_k^{\mathrm{T}} q_k} + \left(1 + \frac{q_k^{\mathrm{T}} q_k}{p_k^{\mathrm{T}} q_k}\right)\frac{p_k p_k^{\mathrm{T}}}{p_k^{\mathrm{T}} q_k}. \tag{12.1.14}$$

Add 1 to k and return to Step 2. If k is an integer multiple of n, return to Step 1.

Combining eq. (12.1.13) and eq. (12.1.14), it is easily seen that

$$d_{k+1} = -g_{k+1} + \frac{q_k p_k^{\mathrm{T}} g_{k+1} + p_k q_k^{\mathrm{T}} g_{k+1}}{p_k^{\mathrm{T}} q_k} - \left(1 + \frac{q_k^{\mathrm{T}} q_k}{p_k^{\mathrm{T}} q_k}\right)\frac{p_k p_k^{\mathrm{T}} g_{k-1}}{p_k^{\mathrm{T}} q_k} \tag{12.1.15}$$

If the line search is exact, then $p_k^{\mathrm{T}} g_{k+1} = 0$ and hence $p_k^{\mathrm{T}} q_k = -p_k^{\mathrm{T}} g_k$. In this case eq. (12.1.15) is equivalent to

$$\begin{aligned} d_{k+1} &= -g_{k+1} + \frac{q_k^{\mathrm{T}} g_{k+1}}{p_k^{\mathrm{T}} q_k} p_k \\ &= -g_{k+1} + \beta_k d_k \end{aligned} \tag{12.1.16}$$

where $\beta_k = \dfrac{q_k q_{k+1}^{\mathrm{T}}}{g_k^{\mathrm{T}} q_k}$.

This coincides exactly with the Polak-Ribiere form of the conjugate gradient method.

12.2 Constrained Minimization Conditions

12.2.1 Constraints

We deal with general nonlinear programming problems of the form

$$\begin{aligned} &\text{minimize} && f(x) \\ &\text{subject to} && h_1(x) = 0,\ g_1(x) \leqslant 0 \\ & && h_2(x) = 0,\ g_2(x) \leqslant 0 \\ & && \vdots \qquad\qquad \vdots \\ & && h_m(x) = 0,\ g_p(x) \leqslant 0 \\ & && x \in \Omega \subset E^n \end{aligned} \tag{12.2.1}$$

where $m \leqslant n$ and the functions $f, h_i (i=1,2,\ldots,m)$ and $g_j (j=1,2,\ldots,p)$ are continuous, and usually assumed to possess continuous second partial derivatives. For notational simplicity, we introduce the vector-valued functions $\boldsymbol{h} = (h_1, h_2, \ldots, h_m)$ and $\boldsymbol{g} = (g_1, g_2, \ldots, g_p)$ and rewrite eq. (12.2.1) as

$$\begin{aligned} &\text{minimize} \quad f(\boldsymbol{x}) \\ &\text{subject to} \quad \boldsymbol{h}(\boldsymbol{x}) = \boldsymbol{0}, \boldsymbol{g}(\boldsymbol{x}) \leqslant \boldsymbol{0} \\ &\qquad\qquad\quad \boldsymbol{x} \in \Omega \end{aligned} \tag{12.2.2}$$

The constraints $\boldsymbol{h}(\boldsymbol{x}) = \boldsymbol{0}, \boldsymbol{g}(\boldsymbol{x}) \leqslant \boldsymbol{0}$ are referred to as functional constraints, while the constraint $\boldsymbol{x} \in \Omega$ is a set constraint. As before we continue to deemphasize the set constraint, assuming in most cases that either Ω is the whole space E^n or that the solution to eq. (12.2.2) is in the interior of Ω. A point $\boldsymbol{x} \in \Omega$ that satisfies all the functional constraints is said to be feasible.

12.2.2 Tangent Plane

A set of equality constraints on E^n

$$\begin{aligned} h_1(\boldsymbol{x}) &= 0 \\ h_2(\boldsymbol{x}) &= 0 \\ &\vdots \\ h_m(\boldsymbol{x}) &= 0 \end{aligned} \tag{12.2.3}$$

defines a subset of E^n which is best viewed as a hypersurface.

Definition: A point $\boldsymbol{x}^*$ satisfying the constraint $\boldsymbol{h}(\boldsymbol{x}^*) = \boldsymbol{0}$ is said to be a *regular point* of the constraint if the gradient vectors $\nabla h_1(\boldsymbol{x}^*), \nabla h_2(\boldsymbol{x}^*), \ldots, \nabla h_m(\boldsymbol{x}^*)$ are linearly independent.

Note that if $\boldsymbol{h}$ is affine, $\boldsymbol{h}(\boldsymbol{x}) = \boldsymbol{A}\boldsymbol{x} + \boldsymbol{b}$, regularity is equivalent to $\boldsymbol{A}$ having rank equal to m, and this condition is independent of $\boldsymbol{x}$.

Theorem: At a regular point $\boldsymbol{x}^*$ of the surface S defined by $\boldsymbol{h}(\boldsymbol{x}) = \boldsymbol{0}$, the tangent plane is equal to $M = \{\boldsymbol{y} : \nabla \boldsymbol{h}(\boldsymbol{x}^*)\boldsymbol{y} = \boldsymbol{0}\}$.

12.2.3 First-order Necessary Conditions (Equality Constraints)

Lemma: Let $\boldsymbol{x}^*$ be a regular point of the constraints $\boldsymbol{h}(\boldsymbol{x}) = \boldsymbol{0}$ and a local extremum point (a minimum or maximum) of f subject to these constraints, then all $\boldsymbol{y} \in E^n$ satisfying

$$\nabla \boldsymbol{h}(\boldsymbol{x}^*)\boldsymbol{y} = \boldsymbol{0} \tag{12.2.4}$$

must also satisfy

$$\nabla f(\boldsymbol{x}^*)\boldsymbol{y} = \boldsymbol{0} \tag{12.2.5}$$

Theorem: Let $\boldsymbol{x}^*$ be a local extremum point of f subject to the constraints $\boldsymbol{h}(\boldsymbol{x}) = \boldsymbol{0}$. Assume further that $\boldsymbol{x}^*$ is a regular point of these constraints, then there is a $\boldsymbol{\lambda} \in E^m$ such that

$$\nabla f(\boldsymbol{x}^*) + \boldsymbol{\lambda}^{\mathrm{T}} \nabla \boldsymbol{h}(\boldsymbol{x}^*) = \boldsymbol{0} \tag{12.2.6}$$

12.2.4 Second-order Conditions

Second-order necessary conditions: Suppose that $\boldsymbol{x}^*$ is a local minimum of f subject to $\boldsymbol{h}(\boldsymbol{x}) = \boldsymbol{0}$ and that $\boldsymbol{x}^*$ is a regular point of these constraints, then there is a $\boldsymbol{\lambda} \in E^m$ such that

$$\nabla f(\boldsymbol{x}^*) + \boldsymbol{\lambda}^{\mathrm{T}} \nabla \boldsymbol{h}(\boldsymbol{x}^*) = \boldsymbol{0} \tag{12.2.7}$$

If we denote by M the tangent plane $M = \{\boldsymbol{y} : \nabla \boldsymbol{h}(\boldsymbol{x}^*)\boldsymbol{y} = \boldsymbol{0}\}$, then the matrix

$$\boldsymbol{L}(\boldsymbol{x}^*) = \boldsymbol{F}(\boldsymbol{x}^*) + \boldsymbol{\lambda}^{\mathrm{T}}\boldsymbol{H}(\boldsymbol{x}^*) \tag{12.2.8}$$

is positive semidefinite on M, that is, $\boldsymbol{y}^{\mathrm{T}}\boldsymbol{L}(\boldsymbol{x}^*)\boldsymbol{y} \geqslant \boldsymbol{0}$ for all $\boldsymbol{y} \in M$.

Second-order sufficiency conditions: Suppose there is a point $\boldsymbol{x}^*$ satisfying $\boldsymbol{h}(\boldsymbol{x}^*) = \boldsymbol{0}$, and a $\boldsymbol{\lambda} \in E^m$ such that

$$\nabla f(\boldsymbol{x}^*) + \boldsymbol{\lambda}^{\mathrm{T}} \nabla \boldsymbol{h}(\boldsymbol{x}^*) = \boldsymbol{0} \tag{12.2.9}$$

Suppose also that the matrix $\boldsymbol{L}(\boldsymbol{x}^*) = \boldsymbol{F}(\boldsymbol{x}^*) + \boldsymbol{\lambda}^{\mathrm{T}}\boldsymbol{H}(\boldsymbol{x}^*)$ is positively definite on $M = \{\boldsymbol{y} : \nabla \boldsymbol{h}(\boldsymbol{x}^*)\boldsymbol{y} = \boldsymbol{0}\}$, that is, for $\boldsymbol{y} \in M, \boldsymbol{y} \neq \boldsymbol{0}$ there holds $\boldsymbol{y}^{\mathrm{T}} \boldsymbol{L}(\boldsymbol{x}^*)\boldsymbol{y} > 0$, then $\boldsymbol{x}^*$ is a strict local minimum of f subject to $\boldsymbol{h}(\boldsymbol{x}) = \boldsymbol{0}$.

12.2.5 Eigenvalues in Tangent Subspace

Given any vector $\boldsymbol{y} \in M$, the vector $\boldsymbol{L}\boldsymbol{y}$ is in E^n but not necessarily in M. We project $\boldsymbol{L}\boldsymbol{y}$ orthogonally back onto M, the result is said to be the restriction of $\boldsymbol{L}$ to M operating on $\boldsymbol{y}$. In this way we obtain a linear transformation from M to M. The transformation is determined somewhat implicitly, however, since we do not have an explicit matrix representation.

A vector $\boldsymbol{y} \in M$ is an eigenvector of $\boldsymbol{L}_M$ if there is a real number λ such that $\boldsymbol{L}_M\boldsymbol{y} = \lambda\boldsymbol{y}$; the corresponding λ is an *eigenvalue* of $\boldsymbol{L}_M$. This coincides with the standard definition. In terms of $\boldsymbol{L}$ we see that $\boldsymbol{y}$ is an eigenvector of $\boldsymbol{L}_M$ if $\boldsymbol{L}\boldsymbol{y}$ can be written as the sum of $\boldsymbol{y}$ and a vector orthogonal to M.

To obtain a matrix representation for $\boldsymbol{L}_M$ it is necessary to introduce a basis in the subspace M. For simplicity it is best to introduce an orthonormal basis, say $\boldsymbol{e}_1, \boldsymbol{e}_2, \ldots, \boldsymbol{e}_{n-m}$. Define the matrix $\boldsymbol{E}$ to be the $n \times (n-m)$ matrix whose columns consist of the vectors $\boldsymbol{e}_i$. Then any vector $\boldsymbol{y}$ in M can be written as $\boldsymbol{y} = \boldsymbol{E}\boldsymbol{z}$ for some $\boldsymbol{z} \in E^{n-m}$ and, of course, $\boldsymbol{L}\boldsymbol{E}\boldsymbol{z}$ represents the action of $\boldsymbol{L}$ on such a vector. To project this result back into M and express the result in terms of the basis $\boldsymbol{e}_1, \boldsymbol{e}_2, \ldots, \boldsymbol{e}_{n-m}$, we merely multiply by $\boldsymbol{E}^{\mathrm{T}}$. Thus $\boldsymbol{E}^{\mathrm{T}}\boldsymbol{L}\boldsymbol{E}\boldsymbol{z}$ is the vector whose components give the representation in terms of the basis; and, correspondingly, the $(n-m) \times (n-m)$ matrix $\boldsymbol{E}^{\mathrm{T}}\boldsymbol{L}\boldsymbol{E}$ is the matrix representation of $\boldsymbol{L}$ restricted to M.

The eigenvalues of $\boldsymbol{L}$ restricted to M can be found by determining the eigenvalues of $\boldsymbol{E}^{\mathrm{T}}\boldsymbol{L}\boldsymbol{E}$. These eigenvalues are independent of the particular orthonormal basis $\boldsymbol{E}$.

Bordered Hessians

The above approach for determining the eigenvalues of $\boldsymbol{L}$ projected onto M is quite direct and relatively simple. There is another approach, however, that is useful in some theoretical arguments and convenient for simple applications. It is based on constructing matrices and determinants of order $n+m$ rather than $n-m$, so dimension is increased.

Let us first characterize all vectors orthogonal to M, where $M = \{\boldsymbol{x} \mid \nabla \boldsymbol{h}\boldsymbol{x} = \boldsymbol{0}\}$. A vector $\boldsymbol{z}$ is orthogonal to M if $\boldsymbol{z}^{\mathrm{T}}\boldsymbol{x} = \boldsymbol{0}$ for all $\boldsymbol{x} \in M$. It is not hard to show that $\boldsymbol{z}$ is orthogonal to M iff $\boldsymbol{z} = \nabla \boldsymbol{h}^{\mathrm{T}}\boldsymbol{w}$ for some $\boldsymbol{w} \in E^m$. The proof that this is sufficient follows the calculation $\boldsymbol{z}^{\mathrm{T}}\boldsymbol{x} = \boldsymbol{w}^{\mathrm{T}} \nabla \boldsymbol{h}\boldsymbol{x} = \boldsymbol{0}$. The proof of necessity follows the Duality Theorem of Linear Programming.

Now we may explicitly characterize an eigenvector of $\boldsymbol{L}_M$. The vector $\boldsymbol{x}$ is such an eigenvector if it satisfies these two conditions: ①$\boldsymbol{x}$ belongs to M, and ②$\boldsymbol{L}\boldsymbol{x} = \boldsymbol{\lambda}\boldsymbol{x} + \boldsymbol{z}$, where $\boldsymbol{z}$ is orthogonal to M. These conditions are equivalent, in view of the characterization of $\boldsymbol{z}$, to

$$\nabla \boldsymbol{h}\boldsymbol{x} = \boldsymbol{0}$$

$$\boldsymbol{L}\boldsymbol{x} = \lambda \boldsymbol{x} + \nabla \boldsymbol{h}^{\mathrm{T}}\boldsymbol{w}$$

This can be regarded as a homogeneous system of $n+m$ linear equations in the unknowns $\boldsymbol{w}, \boldsymbol{x}$. It possesses a nonzero solution iff the determinant of the coefficient matrix is zero. Denoting this determinant $p(\lambda)$, we have

$$p(\lambda) \equiv \det\begin{bmatrix} \boldsymbol{0} & \nabla \boldsymbol{h} \\ -\nabla \boldsymbol{h}^{\mathrm{T}} & \boldsymbol{L} - \lambda \boldsymbol{I} \end{bmatrix} = 0 \qquad (12.2.10)$$

as the condition. The function $p(\lambda)$ is a polynomial in λ of degree $n-m$. It is, as we have derived, the characteristic polynomial of $\boldsymbol{L}_M$.

Bordered hessian test: The matrix $\boldsymbol{L}$ is positively definite on the subspace $M=\{\boldsymbol{x}:\nabla \boldsymbol{h}\boldsymbol{x}=\boldsymbol{0}\}$ iff the last $n-m$ principal minors of $\boldsymbol{B}$ all have sign $(-1)^m$.

For the above example we form

$$\boldsymbol{B}=\det\begin{bmatrix} 0 & 1 & 0 & \vdots & 0 \\ 1 & -1 & 0 & \vdots & 0 \\ 0 & 0 & 1 & \vdots & 1 \\ \cdots & \cdots & \cdots & \cdots & \cdots \\ 0 & 0 & 1 & \cdots & 3 \end{bmatrix}$$

and check the last two principal minors—the one indicated by the dashed lines and the whole determinant. These are -1, -2, which both have sign $(-1)^1$, and hence the criterion is satisfied.

12.2.6 Inequality Constraints

We consider now problems of the form

$$\begin{aligned} &\text{minimize} \quad f(\boldsymbol{x}) \\ &\text{subject to} \quad \boldsymbol{h}(\boldsymbol{x})=\boldsymbol{0} \\ &\qquad\qquad\quad \boldsymbol{g}(\boldsymbol{x})\leqslant\boldsymbol{0} \end{aligned} \tag{12.2.11}$$

We assume that f and $\boldsymbol{h}$ are as before and that $\boldsymbol{g}$ is a p-dimensional function.

First-Order Necessary Conditions

Definition. Let $\boldsymbol{x}^*$ be a point satisfying the constraints

$$\boldsymbol{h}(\boldsymbol{x}^*)=\boldsymbol{0}, \boldsymbol{g}(\boldsymbol{x}^*)\leqslant\boldsymbol{0} \tag{12.2.12}$$

and let J be the set of indices j for which $g_j(\boldsymbol{x}^*)=\boldsymbol{0}$. Then $\boldsymbol{x}^*$ is said to be a *regular point* of the constraints eq. (12.2.12) if the gradient vectors $\nabla h_i(\boldsymbol{x}^*)$, $\nabla g_i(\boldsymbol{x}^*)$, $1\leqslant i\leqslant m, j\in J$ are linearly independent.

Karush-kuhn-tucker conditions: Let $\boldsymbol{x}^*$ be a relative minimum point for the problem

$$\begin{aligned} &\text{minimize} \quad f(\boldsymbol{x}) \\ &\text{subject to} \quad \boldsymbol{h}(\boldsymbol{x})=\boldsymbol{0}, \boldsymbol{g}(\boldsymbol{x})\leqslant 0 \end{aligned} \tag{12.2.13}$$

and suppose $\boldsymbol{x}^*$ is a regular point for the constraints, then there is a vector $\boldsymbol{\lambda}\in E^m$ and a vector $\boldsymbol{\mu}\in E^p$ with $\boldsymbol{\mu}\geqslant\boldsymbol{0}$ such that

$$\nabla f(\boldsymbol{x}^*)+\boldsymbol{\lambda}^{\mathrm{T}}\nabla\boldsymbol{h}(\boldsymbol{x}^*)+\boldsymbol{\mu}^{\mathrm{T}}\nabla\boldsymbol{g}(\boldsymbol{x}^*)=\boldsymbol{0} \tag{12.2.14}$$

$$\boldsymbol{\mu}^{\mathrm{T}}\boldsymbol{g}(\boldsymbol{x}^*) = \boldsymbol{0} \tag{12.2.15}$$

Second-order Conditions

Second-order necessary conditions: Suppose the functions $f, \boldsymbol{g}, \boldsymbol{h} \in C^2$ and if $\boldsymbol{x}^*$ satisfying eq. (12.2.12) is a relative minimum point for problem eq. (12.2.11), then there is a $\boldsymbol{\lambda} \in E^m, \boldsymbol{\mu} \in E^p, \boldsymbol{\mu} \geqslant \boldsymbol{0}$ such that eq. (12.2.14) and eq. (12.2.15) and

$$\boldsymbol{L}(\boldsymbol{x}^*) = \boldsymbol{F}(\boldsymbol{x}^*) + \boldsymbol{\lambda}^{\mathrm{T}}\boldsymbol{H}(\boldsymbol{x}^*) + \boldsymbol{\mu}^{\mathrm{T}}\boldsymbol{G}(\boldsymbol{x}^*) \tag{12.2.16}$$

is positively semidefinite on the tangent subspace of the active constraints at $\boldsymbol{x}^*$.

Second-order sufficiency conditions: Let $f, \boldsymbol{g}, \boldsymbol{h} \in C^2$ and if a point $\boldsymbol{x}^*$ satisfying eq. (12.2.12) is a strict relative minimum point of problem eq. (12.2.11), then there exist $\boldsymbol{\lambda} \in E^m, \boldsymbol{\mu} \in E^p$, such that

$$\boldsymbol{\mu} \geqslant \boldsymbol{0} \tag{12.2.17}$$

$$\boldsymbol{\mu}^{\mathrm{T}}\boldsymbol{g}(\boldsymbol{x}^*) = \boldsymbol{0} \tag{12.2.18}$$

$$\nabla f(\boldsymbol{x}^*) + \boldsymbol{\lambda}^{\mathrm{T}} \nabla \boldsymbol{h}(\boldsymbol{x}^*) + \boldsymbol{\mu}^{\mathrm{T}} \nabla \boldsymbol{g}(\boldsymbol{x}^*) = \boldsymbol{0} \tag{12.2.19}$$

and the Hessian matrix

$$\boldsymbol{L}(\boldsymbol{x}^*) = \boldsymbol{F}(\boldsymbol{x}^*) + \boldsymbol{\lambda}^{\mathrm{T}}\boldsymbol{H}(\boldsymbol{x}^*) + \boldsymbol{\mu}^{\mathrm{T}}\boldsymbol{G}(\boldsymbol{x}^*) \tag{12.2.20}$$

is positively definite on the subspace

$M' = \{\boldsymbol{y} : \nabla \boldsymbol{h}(\boldsymbol{x}^*)\boldsymbol{y} = \boldsymbol{0}, \nabla \boldsymbol{g}_j(\boldsymbol{x}^*)\boldsymbol{y} = \boldsymbol{0}$ for all $j \in J = \{j : g_j(\boldsymbol{x}^*) = \boldsymbol{0}, \mu_j > 0\}\}$.

12.2.7 Zero-order Conditions and Lagrange Multipliers

Zero-order conditions for functionally constrained problems express conditions in terms of Lagrange multipliers without the use of derivatives. As before, the basic constrained problem is

$$\begin{aligned} &\text{minimize} \quad f(\boldsymbol{x}) \\ &\text{subject to} \quad \boldsymbol{h}(\boldsymbol{x}) = \boldsymbol{0}, \boldsymbol{g}(\boldsymbol{x}) \leqslant \boldsymbol{0} \\ &\qquad\qquad \boldsymbol{x} \in \Omega \end{aligned} \tag{12.2.21}$$

where $\boldsymbol{x}$ is a vector in E^n, and $\boldsymbol{h}$ and $\boldsymbol{g}$ are m-dimensional and p-dimensional functions, respectively.

In purest form, zero-order conditions require that the functions that define the objective and the constraints are convex functions and sets. The vector-valued function $\boldsymbol{g}$ consisting of p individual component functions $g_1, g_2, \ldots, g_p$ is said to be *convex* if each of the component functions is convex. The programming problem above is termed as a *convex programming problem* if the functions f and $\boldsymbol{g}$ are convex, the function $\boldsymbol{h}$ is affine (that is, linear plus a constant), and the set $\Omega \subset E^n$ is convex.

Notice that the set defined by each of the inequalities $g_j(\boldsymbol{x}) \leqslant 0$ is convex. This is true also of a set defined by $h_i(\boldsymbol{x}) = 0$. Since the overall constraint set is the intersection of these and Ω, the problem can be regarded as minimize $f(\boldsymbol{x})$, $\boldsymbol{x} \in \Omega_1$, where Ω_1 is a convex subset of Ω.

With this view, one could apply the zero-order conditions to the problem with constraint set Ω_1. However, in the case of functional constraints it is common to keep the structure of the constraints explicit instead of folding them into an amorphous set.

The equality problem is

$$\begin{aligned} \text{minimize} \quad & f(\boldsymbol{x}) \\ \text{subject to} \quad & \boldsymbol{h}(\boldsymbol{x}) = \boldsymbol{0} \\ & \boldsymbol{x} \in \Omega \end{aligned} \tag{12.2.22}$$

Let $Y = E^n$, we have $\boldsymbol{h}(\boldsymbol{x}) \in Y$ for all $\boldsymbol{x}$. For this problem we require a regularity condition.

Definition. An affine function $\boldsymbol{h}$ is *regular* with respect to Ω if the set C in Y defined by $C = \{\boldsymbol{y} : \boldsymbol{h}(\boldsymbol{x}) = \boldsymbol{y}\}$ for some $\boldsymbol{x} \in \Omega$ contains an open sphere around $\boldsymbol{0}$; that is, C contains a set of the form $\{y : |y| < \varepsilon\}$ for some $\varepsilon > 0$. This condition means that $\boldsymbol{h}(\boldsymbol{x})$ can attain $\boldsymbol{0}$ and can vary in arbitrary directions from $\boldsymbol{0}$.

Notice that this condition is similar to the definition of a regular point in the context of first-order conditions. If $\boldsymbol{h}$ has continuous derivatives at a point $\boldsymbol{x}^*$ the earlier regularity condition implies that $\nabla \boldsymbol{h}(\boldsymbol{x}^*)$ is of full rank, then guarantees that there is an $\varepsilon > 0$ such that for any $\boldsymbol{y}$ with $|\boldsymbol{y} - \boldsymbol{h}(\boldsymbol{x}^*)| < \varepsilon$ there is an $\boldsymbol{x}$ such that $\boldsymbol{h}(\boldsymbol{x}) = \boldsymbol{y}$. In other words, there is an open sphere around $\boldsymbol{y}^* = \boldsymbol{h}(\boldsymbol{x}^*)$ that is attainable. In the present situation we assume this attainability directly, at the point $0 \in Y$.

Next we introduce the following important construction.

Definition: The *primal function* associated with problem is $w(\boldsymbol{y}) = \inf\{f(\boldsymbol{x}) : \boldsymbol{h}(\boldsymbol{x}) = \boldsymbol{y}, \boldsymbol{x} \in \Omega\}$ defined for all $\boldsymbol{y} \in C$.

Notice that the primal function is defined by varying the right hand side of the constraint. The original problem corresponds to $\omega(\boldsymbol{0})$.

Proposition 1: Suppose Ω is convex, the function f is convex, and $\boldsymbol{h}$ is affine, then the primal function ω is convex.

Proposition 2: Assume that $\Omega \subset E^n$ is convex, f is a convex function on Ω and $\boldsymbol{h}$ is an m-dimensional affine function on Ω. Assume that $\boldsymbol{h}$ is regular with respect to Ω. If $\boldsymbol{x}^*$ solves (12.2.22), then there is $\boldsymbol{\lambda} \in E^m$ such that $\boldsymbol{x}^*$ solves the Lagrangian problem

$$\text{minimize} \quad f(\boldsymbol{x}) + \boldsymbol{\lambda}^{\mathrm{T}} \boldsymbol{h}(\boldsymbol{x})$$
$$\text{subject to } \boldsymbol{x} \in \Omega$$

Inequality Constraints

We outline the parallel results for the inequality constrained problem

$$\begin{aligned} &\text{minimize} && f(\boldsymbol{x}) \\ &\text{subject to} && \boldsymbol{g}(\boldsymbol{x}) \leqslant \boldsymbol{0} \\ & && \boldsymbol{x} \in \Omega \end{aligned} \qquad (12.2.23)$$

where $\boldsymbol{g}$ is a p-dimensional function.

Let $Z = E^p$ and $D = \{z \in Z : \boldsymbol{g}(\boldsymbol{x}) \leqslant z \text{ for some } \boldsymbol{x} \in \Omega\}$. The regularity condition (or the *Slater condition*) is that there exists a negative $z_1 \in D$.

We again introduce the primal function below.

Definition: The *primal function* associated with problem eq. (12.2.22) is $w(z) = \inf \{f(\boldsymbol{x}) : \boldsymbol{g}(\boldsymbol{x}) \leqslant z, \boldsymbol{x} \in \Omega\}$.

The primal function is again defined by varying the right hand side of the constraint function, using the variable z. Now the primal function in monotonically decreasing with z, since an increase in z enlarges the constraint region.

Proposition 3: Suppose $\Omega \subset E^n$ is convex and f and $\boldsymbol{g}$ are convex functions, then the primal function ω is also convex.

Proposition 4: Assume Ω is a convex subset of E^n and that f and $\boldsymbol{g}$ are convex functions. Assume also that there is a point $\boldsymbol{x}_1 \in \Omega$ such that $\boldsymbol{g}(\boldsymbol{x}_1) < \boldsymbol{0}$, then, if $\boldsymbol{x}^*$ solves (12.2.23), there is a vector $\boldsymbol{\mu} = E^p$ with $\boldsymbol{\mu} \geqslant \boldsymbol{0}$ such that $\boldsymbol{x}^*$ solves the Lagrangian problem

$$\begin{aligned} &\text{minimize} && f(\boldsymbol{x}^*) + \boldsymbol{\mu}^{\mathrm{T}} \boldsymbol{g}(\boldsymbol{x}) \\ &\text{subject to} && \boldsymbol{x} \in \Omega \end{aligned} \qquad (12.2.24)$$

Furthermore, $\boldsymbol{\mu}^{\mathrm{T}} \boldsymbol{g}(\boldsymbol{x}^*) = \boldsymbol{0}$.

12.3 Primal Methods

12.3.1 Feasible Direction Methods

The idea of feasible direction methods is to take steps through the feasible region of the form

$$\boldsymbol{x}_{k+1} = \boldsymbol{x}_k + \alpha_k \boldsymbol{d}_k \tag{12.3.1}$$

where $\boldsymbol{d}_k$ is a direction vector and α_k is a nonnegative scalar. The scalar is chosen to minimize the objective function f with the restriction that the point $\boldsymbol{x}_{k+1}$ and the line segment joining $\boldsymbol{x}_k$ and $\boldsymbol{x}_{k+1}$ are feasible. Thus, in order that the process of minimizing with respect to α is nontrivial, an initial segment of the ray $\boldsymbol{x}_k + \alpha \boldsymbol{d}_k, \alpha > 0$ must be contained in the feasible region. This motivates the use of *feasible directions* for the directions of search. For example, a vector $\boldsymbol{d}_k$ is a feasible direction at $\boldsymbol{x}_k$ if there is an $\bar{\alpha} > 0$ such that $\boldsymbol{x}_k + \alpha \boldsymbol{d}_k$ is feasible for all $\alpha, 0 \leqslant \alpha \leqslant \bar{\alpha}$.

12.3.2 Active Set Methods

Consider the constrained problem

$$\begin{aligned} &\text{minimize} \quad && f(\boldsymbol{x}) \\ &\text{subject to} \quad && \boldsymbol{g}(\boldsymbol{x}) \leqslant \boldsymbol{0} \end{aligned} \tag{12.3.2}$$

The necessary conditions for this problem are

$$\begin{aligned} &\nabla f(\boldsymbol{x}) + \boldsymbol{\lambda}^{\mathrm{T}} \nabla \boldsymbol{g}(\boldsymbol{x}) = \boldsymbol{0} \\ &\boldsymbol{g}(\boldsymbol{x}) \leqslant \boldsymbol{0}. \\ &\boldsymbol{\lambda}^{\mathrm{T}} \nabla \boldsymbol{g}(\boldsymbol{x}) = \boldsymbol{0} \\ &\boldsymbol{\lambda} \geqslant \boldsymbol{0} \end{aligned} \tag{12.3.3}$$

Let A denote the index set of active constraints; that is, A is the set of i such that $g_i(\boldsymbol{x}^*) = 0$. Then the necessary conditions eq. (12.3.3) become

$$\begin{aligned} &\nabla f(\boldsymbol{x}) + \sum_{i \in A} \lambda_i \nabla g_i(\boldsymbol{x}) = \boldsymbol{0} \\ &g_i(\boldsymbol{x}) = 0, \quad i \in A \\ &g_i(\boldsymbol{x}) < 0, \quad i \notin A \\ &\lambda_i \geqslant 0, \quad i \in A \\ &\lambda_i = 0, \quad i \notin A \end{aligned} \tag{12.3.4}$$

Changes in Working Set

Suppose that for a given working set W, the problem with equality constraints

$$\begin{aligned} &\text{minimize} \quad f(\boldsymbol{x}) \\ &\text{subject to} \quad g_i(\boldsymbol{x}) = 0, \quad i \in W \end{aligned}$$

is solved yielding the point $\boldsymbol{x}_W$ that satisfies $g_i(\boldsymbol{x}_W) < 0, i \notin W$. This point satisfies the necessary conditions

$$\nabla f(\boldsymbol{x}_W) + \sum_{i \in W} \lambda_i \nabla g_i(\boldsymbol{x}_W) = \mathbf{0} \tag{12.3.5}$$

If $\lambda_i \geqslant 0$ for all $i \in W$, then the point $\boldsymbol{x}_W$ is a local solution to the original problem.

Active set theorem: Suppose that for every subset W of the constraint indices, the constrained problem

$$\begin{aligned} &\text{minimize} \quad f(\boldsymbol{x}) \\ &\text{subject to} \quad g_i(\boldsymbol{x}) = 0, \quad i \in W \end{aligned} \tag{12.3.6}$$

is well-defined with a unique nondegenerate solution (that is, for all $i \in W, \boldsymbol{\lambda}_i \neq 0$), then the sequence of points generated by the basic active set strategy converges to the solution of the inequality constrained problem eq. (12.3.3).

12.3.3 The Gradient Projection Method

Linear Constraints

Consider first problems of the form

$$\begin{aligned} &\text{minimize} \quad f(\boldsymbol{x}) \\ &\text{subject to} \quad \boldsymbol{a}_i^{\mathrm{T}} \boldsymbol{x} \leqslant b_i, \quad i \in \boldsymbol{I}_1 \\ &\qquad\qquad\quad \boldsymbol{a}_i^{\mathrm{T}} \boldsymbol{x} = b_i, \quad i \in \boldsymbol{I}_2 \end{aligned} \tag{12.3.7}$$

having linear equalities and inequalities.

At a given feasible point $\boldsymbol{x}$, there will be a certain number q of active constraints satisfying $\boldsymbol{a}_i^{\mathrm{T}} \boldsymbol{x} = b_i$ and some inactive constraints $\boldsymbol{a}_i^{\mathrm{T}} \boldsymbol{x} < b_i$. We initially take the working set $W(\boldsymbol{x})$ to be the set of active constraints.

At the feasible point $\boldsymbol{x}$, we seek a feasible direction vector $\boldsymbol{d}$ satisfying $\nabla f(\boldsymbol{x}) \boldsymbol{d} < 0$, so that movement in the direction $\boldsymbol{d}$ will cause a decrease in the function f. Initially, we consider directions satisfying $\boldsymbol{a}_i^{\mathrm{T}} \boldsymbol{d} = 0, i \in W(\boldsymbol{x})$ so that all working constraints remain active.

In summary, the algorithm is as follows: Given a feasible point $\boldsymbol{x}$,

1) Find the subspace of active constraints M, and form $A_q, W(\boldsymbol{x})$;

2) Calculate $\boldsymbol{P} = \boldsymbol{I} - \boldsymbol{A}_q^{\mathrm{T}} (\boldsymbol{A}_q \boldsymbol{A}_q^{\mathrm{T}})^{-1} \boldsymbol{A}_q$ and $\boldsymbol{d} = -\boldsymbol{P} \nabla f(\boldsymbol{x})^{\mathrm{T}}$;

3) If $\boldsymbol{d} \neq \boldsymbol{0}$, find α_1 and α_2 achieving the following, respectively,

$$\max \quad \{\alpha : \boldsymbol{x} + \alpha \boldsymbol{d} \quad \text{is feasible}\}$$
$$\min \quad \{f(\boldsymbol{x} + \alpha \boldsymbol{d}) : 0 \leqslant \alpha \leqslant \alpha_1\}$$

Set $\boldsymbol{x}$ to $\boldsymbol{x} + \alpha_2 \boldsymbol{d}$ and return to 1).

4) If $\boldsymbol{d} = \boldsymbol{0}$, find $\lambda = -(\boldsymbol{A}_q \boldsymbol{A}_q^{\mathrm{T}})^{-1} \boldsymbol{A}_q \nabla f(\boldsymbol{x})^{\mathrm{T}}$.

a) If $\lambda_j \geqslant 0$, for all j corresponding to active inequalities, stop; $\boldsymbol{x}$ satisfies the Karush-Kuhn-Tucker conditions.

b) Otherwise, delete the row from $\boldsymbol{A}_q$ corresponding to the inequality with the most negative component of λ (and drop the corresponding constraint from $W(\boldsymbol{x})$) and return to 2).

12.3.4 Convergence Rate of the Gradient Projection Method

For simplicity we consider first the problem having only equality constraints

$$\begin{aligned} &\text{minimize} \quad f(\boldsymbol{x}) \\ &\text{subject to} \quad \boldsymbol{h}(\boldsymbol{x}) = \boldsymbol{0} \end{aligned} \tag{12.3.8}$$

The constraints define a continuous surface Ω in E^n.

Geodesics: Given the surface $\Omega = \{\boldsymbol{x} : \boldsymbol{h}(\boldsymbol{x}) = \boldsymbol{0}\} \subset E^n$, a smooth curve $\boldsymbol{x}(t) \in \Omega, 0 \leqslant t \leqslant T$ starting at $\boldsymbol{x}(0)$ and terminating at $\boldsymbol{x}(T)$ that minimizes the total arc length $\int_0^T |\dot{\boldsymbol{x}}(t)| \mathrm{d}t$ with respect to all other such curves on Ω is said to be a *geodesic* connecting $\boldsymbol{x}(0)$ and $\boldsymbol{x}(T)$.

It is common to parameterize a geodesic $\boldsymbol{x}(t), 0 \leqslant t \leqslant T$ so that $|\dot{\boldsymbol{x}}(t)| = 1$. The parameter t is then itself the arc length. If the parameter t is also regarded as time, then this parameterization corresponds to moving along the geodesic curve with unit velocity. Parameterized in this way, the geodesic is said to be *normalized*. On any linear subspace of E^n geodesics are straight lines. On a three-dimensional sphere, the geodesics are arcs of great circles.

It can be shown, using the calculus of variations, that any normalized geodesic on Ω satisfies the condition

$$\ddot{\boldsymbol{x}}(t) = \nabla \boldsymbol{h}(\boldsymbol{x}(t)) \boldsymbol{\omega}(t) \tag{12.3.9}$$

for some function $\boldsymbol{\omega}$ taking values in E^m. Geometrically, this condition says that if one moves along the geodesic curve with unit velocity, the acceleration at every point will be orthogonal to the surface. Indeed, this property can be regarded as the fundamental defining characteristic of a geodesic. To stay on the surface Ω, the geodesic must also satisfy the equation

$$\nabla \boldsymbol{h}(\boldsymbol{x}(t))\dot{\boldsymbol{x}}(t)=\boldsymbol{0} \tag{12.3.10}$$

since the velocity vector at every point is tangent to Ω. At a regular point $\boldsymbol{x}_0$ these two differential equations, together with the initial conditions $\boldsymbol{x}(t)=\boldsymbol{x}_0$, $\dot{\boldsymbol{x}}(0)$ specified, and $|\dot{\boldsymbol{x}}(0)|=1$, uniquely specify a curve $\boldsymbol{x}(t)$, $t\geqslant 0$ that can be continued as long as points on the curve are regular. Furthermore, $|\dot{\boldsymbol{x}}(t)|=1$ for $t\geqslant 0$. Hence geodesic curves emanate in every direction from a regular point. Thus, for example, at any point on a sphere there is a unique great circle passing through the point in a given direction.

Lagrangian and Geodesics

Corresponding to any regular point $\boldsymbol{x}\in\Omega$, we may define a corresponding Lagrange multiplier $\boldsymbol{\lambda}(\boldsymbol{x})$ by calculating the projection of the gradient of f onto the tangent subspace at $\boldsymbol{x}$, denoted $M(\boldsymbol{x})$. The matrix that, when operating on a vector, projects it onto $M(\boldsymbol{x})$ is

$$\boldsymbol{P}(\boldsymbol{x})=\boldsymbol{I}-\nabla \boldsymbol{h}(\boldsymbol{x}_k)^{\mathrm{T}}[\nabla \boldsymbol{h}(\boldsymbol{x}_k)\nabla \boldsymbol{h}(\boldsymbol{x}_k)^{\mathrm{T}}]^{-1}\boldsymbol{h}(\boldsymbol{x})$$

and it follows immediately that the projection of $\nabla f(\boldsymbol{x})^{\mathrm{T}}$ onto $M(\boldsymbol{x})$ has the form

$$\boldsymbol{y}(\boldsymbol{x})=[\nabla f(\boldsymbol{x})+\boldsymbol{\lambda}(\boldsymbol{x})^{\mathrm{T}}\nabla \boldsymbol{h}(\boldsymbol{x})]^{\mathrm{T}} \tag{12.3.11}$$

where $\boldsymbol{\lambda}(\boldsymbol{x})$ is given explicitly as

$$\boldsymbol{\lambda}(\boldsymbol{x})^{\mathrm{T}}=-\nabla f(\boldsymbol{x})\nabla \boldsymbol{h}(\boldsymbol{x})^{\mathrm{T}}[\nabla \boldsymbol{h}(\boldsymbol{x})\nabla \boldsymbol{h}(\boldsymbol{x})^{\mathrm{T}}]^{-1} \tag{12.3.12}$$

Thus, in terms of the Lagrangian function $l(\boldsymbol{x},\boldsymbol{\lambda})=f(\boldsymbol{x})+\boldsymbol{\lambda}^{\mathrm{T}}\boldsymbol{h}(\boldsymbol{x})$, the projected gradient is

$$\boldsymbol{y}(\boldsymbol{x})=\boldsymbol{l}_x(\boldsymbol{x},\boldsymbol{\lambda}(\boldsymbol{x}))^{\mathrm{T}} \tag{12.3.13}$$

If a local solution to the original problem occurs at a regular point $\boldsymbol{x}^*\in\Omega$, then as we know

$$\boldsymbol{l}_x(\boldsymbol{x}^*,\boldsymbol{\lambda}(\boldsymbol{x}^*))=\boldsymbol{0} \tag{12.3.14}$$

which states that the projected gradient must vanish at $\boldsymbol{x}^*$. Define $\boldsymbol{L}(\boldsymbol{x})=\boldsymbol{l}_{xx}(\boldsymbol{x},\boldsymbol{\lambda}(\boldsymbol{x}))=\boldsymbol{F}(\boldsymbol{x})+\boldsymbol{\lambda}(\boldsymbol{x})^{\mathrm{T}}\boldsymbol{H}(\boldsymbol{x})$, we also know that at $\boldsymbol{x}^*$ we have the second-order necessary condition that $\boldsymbol{L}(\boldsymbol{x}^*)$ is positive semidefinite on $M(\boldsymbol{x}^*)$; that is, $z^{\mathrm{T}}\boldsymbol{L}(\boldsymbol{x}^*)z\geqslant 0$ for all $z\in M(\boldsymbol{x}^*)$. Equivalently, let

$$\overline{\boldsymbol{L}}(\boldsymbol{x})=\boldsymbol{P}(\boldsymbol{x})\boldsymbol{L}(\boldsymbol{x})\boldsymbol{P}(\boldsymbol{x}) \tag{12.3.15}$$

it follows that $\overline{\boldsymbol{L}}(\boldsymbol{x}^*)$ is positive semidefinite.

We have the following fundamental result along a geodesic.

Proposition 1: Let $\boldsymbol{x}(t), 0 \leqslant t \leqslant T$, be a geodesic on Ω, then

$$\frac{\mathrm{d}}{\mathrm{d}t} f(\boldsymbol{x}(t)) = \boldsymbol{l}_x(\boldsymbol{x}, \boldsymbol{\lambda}(\boldsymbol{x})) \dot{\boldsymbol{x}}(t) \tag{12.3.16}$$

$$\frac{\mathrm{d}^2}{\mathrm{d}t^2} f(\boldsymbol{x}(t)) = \dot{\boldsymbol{x}}(t)^{\mathrm{T}} \boldsymbol{L}(\boldsymbol{x}(t)) \dot{\boldsymbol{x}}(t) \tag{12.3.17}$$

Rate of Convergence

We assume that all functions are three times continuously differentiable, and that every point in a region near the solution $\boldsymbol{x}^*$ is regular and that the method of geodesic descent generates a sequence $\{\boldsymbol{x}_k\}$ converging to $\boldsymbol{x}^*$. We now prove the main theorem regarding the rate of convergence below.

Theorem: Let $\boldsymbol{x}^*$ be a local solution to the problem eq. (12.3.8) and suppose that A and $a > 0$ are, respectively, the largest and smallest eigenvalues of $\boldsymbol{L}(\boldsymbol{x}^*)$ restricted to the tangent subspace $M(\boldsymbol{x}^*)$. If $\{\boldsymbol{x}_k\}$ is a sequence generated by the method of geodesic descent that converges to $\boldsymbol{x}^*$, then the sequence of objective values $\{f(\boldsymbol{x}_k)\}$ converges to $f(\boldsymbol{x}^*)$ linearly with a ratio no greater than $[(A-a)/(A+a)]^2$.

12.3.5 The Reduced Gradient Method

Linear Constraints

Consider the problem

$$\begin{aligned} &\text{minimize} \quad f(\boldsymbol{x}) \\ &\text{subject to} \quad \boldsymbol{Ax} = \boldsymbol{b}, \boldsymbol{x} \geqslant \boldsymbol{0} \end{aligned} \tag{12.3.18}$$

where $\boldsymbol{x} \in E^n, \boldsymbol{b} \in E^m, \boldsymbol{A}$ is an $m \times n$ matrix, and f is a function in C^2. The constraints are expressed in the format of the standard form of linear programming. For simplicity of notation it is assumed that each variable is required to be non-negative—if some variables were free, the procedure (but not the notation) would be somewhat simplified.

We invoke the *non-degeneracy assumptions* that every collection of m columns from $\boldsymbol{A}$ is linearly independent and every basic solution to the constraints has m strictly positive variables. With these assumptions any feasible solution will have at most $n-m$ variables taking the value zero. Given a vector $\boldsymbol{x}$ satisfying the constraints, we partition the variables into two groups: $\boldsymbol{x} = (\boldsymbol{y}, \boldsymbol{z})$ where $\boldsymbol{y}$ has dimension m and $\boldsymbol{z}$ has dimension $n-m$. This partition is formed in such a way that all variables in $\boldsymbol{y}$ are strictly positive (for simplicity of notation we indicate the basic variables as being the first m components of $\boldsymbol{x}$ but, of course, in general this will not be so). With respect to the partition, the original problem can be expressed as

$$\text{minimize} \quad f(\boldsymbol{y},\boldsymbol{z}) \tag{12.3.19a}$$

$$\text{subject to} \quad \boldsymbol{By}+\boldsymbol{Cz}=\boldsymbol{b} \tag{12.3.19b}$$

$$\boldsymbol{y}\geqslant\boldsymbol{0}, \quad \boldsymbol{z}\geqslant\boldsymbol{0} \tag{12.3.19c}$$

where, of course, $\boldsymbol{A}=[\boldsymbol{B},\boldsymbol{C}]$. We can regard $\boldsymbol{z}$ as consisting of the independent variables and $\boldsymbol{y}$ the dependent variables, since if $\boldsymbol{z}$ is specified, eq. (12.3.19b) can be uniquely solved for $\boldsymbol{y}$. Furthermore, a small change Δz from the original value that leaves $\boldsymbol{z}+\boldsymbol{\Delta z}$ nonnegative will, upon solution of eq. (12.3.19b), yield another feasible solution, since $\boldsymbol{y}$ was originally taken to be strictly positive and thus $\boldsymbol{y}+\boldsymbol{\Delta y}$ will also be positive for small $\boldsymbol{\Delta y}$. We may therefore move from one feasible solution to another by selecting a $\boldsymbol{\Delta z}$ and moving $\boldsymbol{z}$ on the line $\boldsymbol{z}+\alpha\boldsymbol{\Delta z}$, $\alpha\geqslant 0$. Accordingly, $\boldsymbol{y}$ will move along a corresponding line $\boldsymbol{y}+\alpha\boldsymbol{\Delta y}$. If in moving this way some variable becomes zero, a new inequality constraint becomes active. If some independent variable becomes zero, a new direction $\boldsymbol{\Delta z}$ must be chosen. If a dependent (basic) variable becomes zero, the partition must be modified. The zero-valued basic variable is declared independent and one of the strictly positive independent variables is made dependent. Operationally, this interchange will be associated with a pivot operation.

The idea of the reduced gradient method is to consider, at each stage, the problem only in terms of the independent variables. Since the vector of dependent variables $\boldsymbol{y}$ is determined through the constraints eq. (12.3.19b) from the vector of independent variables $\boldsymbol{z}$, the objective function can be considered to be a function of $\boldsymbol{z}$ only. Hence a simple modification of steepest descent, accounting for the constraints, can be executed. The gradient with respect to the independent variables $\boldsymbol{z}$ (the *reduced gradient*) is found by evaluating the gradient of $f(\boldsymbol{B}^{-1}\boldsymbol{b}-\boldsymbol{B}^{-1}\boldsymbol{Cz},\boldsymbol{z})$. It is equal to

$$\boldsymbol{r}^{\mathrm{T}}=\nabla_z f(\boldsymbol{y},\boldsymbol{z})-\nabla_y f(\boldsymbol{y},\boldsymbol{z})\boldsymbol{B}^{-1}\boldsymbol{C} \tag{12.3.20}$$

It is easy to see that a point $(\boldsymbol{y},\boldsymbol{z})$ satisfies the first-order necessary conditions for optimality iff

$$r_i=0 \quad \text{for all} \quad z_i>0$$

$$r_i\geqslant 0 \quad \text{for all} \quad z_i=0$$

In the active set form of the reduced gradient method the vector z is moved in the direction of the reduced gradient on the working surface. Thus at each step, a direction of the form

$$\Delta z_i = \begin{cases} -r_i, i \notin W(z) \\ 0, i \in W(z) \end{cases}$$

is determined and a descent is made in this direction. The working set is augmented whenever a new variable reaches zero; if it is a basic variable, a new partition is also formed. If a point is found where $r_i = 0$ for all $i \notin W(z)$ (representing a vanishing reduced gradient on the working surface) but $r_j < 0$ for some $j \in W(z)$, then j is deleted from $W(z)$ as in the standard active set strategy.

The basic procedures of the reduced gradient method are as follows:

Step 1. Let $\Delta z_i = \begin{cases} -r_i, \text{if } r_i < 0 \quad \text{or } z_i > 0 \\ 0, \text{otherwise} \end{cases}$

Step 2. If $\boldsymbol{\Delta z}$ is zero, stop; the current point is a solution. Otherwise, find $\boldsymbol{\Delta y} = -\boldsymbol{B}^{-1}\boldsymbol{C}\boldsymbol{\Delta z}$.

Step 3. Find $\alpha_1, \alpha_2, \alpha_3$ achieving, respectively,

max $\{\alpha : \boldsymbol{y} + \alpha\boldsymbol{\Delta y} \geqslant 0\}$

max $\{\alpha : \boldsymbol{z} + \alpha\boldsymbol{\Delta z} \geqslant 0\}$

min $\{f(\boldsymbol{x} + \alpha\boldsymbol{\Delta x}) : 0 \leqslant \alpha \leqslant \alpha_1, 0 \leqslant \alpha \leqslant \alpha_2\}$

Let $\bar{\boldsymbol{x}} = \boldsymbol{x} + \alpha_3\boldsymbol{\Delta x}$.

Step 4. If $\alpha_3 \leqslant \alpha_1$, return to Step 1. Otherwise, declare the vanishing variable in the dependent set independent and declare a strictly positive variable in the independent set dependent. Update $\boldsymbol{B}$ and $\boldsymbol{C}$.

12.4 Penalty and Barrier Methods

12.4.1 Penalty Methods

Consider the problem

$$\begin{aligned} &\text{minimize} \quad f(\boldsymbol{x}) \\ &\text{subject to} \quad \boldsymbol{x} \in S \end{aligned} \tag{12.4.1}$$

where f is a continuous function on E^n and S is a constraint set in E^n. In most applications S is defined implicitly by a number of functional constraints, but in this section the more general description in eq. (12.4.1) can be handled. The idea of a penalty function method is to replace problem eq. (12.4.1) by an unconstrained problem of the form

$$\text{minimize} \quad f(\boldsymbol{x}) + c\boldsymbol{P}(\boldsymbol{x}) \tag{12.4.2}$$

where c is a positive constant and $\boldsymbol{P}$ is a function on E^n satisfying: (i) $\boldsymbol{P}$ is continuous, (ii) $\boldsymbol{P}(\boldsymbol{x}) \geqslant 0$ for all $\boldsymbol{x} \in E^n$, and (iii) $\boldsymbol{P}(\boldsymbol{x}) = 0$ iff $\boldsymbol{x} \in S$.

The Penalty Function Method

The basic procedure for solving problem eq. (12.4.1) by the penalty function method is as follows.

Step 1. Let $\{c_k\}$, $k = 1, 2, \ldots$, be a sequence tending to infinity such that for each k, $c_k \geqslant 0$, $c_{k+1} > c_k$. Define the function

$$q(c, \boldsymbol{x}) = f(\boldsymbol{x}) + c\boldsymbol{P}(\boldsymbol{x}) \tag{12.4.3}$$

For each k, a solution point $\boldsymbol{x}_k$ to the problem

$$\text{minimize} \quad q(c_k, \boldsymbol{x}) \tag{12.4.4}$$

is the answer.

Step 2. For the definition of $\boldsymbol{x}_k$ and the inequality $c_{k+1} > c_k$, we have the following sequence

$$q(c_k, \boldsymbol{x}_k) \leqslant q(c_{k+1}, \boldsymbol{x}_{k+1}) \tag{12.4.5}$$

$$\boldsymbol{P}(\boldsymbol{x}_k) \geqslant \boldsymbol{P}(\boldsymbol{x}_{k+1}) \tag{12.4.6}$$

$$f(\boldsymbol{x}_k) \leqslant f(\boldsymbol{x}_{k+1}) \tag{12.4.7}$$

Step 3. Let $\boldsymbol{x}^*$ be a solution to problem (12.4.1), then for each k,

$$f(\boldsymbol{x}^*) \geqslant q(c_k, \boldsymbol{x}_k) \geqslant f(\boldsymbol{x}_k)$$

Step 4. Any limit point of the sequence $\{\boldsymbol{x}_k\}$ is a solution to eq. (12.4.1).

12.4.2 Barrier Methods

Barrier methods are applicable to problems of the form

$$\begin{aligned} &\text{minimize} \quad f(\boldsymbol{x}) \\ &\text{subject to} \quad \boldsymbol{x} \in S \end{aligned} \tag{12.4.8}$$

where the constraint set S has a nonempty interior that is arbitrarily close to any point of S. Intuitively, what this means is that the set has an interior and it is possible to get to any boundary point by approaching it from the interior. This kind of set often arises in conjunction with inequality constraints, where S takes the form

$$S = \{x : g_i(x) \leqslant 0, i = 1,2,\ldots,p\}$$

Barrier methods are also termed as *interior methods*. They work by establishing a barrier on the boundary of the feasible region that prevents a search procedure from leaving the region. A *barrier function* is a function B defined on the interior of S such that: (i) B is continuous, (ii) $B(x) \geqslant 0$, (iii) $B(x) \to \infty$ as x approaches the boundary of S.

The Barrier Method

The barrier method is quite analogous to the penalty method. Let c_k be a sequence tending to infinity such that for each k, $k = 1,2,\ldots, c_k \geqslant 0, c_{k+1} > c_k$. Define the function $r(c,x) = f(x) + \frac{1}{c}B(x)$, and for each k, solve the problem

$$\text{minimize} \quad r(c_k, x)$$

$$\text{subject to} \quad x \in \text{interior of } S$$

obtaining the point x_k.

The main result is as follows.

Theorem: Any limit point of a sequence $\{x_k\}$ generated by the barrier method is a solution to problem eq. (12.4.8).

12.4.3 Properties of Penalty and Barrier Functions

We consider problems of the form

$$\begin{aligned} &\text{minimize} \quad f(x) \\ &\text{subject to} \quad g_i(x) \leqslant 0, \quad i = 1,2,\ldots,p \end{aligned} \tag{12.4.9}$$

Penalty Functions

A penalty function for a problem expressed in the form eq. (12.4.9) will most naturally be expressed in terms of the auxiliary constraint functions

$$g_i^+(x) \equiv \max[0, g_i(x)], \quad i = 1,2,\ldots,p \tag{12.4.10}$$

This is because in the interior of the constraint region $P(x) \equiv 0$ and hence P should be a function only of violated constraints. Denoting by $g^+(x)$ the p-dimensional vector made up of $g_i^+(x)$, we consider the general class of penalty functions

$$P(x) = \gamma(g^+(x)) \tag{12.4.11}$$

where γ is a continuous function from E^p to the real numbers, defined in such a way that $\boldsymbol{P}$ satisfies the requirements demanded of a penalty function.

Lagrange Multipliers

In the penalty method we solve, for various c_k, the unconstrained problem

$$\text{minimize} \quad f(\boldsymbol{x}) + c_k \boldsymbol{P}(\boldsymbol{x}) \tag{12.4.12}$$

Most algorithms require that the objective function have continuous first partial derivatives. Since we shall, as usual, assume that both f and $\boldsymbol{g} \in C^1$, it is natural to require, then, the penalty function $\boldsymbol{P} \in C^1$. We define

$$\nabla g_i^+(\boldsymbol{x}) = \begin{cases} \nabla g_i(\boldsymbol{x}) & \text{if} \quad g_i(\boldsymbol{x}) \geqslant 0 \\ 0 & \text{if} \quad g_i(\boldsymbol{x}) < 0 \end{cases} \tag{12.4.13}$$

and, of course, $\nabla \boldsymbol{g}^+(\boldsymbol{x})$ is the $m \times n$ matrix whose rows are ∇g_i^+. Unfortunately, $\nabla \boldsymbol{g}^+$ is usually discontinuous at points where $g_i^+(\boldsymbol{x}) = 0$ for some $i = 1, 2, \ldots, p$, and thus some restrictions must be placed on γ in order to guarantee $\boldsymbol{P} \in C^1$. We assume that $\boldsymbol{\gamma} \in \mathrm{C}^1$ *and that if* $\boldsymbol{y} = (y_1, y_2, \ldots, y_n)$, $\nabla \boldsymbol{\gamma}(\boldsymbol{y}) = (\nabla \gamma_1, \nabla \gamma_2, \ldots, \nabla \gamma_n)$, then

$$y_i = 0 \quad \text{implies} \quad \nabla \gamma_i = 0 \tag{12.4.14}$$

(for instance, this condition is satisfied only for $\varepsilon > 1$). With this assumption, the derivative of $\boldsymbol{\gamma}(\boldsymbol{g}^+(\boldsymbol{x}))$ with respect to $\boldsymbol{x}$ is continuous and can be written as $\nabla \boldsymbol{\gamma}(\boldsymbol{g}^+(\boldsymbol{x})) \nabla \boldsymbol{g}(\boldsymbol{x})$. In this result, $\nabla \boldsymbol{g}(\boldsymbol{x})$ legitimately replaces the discontinuous $\nabla \boldsymbol{g}^+(\boldsymbol{x})$, because it is premultiplied by $\nabla \boldsymbol{\gamma}(\boldsymbol{g}^+(\boldsymbol{x}))$. Of course, these considerations are necessary only for inequality constraints. If equality constraints are treated directly, the situation is far simpler.

In view of this assumption, problem eq. (12.4.12) will have its solution at a point $\boldsymbol{x}_k$ satisfying

$$\nabla f(\boldsymbol{x}_k) + \boldsymbol{c}_k \nabla \boldsymbol{\gamma}(\boldsymbol{g}^+(\boldsymbol{x}_k)) \nabla \boldsymbol{g}(\boldsymbol{x}_k) = 0$$

which can be written as

$$\nabla f(\boldsymbol{x}_k) + \boldsymbol{\lambda}_k^{\mathrm{T}} \nabla \boldsymbol{g}(\boldsymbol{x}_k) = 0 \tag{12.4.15}$$

where

$$\boldsymbol{\lambda}_k^{\mathrm{T}} \equiv \boldsymbol{c}_k \nabla \boldsymbol{\gamma}(\boldsymbol{g}^+(\boldsymbol{x}_k)) \tag{12.4.16}$$

Thus, associated with every c is a Lagrange multiplier vector that is determined after the unconstrained minimization is performed.

If a solution $\boldsymbol{x}^*$ to the original problem eq. (12.4.9) is a regular point of the constraints, then there is a unique Lagrange multiplier vector $\boldsymbol{\lambda}^*$ associated with the solution. The result stated below says that $\boldsymbol{\lambda}_k \to \boldsymbol{\lambda}^*$.

Proposition: Suppose that the penalty function method is applied to problem eq. (12.4.9) using a penalty function of the form eq. (12.4.11) with $\gamma \in C^1$ and satisfying eq. (12.4.14). Corresponding to the sequence $\{\boldsymbol{x}_k\}$ generated by this method, define $\boldsymbol{\lambda}_k^{\mathrm{T}} = c_k \nabla \gamma(\boldsymbol{g}^+(\boldsymbol{x}_k))$. If $\boldsymbol{x}_k \to \boldsymbol{x}^*$, a solution to eq. (12.4.9), and this solution is a regular point, then $\boldsymbol{\lambda}_k \to \boldsymbol{\lambda}^*$, the Lagrange multiplier associated with problem eq. (12.4.9).

The Hessian Matrix

Since the penalty function method must, for various (large) values of c, solve the unconstrained problem

$$\text{minimize} \quad f(\boldsymbol{x}) + c\boldsymbol{P}(\boldsymbol{x}) \tag{12.4.17}$$

it is important, in order to evaluate the difficulty of such a problem, to determine the eigenvalue structure of the Hessian of this modified objective function. We show here that the structure becomes increasingly unfavorable as c increases.

Here we only require the function $\boldsymbol{P} \in C^1$, not $\boldsymbol{P} \in C^2$. In particular, the most popular penalty function $\boldsymbol{P}(\boldsymbol{x}) = \frac{1}{2}|\boldsymbol{g}^+(\boldsymbol{x})|^2$ is a discontinuity in its second derivative at any point where a component of $\boldsymbol{g}$ is zero. This is clearly a serious drawback, since it means the Hessian is discontinuous at the boundary of the constraint region—right where, in general, the solution is expected to lie. However, the above penalty method generates points that approach a boundary solution from outside the constraint region. the sequence will, as $\boldsymbol{x}_k \to \boldsymbol{x}^*$, be at points where the Hessian is well-defined. Hence the standard type of analysis will be applicable to the tail of such a sequence.

Defining $q(c, \boldsymbol{x}) = f(\boldsymbol{x}) + c\gamma(\boldsymbol{g}^+(\boldsymbol{x}))$, we have for the Hessian, $\boldsymbol{Q}$, of q (with respect to $\boldsymbol{x}$)

$$\boldsymbol{Q}(c, \boldsymbol{x}) = \boldsymbol{F}(\boldsymbol{x}) + c\,\nabla\gamma(\boldsymbol{g}^+(\boldsymbol{x}))\boldsymbol{G}(\boldsymbol{x}) + c\,\nabla\boldsymbol{g}^+(\boldsymbol{x})^{\mathrm{T}}\boldsymbol{\Gamma}(\boldsymbol{g}^+(\boldsymbol{x}))\nabla\boldsymbol{g}^+(\boldsymbol{x})$$

where $\boldsymbol{F}$, $\boldsymbol{G}$, and $\boldsymbol{\Gamma}$ are, respectively, the Hessians of f, $\boldsymbol{g}$, and γ. For a fixed c_k, we use the definition of $\boldsymbol{\lambda}_k$ given by eq. (12.4.16) and introduce the following definition

$$\boldsymbol{L}_k(\boldsymbol{x}_k) = \boldsymbol{F}(\boldsymbol{x}_k) + \boldsymbol{\lambda}_k^{\mathrm{T}}\boldsymbol{G}(\boldsymbol{x}_k) \tag{12.4.18}$$

which is the Hessian of the corresponding Lagrangian. Then we have

$$\boldsymbol{Q}(c_k, \boldsymbol{x}_k) = \boldsymbol{L}_k(\boldsymbol{x}_k) + c_k\,\nabla\boldsymbol{g}^+(\boldsymbol{x}_k)^{\mathrm{T}}\boldsymbol{\Gamma}(\boldsymbol{g}^+(\boldsymbol{x}_k))\nabla\boldsymbol{g}^+(\boldsymbol{x}_k) \tag{12.4.19}$$

which is the desired expression.

The first term on the right side of eq. (12.4.19) converges to the Hessian of the Lagrangian of the original constrained problem as $\boldsymbol{x}_k \to \boldsymbol{x}^*$, and hence has a limit that is independent of c_k. The second term is a matrix having rank equal to the rank of the active constraints and having a magnitude tending to infinity.

Lemma 1: Let $\boldsymbol{A}(c)$ be a symmetric matrix written in partitioned form

$$\boldsymbol{A}(c) = \begin{bmatrix} \boldsymbol{A}_1(c) & \boldsymbol{A}_2(c) \\ \boldsymbol{A}_2^{\mathrm{T}}(c) & \boldsymbol{A}_3(c) \end{bmatrix} \tag{12.4.20}$$

where $\boldsymbol{A}_1(c)$ tends to a positive definite matrix $\boldsymbol{A}_1$, $\boldsymbol{A}_2(c)$ tends to a finite matrix, and $\boldsymbol{A}_3(c)$ is a positive definite matrix tending to infinity with c (that is, for any $s>0$, $\boldsymbol{A}_3(c)\to s\boldsymbol{I}$ is positive definite for sufficiently large c), then

$$\boldsymbol{A}^{-1}(c) \to \begin{bmatrix} \boldsymbol{A}_1^{-1} & \boldsymbol{0} \\ \boldsymbol{0} & \boldsymbol{0} \end{bmatrix} \quad \text{as} \quad c \to \infty$$

Barrier Functions

Essentially the same story holds for barrier function. If we consider for Problem eq. (12.4.9) barrier functions of the form

$$B(\boldsymbol{x}) = \eta(\boldsymbol{g}(\boldsymbol{x})) \tag{12.4.21}$$

then Lagrange multipliers and ill-conditioned Hessians are again inevitable. Rather than parallel the earlier analysis of penalty functions, we illustrate the conclusions with two examples.

The Central Path

The definition of the central path associated with linear programs is easily extended to general nonlinear programs. For example, consider the problem

$$\begin{aligned} &\text{minimize} \quad f(\boldsymbol{x}) \\ &\text{subject to} \quad \boldsymbol{h}(\boldsymbol{x}) = \boldsymbol{0} \\ &\boldsymbol{g}(\boldsymbol{x}) \leqslant \boldsymbol{0} \end{aligned}$$

Assume that $\mathring{F} = \{\boldsymbol{x}: \boldsymbol{h}(\boldsymbol{x}) = \boldsymbol{0}, \boldsymbol{g}(\boldsymbol{x}) < \boldsymbol{0}\} \neq \phi$, then we use the logarithmic barrier function to define the problems

$$\begin{aligned} &\text{minimize} \quad f(\boldsymbol{x}) - \mu \sum_{i=1}^{p} \log[-g_i(\boldsymbol{x})] \\ &\text{subject to} \quad \boldsymbol{h}(\boldsymbol{x}) = \boldsymbol{0} \end{aligned}$$

The solution $\boldsymbol{x}_\mu$ parameterized by $\mu \to 0$ is the central path.

The necessary conditions for the problem can be written as

$$\nabla f(\boldsymbol{x}_\mu) + \boldsymbol{\lambda}^{\mathrm{T}} \nabla \boldsymbol{g}(\boldsymbol{x}_\mu) + \boldsymbol{y}^{\mathrm{T}} \boldsymbol{h}(\boldsymbol{x}_\mu) = \boldsymbol{0}$$
$$\boldsymbol{h}(\boldsymbol{x}_\mu) = \boldsymbol{0}$$
$$\lambda_i g_i(\boldsymbol{x}_\mu) = -\mu; \quad i = 1, 2, \ldots, p$$

where $\boldsymbol{y}$ is the Lagrange multiplier vector for the constraint $\boldsymbol{h}(\boldsymbol{x}_\mu) = \boldsymbol{0}$.

Geometric Interpretation—The Primal Function

Let us again consider the problem

$$\begin{aligned} &\text{minimize} \quad f(\boldsymbol{x}) \\ &\text{subject to} \quad \boldsymbol{h}(\boldsymbol{x}) = \boldsymbol{0} \end{aligned} \tag{12.4.22}$$

where $\boldsymbol{h}(\boldsymbol{x}) \in E^m$. We assume that the solution point $\boldsymbol{x}^*$ of eq. (12.4.22) is a regular point and that the second-order sufficiency conditions are satisfied. Corresponding to this problem we introduce the following definition:

Definition: Corresponding to the constrained minimization problem eq. (12.4.22), the primal function ω is defined on E^m in a neighborhood of $\boldsymbol{0}$ to be

$$\omega(\boldsymbol{y}) = \min\{f(\boldsymbol{x}) : \boldsymbol{h}(\boldsymbol{x}) = \boldsymbol{y}\} \tag{12.4.23}$$

The primal function gives the optimal value of the objective for various values of the right-hand side. In particular $\omega(\boldsymbol{0})$ gives the value of the original problem. Strictly speaking, the minimum in the definition eq. (12.4.23) must be specified as a local minimum, in a neighborhood of $\boldsymbol{x}^*$. The existence of $\omega(\boldsymbol{y})$ then follows directly from the Sensitivity Theorem. Furthermore, from that theorem it follows that $\nabla \omega(\boldsymbol{0}) = -\boldsymbol{\lambda}^{*\mathrm{T}}$.

Now consider the penalty problem and note the following relations:

$$\begin{aligned} \min\{f(\boldsymbol{x}) + \frac{1}{2}c|\boldsymbol{h}(\boldsymbol{x})|^2\} &= \min_{x,y}\{f(\boldsymbol{x}) + \frac{1}{2}c|\boldsymbol{y}|^2 : \boldsymbol{h}(\boldsymbol{x}) = \boldsymbol{y}\} \\ &= \min_y\{\omega(\boldsymbol{y}) + \frac{1}{2}c|\boldsymbol{y}|^2\} \end{aligned} \tag{12.4.24}$$

12.5 Dual and Cutting Plane Methods

12.5.1 Global Duality

Here we shall first consider a problem with inequality constraints. In particular, consider the problem

$$\begin{aligned} &\text{minimize} \quad f(\boldsymbol{x}) \\ &\text{subject to} \quad \boldsymbol{g}(\boldsymbol{x}) \leqslant \boldsymbol{0} \\ &\qquad\qquad\quad \boldsymbol{x} \in \Omega \end{aligned} \tag{12.5.1}$$

where $\Omega \subset E^n$ is a convex set, and the functions f and $\boldsymbol{g}$ are defined on Ω. The function $\boldsymbol{g}$ is p-dimensional. The problem is not necessarily convex, but we assume that there is a feasible point. Recall that the primal function associated with eq. (12.5.1) is defined for $z \in E^p$ as

$$\omega(z) = \inf\{f(\boldsymbol{x}) : \boldsymbol{g}(\boldsymbol{x}) \leqslant z, \boldsymbol{x} \in \Omega\} \tag{12.5.2}$$

defined by letting the right hand side of inequality constraint take on arbitrary values. It is understood that eq. (12.5.2) is defined on the set $D = \{z : \boldsymbol{g}(\boldsymbol{x}) \leqslant z,$ for some $\boldsymbol{x} \in \Omega\}$.

If problem eq. (12.5.1) has a solution $\boldsymbol{x}^*$ with value $f^* = f(\boldsymbol{x}^*)$, then f^* is the point on the vertical axis in E^{p+1} where the primal function passes through the axis. If eq. (12.5.1) does not have a solution, then $f^* = \inf\{f(\boldsymbol{x}) : \boldsymbol{g}(\boldsymbol{x}) \leqslant \boldsymbol{0}, \boldsymbol{x} \in \Omega\}$ is the intersection point.

In general, we define the *dual function* on the positive cone in E^p as

$$\phi(\boldsymbol{\mu}) = \inf\{f(\boldsymbol{x}) + \boldsymbol{\mu}^{\mathrm{T}} \boldsymbol{g}(\boldsymbol{x}) : \boldsymbol{x} \in \Omega\} \tag{12.5.3}$$

Clearly, ϕ may not be finite throughout the positive orthant E_+^p but the region where it is finite is convex.

Proposition 1: The dual function is concave on the region where it is finite.

Weak duality proposition: $\phi^* \leqslant f^*$.

We shall state the result for the more general problem that includes equality constraints of the form $\boldsymbol{h}(\boldsymbol{x}) = \boldsymbol{0}$ as follows

$$\begin{aligned} &\text{maximize} \quad f(\boldsymbol{x}) \\ &\text{subject to} \quad \boldsymbol{h}(\boldsymbol{x}) = \boldsymbol{0}, \quad \boldsymbol{g}(\boldsymbol{x}) \leqslant \boldsymbol{0} \\ &\boldsymbol{x} \in \Omega \end{aligned} \tag{12.5.4}$$

where $\boldsymbol{h}$ is affine of dimension m, $\boldsymbol{g}$ is convex of dimension p, and Ω is a convex set.

In this case the dual is $\phi(\boldsymbol{\lambda}, \boldsymbol{\mu}) = \inf\{f(\boldsymbol{x}) + \boldsymbol{\lambda}^{\mathrm{T}} \boldsymbol{h}(\boldsymbol{x}) + \boldsymbol{\mu}^{\mathrm{T}} \boldsymbol{g}(\boldsymbol{x}) : \boldsymbol{x} \in \Omega\}$, and $\phi^* = \sup\{\phi(\boldsymbol{\lambda}, \boldsymbol{\mu}) : \boldsymbol{\lambda} \in E^m, \boldsymbol{\mu} \in E^p, \boldsymbol{\mu} \geqslant \boldsymbol{0}\}$.

Strong duality theorem: Suppose in the problem eq. (12.5.4), $\boldsymbol{h}$ is regular with respect to Ω and there is a point $\boldsymbol{x}_1 \in \Omega$ with that $\boldsymbol{h}(\boldsymbol{x}) = \boldsymbol{0}$ and $\boldsymbol{g}(\boldsymbol{x}) < \boldsymbol{0}$, the problem has solution $\boldsymbol{x}^*$ with value $f(\boldsymbol{x}^*) = f^*$, then for every $\boldsymbol{\lambda}$ and $\boldsymbol{\mu} \geqslant \boldsymbol{0}$ there holds $\phi^* \leqslant f^*$. Furthermore, there are $\boldsymbol{\lambda}, \boldsymbol{\mu} \geqslant \boldsymbol{0}$ such that $\phi(\boldsymbol{\lambda}, \boldsymbol{\mu}) = f^*$

and hence $\phi^* = f^*$. Moreover, the $\boldsymbol{\lambda}$ and $\boldsymbol{\mu}$ above are Lagrange multipliers for the problem.

12.5.2 Local Duality

Here we again consider nonlinear programming problems of the form

$$\begin{aligned} &\text{minimize} \quad f(\boldsymbol{x}) \\ &\text{subject to} \quad \boldsymbol{h}(\boldsymbol{x}) = \boldsymbol{0} \end{aligned} \tag{12.5.5}$$

where $\boldsymbol{x} \in E^n$, $\boldsymbol{h}(\boldsymbol{x}) \in E^n$ and $f, \boldsymbol{h} \in C^2$, and global convexity is not necessary here. We focus attention on a local solution $\boldsymbol{x}^*$ of eq. (12.5.5). Assuming that $\boldsymbol{x}^*$ is a regular point of the constraints, then, as we know, there will be a corresponding Lagrange multiplier (row) vector $\boldsymbol{\lambda}^*$ such that

$$\nabla f(\boldsymbol{x}^*) + (\boldsymbol{\lambda}^*)^{\mathrm{T}} \nabla \boldsymbol{h}(\boldsymbol{x}^*) = \boldsymbol{0} \tag{12.5.6}$$

and the Hessian of the Lagrangian

$$\boldsymbol{L}(\boldsymbol{x}^*) = \boldsymbol{F}(\boldsymbol{x}^*) + (\boldsymbol{\lambda}^*)^{\mathrm{T}} \boldsymbol{H}(\boldsymbol{x}^*) \tag{12.5.7}$$

must be positive semidefinite on the tangent subspace

$$M = \{\boldsymbol{x} : \nabla \boldsymbol{h}(\boldsymbol{x}^*)\boldsymbol{x} = \boldsymbol{0}\}$$

At this point we introduce the special local convexity assumption necessary for the development of the local duality theory. Specifically, we assume that the Hessian $\boldsymbol{L}(\boldsymbol{x}^*)$ is positive definite. Of course, it should be emphasized that by this we mean $\boldsymbol{L}(\boldsymbol{x}^*)$ is positive definite on the whole space E^n, not just on the subspace M. The assumption guarantees that the Lagrangian $l(\boldsymbol{x}) = f(\boldsymbol{x}) + (\boldsymbol{\lambda}^*)^{\mathrm{T}} \boldsymbol{h}(\boldsymbol{x})$ is locally convex at $\boldsymbol{x}^*$.

With this assumption, the point $\boldsymbol{x}^*$ is not only a local solution to the constrained problem eq. (12.5.5); it is also a local solution to the unconstrained problem

$$\text{minimize} \quad f(\boldsymbol{x}) + (\boldsymbol{\lambda}^*)^{\mathrm{T}} \boldsymbol{h}(\boldsymbol{x}) \tag{12.5.8}$$

since it satisfies the first- and second-order sufficiency conditions for a local minimum point. Furthermore, for any $\boldsymbol{\lambda}$ sufficiently close to $\boldsymbol{\lambda}^*$ the function $f(\boldsymbol{x}) + \boldsymbol{\lambda}^{\mathrm{T}} \boldsymbol{h}(\boldsymbol{x})$ will have a local minimum point at a point $\boldsymbol{x}$ near $\boldsymbol{x}^*$. This follows by noting that, by the Implicit Function Theorem, the equation

$$\nabla f(\boldsymbol{x}) + \boldsymbol{\lambda}^{\mathrm{T}} \nabla \boldsymbol{h}(\boldsymbol{x}) = \boldsymbol{0} \tag{12.5.9}$$

has a solution $\boldsymbol{x}$ near $\boldsymbol{x}^*$ when $\boldsymbol{\lambda}$ is near $\boldsymbol{\lambda}^*$, because $\boldsymbol{L}^*$ is non-singular; and by the fact that, at this solution $\boldsymbol{x}$, the Hessian $\boldsymbol{F}(\boldsymbol{x}) + \boldsymbol{\lambda}^{\mathrm{T}} \boldsymbol{H}(\boldsymbol{x})$ is positive definite. Thus locally there is a unique correspondence between $\boldsymbol{\lambda}$ and $\boldsymbol{x}$ through the solution of the unconstrained problem

$$\text{minimize} \quad f(\boldsymbol{x}) + \boldsymbol{\lambda}^{\mathrm{T}}\boldsymbol{h}(\boldsymbol{x}) \tag{12.5.10}$$

Furthermore, this correspondence is continuously differentiable.

Near $\boldsymbol{\lambda}^*$ we define the *dual function* ϕ by the equation

$$\phi(\boldsymbol{\lambda}) = \text{minimum}[f(\boldsymbol{x}) + \boldsymbol{\lambda}^{\mathrm{T}}\boldsymbol{h}(\boldsymbol{x})] \tag{12.5.11}$$

where it is understood that the minimum is taken locally with respect to $\boldsymbol{x}$ near $\boldsymbol{x}^*$. We are then able to show (and will do so below) that locally the original constrained problem eq. (12.5.5) is equivalent to unconstrained local maximization of the dual function ϕ with respect to $\boldsymbol{\lambda}$. Hence we establish an equivalence between a constrained problem in $\boldsymbol{x}$ and an unconstrained problem in $\boldsymbol{\lambda}$.

To establish the duality relation we must prove two important lemmas. In the statements below we denote by $\boldsymbol{x}(\boldsymbol{\lambda})$ the unique solution to eq. (12.5.10) in the neighborhood of $\boldsymbol{x}^*$.

Lemma 1: The dual function ϕ has gradient

$$\nabla\phi(\boldsymbol{\lambda}) = \boldsymbol{h}(\boldsymbol{x}(\boldsymbol{\lambda}))^{\mathrm{T}} \tag{12.5.12}$$

Lemma 2: The Hessian of the dual function is

$$\phi(\boldsymbol{\lambda}) = -\nabla\boldsymbol{h}(\boldsymbol{x}(\boldsymbol{\lambda}))\boldsymbol{L}^{-1}(\boldsymbol{x}(\boldsymbol{\lambda}),\boldsymbol{\lambda})\nabla\boldsymbol{h}(\boldsymbol{x}(\boldsymbol{\lambda}))^{\mathrm{T}} \tag{12.5.13}$$

Local Duality Theorem: Suppose that the problem

$$\begin{aligned} &\text{minimize} \quad f(\boldsymbol{x}) \\ &\text{subject to} \quad \boldsymbol{h}(\boldsymbol{x}) = \boldsymbol{0} \end{aligned} \tag{12.5.14}$$

has a local solution at $\boldsymbol{x}^*$ with corresponding value r^*, Lagrange multiplier $\boldsymbol{\lambda}^*$, that $\boldsymbol{x}^*$ is a regular point of the constraints and that the corresponding Hessian of the Lagrangian $\boldsymbol{L}^* = \boldsymbol{L}(\boldsymbol{x}^*)$ is positive definite, then the dual problem

$$\text{maximize} \quad \phi(\boldsymbol{\lambda}) \tag{12.5.15}$$

has a local solution at $\boldsymbol{\lambda}^*$ with corresponding value r^* and $\boldsymbol{x}^*$ as the point corresponding to $\boldsymbol{\lambda}^*$ in the definition of ϕ.

Inequality Constraints

For problems simultaneously having inequality constraints, equality constraints, we only need to have minor modifications. Consider the problem

$$\begin{aligned} &\text{minimize} \quad f(\boldsymbol{x}) \\ &\text{subject to} \quad \boldsymbol{h}(\boldsymbol{x}) = \boldsymbol{0} \\ &\qquad \boldsymbol{g}(\boldsymbol{x}) \leqslant \boldsymbol{0} \end{aligned} \tag{12.5.16}$$

where $\boldsymbol{g}(\boldsymbol{x}) \in E^p$, $\boldsymbol{g} \in C^2$ and others defined as above. Suppose $\boldsymbol{x}^*$ is a local solution of eq. (12.5.16) and is a regular point of the constraints. Then, as we know, there are Lagrange multipliers $\boldsymbol{\lambda}^*$ and $\boldsymbol{\mu}^* \geqslant \mathbf{0}$ such that

$$\nabla f(\boldsymbol{x}^*) + (\boldsymbol{\lambda}^*)^{\mathrm{T}} \nabla \boldsymbol{h}(\boldsymbol{x}^*) + (\boldsymbol{\mu}^*)^{\mathrm{T}} \nabla \boldsymbol{g}(\boldsymbol{x}^*) = \mathbf{0} \tag{12.5.17}$$

$$(\boldsymbol{\mu}^*)^{\mathrm{T}} \boldsymbol{g}(\boldsymbol{x}^*) = \mathbf{0} \tag{12.5.18}$$

We impose the local convexity assumptions that the Hessian of the Lagrangian

$$\boldsymbol{L}(\boldsymbol{x}^*) = \boldsymbol{F}(\boldsymbol{x}^*) + (\boldsymbol{\lambda}^*)^{\mathrm{T}} \boldsymbol{H}(\boldsymbol{x}^*) + (\boldsymbol{\mu}^*)^{\mathrm{T}} \boldsymbol{G}(\boldsymbol{x}^*) \tag{12.5.19}$$

is positive definite (on the whole space).

For $\boldsymbol{\lambda}$ and $\boldsymbol{\mu} \geqslant \mathbf{0}$ near $\boldsymbol{\lambda}^*$ and $\boldsymbol{\mu}^*$ we define the dual function

$$\phi(\boldsymbol{\lambda}, \boldsymbol{\mu}) = \min[f(\boldsymbol{x}) + \boldsymbol{\lambda}^{\mathrm{T}} \boldsymbol{h}(\boldsymbol{x}) + \boldsymbol{\mu}^{\mathrm{T}} \boldsymbol{g}(\boldsymbol{x})] \tag{12.5.20}$$

where the minimum is taken locally near $\boldsymbol{x}^*$. Then, it is easy to show, paralleling the development above for equality constraints, that ϕ achieves a local maximum with respect to $\boldsymbol{\lambda}, \boldsymbol{\mu} \geqslant \mathbf{0}$ at $\boldsymbol{\lambda}^*, \boldsymbol{\mu}^*$.

Partial Duality

It is not necessary to include the Lagrange multipliers of all the constraints of a problem in the definition of the dual function. In general, if the local convexity assumption holds, local duality can be defined with respect to any subset of functional constraints. Thus, for example, in the problem

$$\begin{aligned} \text{minimize} \quad & f(\boldsymbol{x}) \\ \text{subject to} \quad & \boldsymbol{h}(\boldsymbol{x}) = \mathbf{0} \\ & \boldsymbol{g}(\boldsymbol{x}) \leqslant \mathbf{0} \end{aligned} \tag{12.5.21}$$

we might define the dual function with respect to only the equality constraints. In this case we would define

$$\phi(\boldsymbol{\lambda}) = \min_{\boldsymbol{g}(\boldsymbol{x}) \leqslant \mathbf{0}} \{f(\boldsymbol{x}) + \boldsymbol{\lambda}^{\mathrm{T}} \boldsymbol{h}(\boldsymbol{x})\} \tag{12.5.22}$$

where the minimum is taken locally near the solution $\boldsymbol{x}^*$ but constrained by the remaining constraints $\boldsymbol{g}(\boldsymbol{x}) \leqslant \mathbf{0}$. Again, the dual function defined in this way will achieve a local maximum at the optimal Lagrange multiplier $\boldsymbol{\lambda}^*$.

12.5.3 Dual Canonical Convergence Rate

The method of steepest ascent, and other gradient-based algorithms, when applied to the dual problem will have a convergence rate governed by the eigenvalue structure of the Hessian of the dual function ϕ. At the Lagrange multiplier $\boldsymbol{\lambda}^*$ corresponding to a solution $\boldsymbol{x}^*$, this Hessian is $\boldsymbol{\Phi} = -\nabla \boldsymbol{h}(\boldsymbol{x}^*)(\boldsymbol{L}^*)^{-1} \nabla \boldsymbol{h}(\boldsymbol{x}^*)^{\mathrm{T}}$.

This expression shows that $\boldsymbol{\Phi}$ is in some sense a restriction of the matrix $(\boldsymbol{L}^*)^{-1}$ to the subspace spanned by the gradients of the constraint functions, which is the orthogonal complement of the tangent subspace M. This restriction is not the orthogonal restriction of $(\boldsymbol{L}^*)^{-1}$ onto the complement of M since the particular representation of the constraints affects the structure of the Hessian. We see, however, that while the convergence of primal methods is governed by the restriction of $\boldsymbol{L}^*$ to M, the convergence of dual methods is governed by a restriction of $(\boldsymbol{L}^*)^{-1}$ to the orthogonal complement of M.

The *dual canonical convergence rate* associated with the original constrained problem, which is the rate of convergence of steepest ascent applied to the dual, is $(B-b)^2/(B+b)^2$ where b and B are, respectively, the smallest and largest eigenvalues of

$$-\boldsymbol{\Phi} = \nabla \boldsymbol{h}(\boldsymbol{x}^*)(\boldsymbol{L}^*)^{-1}\nabla \boldsymbol{h}(\boldsymbol{x}^*)^{\mathrm{T}}$$

For locally convex programming problems, this rate is as important as the primal canonical rate.

12.5.4 Separable Problems

A structure that arises frequently in mathematical programming applications is that of the separable problem:

$$\text{minimize} \quad \sum_{i=1}^{q} f_i(x_i) \tag{12.5.23}$$

$$\text{subject to} \quad \sum_{i=1}^{q} h_i(x_i) = 0 \tag{12.5.24}$$

$$\sum_{i=1}^{q} g_i(x_i) \leqslant 0 \tag{12.5.25}$$

In this formulation the components of the n-vector $\boldsymbol{x}$ are partitioned into q disjoint groups, $\boldsymbol{x} = (x_1, x_2, \ldots, x_q)$ where the groups may or may not have the same number of components. Both the objective function and the constraints separate into sums of functions of the individual groups. For each i, the functions, f_i, h_i, and g_i are twice continuously differentiable functions of dimensions $1, m$, and p, respectively.

12.5.5 Decomposition

Separable problems are ideally suited to dual methods, because the required unconstrained minimization decomposes into small subproblems. To see this we recall that the generally most difficult aspect of a dual method is evaluation of the dual function. For a separable problem, if we associate $\boldsymbol{\lambda}$ with the equality constraints eq. (12.5.24) and $\boldsymbol{\mu} \geqslant \mathbf{0}$ with the inequality constraints eq. (12.5.25), the required dual function is

$$\phi(\boldsymbol{\lambda},\boldsymbol{\mu}) = \min \sum_{i=1}^{q} f_i(x_i) + \boldsymbol{\lambda}^{\mathrm{T}} h_i(x_i) + \boldsymbol{\mu}^{\mathrm{T}} g_i(x_i)$$

This minimization problem decomposes into the q separate problems

$$\min_{x_i} f_i(x_i) + \boldsymbol{\lambda}^{\mathrm{T}} h_i(x_i) + \boldsymbol{\mu}^{\mathrm{T}} g_i(x_i)$$

The solution of these subproblems can usually be accomplished relatively efficiently, since they are of smaller dimension than the original problem.

12.5.6 The Dual Viewpoint

As we observed earlier, the constrained problem

$$\begin{aligned} &\text{minimize} \quad f(\boldsymbol{x}) \\ &\text{subject to} \quad \boldsymbol{h}(\boldsymbol{x}) = \mathbf{0} \end{aligned} \tag{12.5.26}$$

is equivalent to the problem

$$\begin{aligned} &\text{minimize} \quad f(\boldsymbol{x}) + \frac{1}{2} c |\boldsymbol{h}(\boldsymbol{x})|^2 \\ &\text{subject to} \quad \boldsymbol{h}(\boldsymbol{x}) = \mathbf{0} \end{aligned} \tag{12.5.27}$$

in the sense that the solution points, the optimal values, and the Lagrange multipliers are the same for both problems. However, as spelled out by Proposition 1 of the previous section, whereas problem eq. (12.5.26) may not be locally convex, problem eq. (12.5.27) is locally convex for sufficiently large c; specifically, the Hessian of the Lagrangian is positive definite at the solution pair $\boldsymbol{x}^*, \boldsymbol{\lambda}^*$. Thus local duality theory is applicable to problem eq. (12.5.27) for sufficiently large c.

To apply the dual method to eq. (12.5.27), we define the dual function

$$\phi(\boldsymbol{\lambda}) = \min \{ f(\boldsymbol{x}) + \boldsymbol{\lambda}^{\mathrm{T}} \boldsymbol{h}(\boldsymbol{x}) + \frac{1}{2} c |\boldsymbol{h}(\boldsymbol{x})|^2 \} \tag{12.5.28}$$

in a region near $\boldsymbol{x}^*, \boldsymbol{\lambda}^*$. If $\boldsymbol{x}(\boldsymbol{\lambda})$ is the vector minimizing the right-hand side of eq. (12.5.27), $\boldsymbol{h}(\boldsymbol{x}(\boldsymbol{\lambda}))$ is the gradient of ϕ. Thus the iterative process

$$\lambda_{k+1} = \lambda_k + c h_{\boldsymbol{k}}(\boldsymbol{x}(\lambda))$$

used in the basic augmented Lagrangian method is seen to be a steepest ascent iteration for maximizing the dual function ϕ. It is a simple form of steepest ascent, using a constant stepsize c.

The rate of convergence of the optimal steepest ascent method (where the steplength is selected to maximize ϕ in the gradient direction) is determined by the eigenvalues of the Hessian of ϕ. The Hessian of ϕ is found from eq. (12.5.13) to be

$$\nabla \boldsymbol{h}(\boldsymbol{x}(\boldsymbol{\lambda}))[\boldsymbol{L}(\boldsymbol{x}(\boldsymbol{\lambda}),\boldsymbol{\lambda})+c\,\nabla \boldsymbol{h}(\boldsymbol{x}(\boldsymbol{\lambda}))^{\mathrm{T}}\nabla \boldsymbol{h}(\boldsymbol{x}(\boldsymbol{\lambda}))]^{-1}\nabla \boldsymbol{h}(\boldsymbol{x})^{\mathrm{T}} \tag{12.5.29}$$

The eigenvalues of this matrix at the solution point $\boldsymbol{x}^*, \boldsymbol{\lambda}^*$ determine the convergence rate of the method of steepest ascent.

To analyze these eigenvalues, we make use of the matrix identity

$$c\boldsymbol{B}(\boldsymbol{A}+c\boldsymbol{B}^{\mathrm{T}}\boldsymbol{B})^{-1}\boldsymbol{B}^{\mathrm{T}}=\boldsymbol{I}-(\boldsymbol{I}+c\boldsymbol{B}\boldsymbol{A}^{-1}\boldsymbol{B}^{\mathrm{T}})^{-1}$$

which is a generalization of the Sherman-Morrison formula. It is easily seen from the above identity that the matrices $\boldsymbol{B}(\boldsymbol{A}+c\boldsymbol{B}^{\mathrm{T}}\boldsymbol{B})^{-1}\boldsymbol{B}^{\mathrm{T}}$ and $(\boldsymbol{B}\boldsymbol{A}^{-1}\boldsymbol{B}^{\mathrm{T}})$ have identical eigenvectors. One way to see this is to multiply both sides of the identity by $(\boldsymbol{I}+c\boldsymbol{B}\boldsymbol{A}^{-1}\boldsymbol{B}^{\mathrm{T}})$ on the right to obtain

$$c\boldsymbol{B}(\boldsymbol{A}+c\boldsymbol{B}^{\mathrm{T}}\boldsymbol{B})^{-1}\boldsymbol{B}^{\mathrm{T}}(\boldsymbol{I}+c\boldsymbol{B}\boldsymbol{A}^{-1}\boldsymbol{B}^{\mathrm{T}})=c\boldsymbol{B}\boldsymbol{A}^{-1}\boldsymbol{B}^{\mathrm{T}}$$

Suppose both sides are applied to an eigenvector $\boldsymbol{e}$ of $\boldsymbol{B}\boldsymbol{A}^{-1}\boldsymbol{B}^{\mathrm{T}}$ having eigenvalue w. Then we obtain

$$c\boldsymbol{B}(\boldsymbol{A}+c\boldsymbol{B}^{\mathrm{T}}\boldsymbol{B})^{-1}\boldsymbol{B}^{\mathrm{T}}(1+cw)\boldsymbol{e}=cw\boldsymbol{e}$$

It follows that $\boldsymbol{e}$ is also an eigenvector of $\boldsymbol{B}(\boldsymbol{A}+c\boldsymbol{B}^{\mathrm{T}}\boldsymbol{B})^{-1}\boldsymbol{B}^{\mathrm{T}}$, and if v is the corresponding eigenvalue, the relation

$$cv(1+cw)=cw$$

must hold. Therefore, the eigenvalues are related by

$$v=\frac{w}{1+cw} \tag{12.5.30}$$

The above relations apply directly to the Hessian eq. (12.5.29) through the associations $\boldsymbol{A}=\boldsymbol{L}(\boldsymbol{x}^*,\boldsymbol{\lambda}^*)$ and $\boldsymbol{B}=\nabla \boldsymbol{h}(\boldsymbol{x}^*)$. Note that the matrix $\nabla \boldsymbol{h}(\boldsymbol{x}^*)\boldsymbol{L}(\boldsymbol{x}^*,\boldsymbol{\lambda}^*)^{-1}\nabla \boldsymbol{h}(\boldsymbol{x}^*)^{\mathrm{T}}$, corresponding to $\boldsymbol{B}\boldsymbol{A}^{-1}\boldsymbol{B}^{\mathrm{T}}$ above, is the Hessian of the dual function of the original problem eq. (12.5.26), whose eigenvalues determine the rate of convergence for the ordinary dual method. Let w and W be the smallest and largest eigenvalues of this matrix. From eq. (12.5.30) it follows that the ratio of the smallest to largest eigenvalues of the Hessian of the dual for the augmented problem is

$$\frac{\frac{1}{W}+c}{\frac{1}{w}+c}$$

This shows explicitly how the rate of convergence of the multiplier method depends on c. As c goes to infinity, the ratio of eigenvalues goes to unity, implying arbitrarily fast convergence.

Other unconstrained optimization techniques may be applied to the maximization of the dual function defined by the augmented Lagrangian; conjugate gradient methods, Newton's method, and quasi-Newton methods can all be used. The use of Newton's method requires evaluation of the Hessian matrix eq. (12.5.29). For some problems this may be feasible, but for others some sort of approximation is desirable. One approximation is obtained by noting that for large values of c, the Hessian eq. (12.5.29) is approximately equal to $(1/c)\boldsymbol{I}$. Using this value for the Hessian and $\boldsymbol{h}(\boldsymbol{x}(\boldsymbol{\lambda}))$ for the gradient, we are led to the iterative scheme

$$\lambda_{k+1}=\lambda_k+c\boldsymbol{h}(\boldsymbol{x}(\lambda_k))$$

which is exactly the simple method of multipliers originally proposed.

12.5.7 Cutting Plane Methods

Cutting plane methods are applied to problems having the general form

$$\begin{aligned}&\text{minimize} \quad \boldsymbol{c}^{\mathrm{T}}\boldsymbol{x}\\&\text{subject to} \quad \boldsymbol{x}\in S\end{aligned} \tag{12.5.31}$$

where $S\subset E^n$ is a closed convex set. Problems that involve minimization of a convex function over a convex set, such as the problem

$$\begin{aligned}&\text{minimize} \quad f(\boldsymbol{y})\\&\text{subject to} \quad \boldsymbol{y}\in R\end{aligned} \tag{12.5.32}$$

where $R\subset E^{n-1}$ is a convex set and f is a convex function, can be easily converted to the form eq. (12.5.31) by writing eq. (12.5.32) equivalently as

$$\begin{aligned}&\text{minimize} \quad r\\&\text{subject to} \quad f(\boldsymbol{y})-r\leqslant 0\\&\boldsymbol{y}\in R\end{aligned} \tag{12.5.33}$$

which, with $\boldsymbol{x}=(r,\boldsymbol{y})$, is a special case of eq. (12.5.31).

General Form of Algorithm

The general form of a cutting-plane algorithm for problem eq. (12.5.31) is as follows:

Given a polytope $P_k \supset S$,

Step 1. Minimize $\boldsymbol{c}^{\mathrm{T}}\boldsymbol{x}$ over P_k obtaining a point $\boldsymbol{x}_k$ in P_k. If $\boldsymbol{x}_k \in S$, stop; $\boldsymbol{x}_k$ is optimal. Otherwise,

Step 2. Find a hyperplane $\boldsymbol{H}_k$ separating the point $\boldsymbol{x}_k$ from S, that is, find $\boldsymbol{a}_k \in E^n, b_k \in E^1$ such that $S \subset \{\boldsymbol{x}: \boldsymbol{a}_k^{\mathrm{T}}\boldsymbol{x} \leqslant b_k\}, \boldsymbol{x}_k \in \{\boldsymbol{x}: \boldsymbol{a}_k^{\mathrm{T}}\boldsymbol{x} > b_k\}$. Update P_k to obtain P_{k+1} including a constraint $\boldsymbol{a}_k^{\mathrm{T}}\boldsymbol{x} \leqslant b_k$.

Duality

The general cutting plane algorithm can be regarded as an extended application of duality in linear programming, and although this viewpoint does not particularly aid in the analysis of the method, it reveals the basic interconnection between cutting plane and dual methods. The foundation of this viewpoint is the fact that S can be written as the intersection of all the half-spaces that contain it; thus

$$S = \{\boldsymbol{x}: \boldsymbol{a}_k^{\mathrm{T}}\boldsymbol{x} \leqslant b_i, i \in I\},$$

where I is an (infinite) index set corresponding to all half-spaces containing S. With S viewed in this way problem eq. (12.5.31) can be thought of as an (infinite) linear programming problem.

Corresponding to this linear program there is (at least formally) the dual problem

$$\begin{aligned} \text{maximize} \quad & \sum_{i \in I} \lambda_i b_i \\ \text{subject to} \quad & \sum_{i \in I} \lambda_i \boldsymbol{a}_i = \boldsymbol{c} \\ & \lambda_i \geqslant 0, \quad i \in I \end{aligned} \tag{12.5.34}$$

Selecting a finite subset of I, say $\bar{I}$, and forming

$$P = \{\boldsymbol{x}: \boldsymbol{a}_k^{\mathrm{T}}\boldsymbol{x} \leqslant b_i, i \in \bar{I}\}$$

gives a polytope that contains S. Minimizing $\boldsymbol{c}^{\mathrm{T}}\boldsymbol{x}$ over this polytope yields a point and a corresponding subset of active constraints I_A. The dual problem with the additional restriction $\lambda_i = 0$ for $i \notin I_A$ will then have a feasible solution, but this solution will in general not be optimal. Thus, a solution to a polytope problem corresponds to a feasible but non-optimal solution to the dual. For this reason, the cutting plane method can be regarded as working toward optimality of the (infinite dimensional) dual.

12.5.8 Kelley's Convex Cutting Plane Algorithm

The convex cutting plane method was developed to solve convex programming problems of the form

$$\begin{aligned} &\text{minimize} \quad f(\boldsymbol{x}) \\ &\text{subject to} \quad g_i(\boldsymbol{x}) \leqslant 0, \quad i=1,2,\ldots,p \end{aligned} \tag{12.5.35}$$

where $\boldsymbol{x} \in E^n$ and f and $g_i, i=1,2,\ldots,p$ are differentiable convex functions. As indicated in the last section, it is sufficient to consider the case where the objective function is linear; thus, we consider the problem

$$\begin{aligned} &\text{minimize} \quad \boldsymbol{c}^{\mathrm{T}}\boldsymbol{x} \\ &\text{subject to} \quad \boldsymbol{g}(\boldsymbol{x}) \leqslant \boldsymbol{0} \end{aligned} \tag{12.5.36}$$

where $\boldsymbol{x} \in E^n$ and $\boldsymbol{g}(\boldsymbol{x}) \in E^p$ is convex and differentiable.

For $\boldsymbol{g}$ convex and differentiable, we have the fundamental inequality

$$\boldsymbol{g}(\boldsymbol{x}) \geqslant \boldsymbol{g}(\boldsymbol{w}) + \nabla \boldsymbol{g}(\boldsymbol{w})(\boldsymbol{x}-\boldsymbol{w}) \tag{12.5.37}$$

for any $\boldsymbol{x}$, $\boldsymbol{w}$. We use this equation to determine the separating hyperplane. Specifically, the algorithm is as follows:

Let $S = \{\boldsymbol{x}:\boldsymbol{g}(\boldsymbol{x}) \leqslant \boldsymbol{0}\}$ and let P be an initial polytope containing S and such that $\boldsymbol{c}^{\mathrm{T}}\boldsymbol{x}$ is bounded on P. Then,

Step 1. Minimize $\boldsymbol{c}^{\mathrm{T}}\boldsymbol{x}$ over P obtaining the point $\boldsymbol{x} = \boldsymbol{w}$. If $\boldsymbol{g}(\boldsymbol{w}) \leqslant \boldsymbol{0}$, stop; $\boldsymbol{w}$ is an optimal solution. Otherwise,

Step 2. Let i be an index maximizing $g_i(\boldsymbol{w})$. Clearly $g_i(\boldsymbol{w}) > 0$. Define the new approximating polytope to be the old one intersected with the half-space

$$\{\boldsymbol{x}:g_i(\boldsymbol{w}) + \nabla g_i(\boldsymbol{w})^{\mathrm{T}}(\boldsymbol{x}-\boldsymbol{w}) \leqslant 0\} \tag{12.5.38}$$

Return to Step 1.

The set defined by eq. (12.5.38) is actually a half-space if $\nabla g_i(\boldsymbol{w}) \neq 0$.

Convergence

Theorem: Let the convex functions $g_i, i=1,2,\ldots,p$ be continuously differentiable, and suppose the convex cutting plane algorithm generates the sequence of points $\{w_k\}$. Any limit point of this sequence is a solution to problem eq. (12.5.36).

12.5.9 Modifications

The Supporting Hyperplane Algorithm

The convexity requirements are less severe for this algorithm. It is applicable to problems of the form

$$\begin{aligned}&\text{minimize} \quad \boldsymbol{c}^{\mathrm{T}}\boldsymbol{x}\\&\text{subject to} \quad \boldsymbol{g}(\boldsymbol{x}) \leqslant \boldsymbol{0}\end{aligned}$$

where $\boldsymbol{x} \in E^n$, $\boldsymbol{g}(\boldsymbol{x}) \in E^p$, the $g_i, i = 1, 2, \ldots, p$ are continuously differentiable, and the constraint region S defined by the inequalities is convex. Note that convexity of the functions themselves is not necessary. We also assume the existence of a point interior to the constraint region, that is, we assume the existence of a point $\boldsymbol{y}$ such that $\boldsymbol{g}(\boldsymbol{y}) < \boldsymbol{0}$, and we assume that on the constraint boundary $g_i(\boldsymbol{x}) = 0$ implies $\nabla g_i(\boldsymbol{x}) \neq \boldsymbol{0}$.

The basic algorithm is as follows:

Start with an initial polytope P containing S and such that $\boldsymbol{c}^{\mathrm{T}}\boldsymbol{x}$ is bounded below on S. Then,

Step 1. Determine $\boldsymbol{w} = \boldsymbol{x}$ to minimize $\boldsymbol{c}^{\mathrm{T}}\boldsymbol{x}$ over P. If $\boldsymbol{w} \in S$, stop. Otherwise,

Step 2. Find the point $\boldsymbol{u}$ on the line joining $\boldsymbol{y}$ and $\boldsymbol{w}$ that lies on the boundary of S. Let i be an index for which $g_i(\boldsymbol{u}) = 0$ and define the half-space $H = \{\boldsymbol{x}: \nabla g_i(\boldsymbol{u})^{\mathrm{T}}(\boldsymbol{x} - \boldsymbol{u}) \leqslant 0\}$. Update P by intersecting with H. Return to Step 1.

12.6 Primal-dual Methods

12.6.1 The Standard Problem

Consider again the standard nonlinear program

$$\begin{aligned}&\text{minimize} \quad f(\boldsymbol{x})\\&\text{subject to} \quad \boldsymbol{h}(\boldsymbol{x}) = \boldsymbol{0}\\&\qquad\qquad\quad \boldsymbol{g}(\boldsymbol{x}) \leqslant \boldsymbol{0}\end{aligned} \tag{12.6.1}$$

The first-order necessary conditions for optimality are, as we know,

$$\begin{aligned}\nabla f(\boldsymbol{x}) + \boldsymbol{\lambda}^{\mathrm{T}} \nabla \boldsymbol{h}(\boldsymbol{x}) + \boldsymbol{\mu}^{\mathrm{T}} \nabla \boldsymbol{g}(\boldsymbol{x}) &= \boldsymbol{0}\\ \boldsymbol{h}(\boldsymbol{x}) &= \boldsymbol{0}\\ \boldsymbol{g}(\boldsymbol{x}) &\leqslant \boldsymbol{0}\\ \boldsymbol{\mu}^{\mathrm{T}}\boldsymbol{g}(\boldsymbol{x}) &= 0\end{aligned} \tag{12.6.2}$$

The last requirement is the complementary slackness condition. If it is known which of the inequality constraints is active at the solution, these active constraints can be rolled into the equality constraints $\boldsymbol{h}(\boldsymbol{x}) = \mathbf{0}$ and the inactive inequalities along with the complementary slackness condition dropped, to obtain a problem with equality constraints only. This indeed is the structure of the problem near the solution.

If in this structure the vector $\boldsymbol{x}$ is n-dimensional and $\boldsymbol{h}$ is m-dimensional, then $\boldsymbol{\lambda}$ will also be m-dimensional. The system eq. (12.6.1) will, in this reduced form, consist of $n+m$ equations and $n+m$ unknowns, which is an indication that the system may be well defined, and hence that there is a solution for the pair $(\boldsymbol{x}, \boldsymbol{\lambda})$. In essence, primal-dual methods amount to solving this system of equations, and use additional strategies to account for inequality constraints. In view of the above observation it is natural to consider whether in fact the system of necessary conditions is in fact well conditioned, possessing a unique solution $(\boldsymbol{x}, \boldsymbol{\lambda})$. We investigate this question by considering a linearized version of the conditions.

A useful and somewhat more generally useful approach is to consider the quadratic program

$$\begin{aligned} &\text{minimize} \quad \frac{1}{2}\boldsymbol{x}^{\mathrm{T}}\boldsymbol{Q}\boldsymbol{x} + \boldsymbol{c}^{\mathrm{T}}\boldsymbol{x} \\ &\text{subject to} \quad \boldsymbol{A}\boldsymbol{x} = \boldsymbol{b} \end{aligned} \tag{12.6.3}$$

where $\boldsymbol{x}$ is n-dimensional and $\boldsymbol{b}$ is m-dimensional.

The first-order conditions for this problem are

$$\begin{aligned} \boldsymbol{Q}\boldsymbol{x} + \boldsymbol{A}^{\mathrm{T}}\boldsymbol{\lambda} + \boldsymbol{c} &= \mathbf{0} \\ \boldsymbol{A}\boldsymbol{x} - \boldsymbol{b} &= \mathbf{0} \end{aligned} \tag{12.6.4}$$

These correspond to the necessary conditions eq. (12.6.2) for equality constraints only. The following proposition gives conditions under which the system is nonsingular.

Proposition: Let $\boldsymbol{Q}$ and $\boldsymbol{A}$ be $n \times n$ and $m \times n$ matrices, respectively. Suppose that $\boldsymbol{A}$ has rank m and that $\boldsymbol{Q}$ is positive definite on the subspace $M = \{\boldsymbol{x} : \boldsymbol{A}\boldsymbol{x} = \mathbf{0}\}$. Then the matrix

$$\begin{bmatrix} \boldsymbol{Q} & \boldsymbol{A}^{\mathrm{T}} \\ \boldsymbol{A} & \mathbf{0} \end{bmatrix} \tag{12.6.5}$$

is nonsingular.

12.6.2 Strategies

Descent Measures

A fundamental concept that we have frequently used is that of assuring that progress is made at each step of an iterative algorithm. It is this that is used to guarantee global convergence. In primal methods this measure of descent is the objective function. Even the simplex method of linear programming is founded on this idea of making progress with respect to the objective function. For primal minimization methods, one typically arranges that the objective function decreases at each step.

The objective function is not the only possible way to measure progress. We have, for example, when minimizing a function f, considered the quantity $(1/2)|\nabla f(\boldsymbol{x})|^2$ seeking to monotonically reduce it to zero.

In general, a function used to measure progress is termed as a merit function. Typically, it is defined so as to decrease as progress is made toward the solution of a minimization problem, but the sign may be reversed in some definitions. For primal-dual methods, the merit function may depend on both $\boldsymbol{x}$ and $\boldsymbol{\lambda}$. One especially useful merit function for equality constrained problems is

$$m(\boldsymbol{x},\boldsymbol{\lambda}) = \frac{1}{2}|\nabla f(\boldsymbol{x}) + \boldsymbol{\lambda}^{\mathrm{T}} \nabla \boldsymbol{h}(\boldsymbol{x})|^2 + \frac{1}{2}|\boldsymbol{h}(\boldsymbol{x})|^2$$

It is examined in the next section.

Active Set Methods

Inequality constraints can be treated using active set methods that treat the active constraints as equality constraints, at least for the current iteration. However, in primal-dual methods, both $\boldsymbol{x}$ and $\boldsymbol{\lambda}$ are changed. We shall consider variations of steepest descent, conjugate directions, and Newton's method where movement is made in the $(\boldsymbol{x},\boldsymbol{\lambda})$ space.

Penalty Functions

In some primal-dual methods, a penalty function can serve as a merit function, even though the penalty function depends only on $\boldsymbol{x}$. This is particularly attractive for recursive quadratic programming methods where a quadratic program is solved at each stage to determine the direction of change in the pair $(\boldsymbol{x},\boldsymbol{\lambda})$.

Interior (Barrier) Methods

Barrier methods lead to methods that move within the relative interior of the inequality constraints. This approach leads to the concept of the primal-dual central path. These methods are used for semidefinite programming since these problems are characterized as possessing a special form of inequality constraint.

12.6.3 A Simple Merit Function

It is very natural, when considering the system of necessary conditions, to form the function

$$m(\boldsymbol{x},\boldsymbol{\lambda}) = \frac{1}{2}|\nabla f(\boldsymbol{x}) + \boldsymbol{\lambda}^{\mathrm{T}} \nabla \boldsymbol{h}(\boldsymbol{x})|^2 + \frac{1}{2}|\boldsymbol{h}(\boldsymbol{x})|^2 \qquad (12.6.6)$$

and use it as a measure of how close a point$(\boldsymbol{x},\boldsymbol{\lambda})$ is to a solution.

It must be noted, however, that the function $m(\boldsymbol{x},\boldsymbol{\lambda})$ is not always well-behaved; it may have local minima, and these are of no value in a search for a solution. The following theorem gives the conditions under which the function $m(\boldsymbol{x},\boldsymbol{\lambda})$ can serve as a well-behaved merit function. Basically, the main requirement is that the Hessian of the Lagrangian be positive definite. As usual, we define $l(\boldsymbol{x},\boldsymbol{\lambda}) = f(\boldsymbol{x}) + \boldsymbol{\lambda}^{\mathrm{T}}\boldsymbol{h}(\boldsymbol{x})$.

Theorem. Let f and $\boldsymbol{h}$ respectively be twice continuously differentiable functions on E^n of dimension l and m, and that $\boldsymbol{x}^*$ and $\boldsymbol{\lambda}^*$ satisfy the first-order necessary conditions for a local minimum of $m(\boldsymbol{x},\boldsymbol{\lambda}) = \frac{1}{2}|\nabla f(\boldsymbol{x}) + \boldsymbol{\lambda}^{\mathrm{T}} \nabla \boldsymbol{h}(\boldsymbol{x})|^2 + \frac{1}{2}|\boldsymbol{h}(\boldsymbol{x})|^2$ with respect to $\boldsymbol{x}$ and $\boldsymbol{\lambda}$, and also that the rank of $\nabla \boldsymbol{h}(\boldsymbol{x}^*)$ is m and the Hessian matrix $\boldsymbol{L}(\boldsymbol{x}^*,\boldsymbol{\lambda}^*) = \boldsymbol{F}(\boldsymbol{x}^*) + \boldsymbol{\lambda}^{*\mathrm{T}}\boldsymbol{H}(\boldsymbol{x}^*)$ is positive definite at $\boldsymbol{x}^*,\boldsymbol{\lambda}^*$, then $\boldsymbol{x}^*,\boldsymbol{\lambda}^*$, are a global minimum solution of $m(\boldsymbol{x},\boldsymbol{\lambda})$, are not necessarily unique.

12.6.4 Basic Primal-dual Methods

First-order Method

We consider first a simple straightforward approach, which only uses a first-order approximation to the primal-dual equations and in a sense, which parallels to the steepest descent method. It is defined by

$$\boldsymbol{x}_{k+1} = \boldsymbol{x}_k - \alpha_k \nabla l(\boldsymbol{x}_k,\boldsymbol{\lambda}_k)^{\mathrm{T}}$$
$$\boldsymbol{\lambda}_{k+1} = \boldsymbol{\lambda}_k - \alpha_k \boldsymbol{h}(\boldsymbol{x}_k) \qquad (12.6.7)$$

where α_k is not yet determined. This is based on the error in satisfying eq. (12.6.

2). Assume that the Hessian of the Lagrangian $\boldsymbol{L}(\boldsymbol{x},\boldsymbol{\lambda})$ is positive definite in some compact region of interest, and consider the simple merit function

$$m(\boldsymbol{x},\boldsymbol{\lambda}) = \frac{1}{2}|\nabla \boldsymbol{l}(\boldsymbol{x},\boldsymbol{\lambda})|^2 + \frac{1}{2}|\boldsymbol{h}(\boldsymbol{x})|^2 \tag{12.6.8}$$

discussed above. We would like to determine whether the direction of change in eq. (12.6.7) is a descent direction with respect to this merit function. The gradient of the merit function has components corresponding to $\boldsymbol{x}$ and $\boldsymbol{\lambda}$ of

$$\nabla \boldsymbol{l}(\boldsymbol{x},\boldsymbol{\lambda})\boldsymbol{L}(\boldsymbol{x},\boldsymbol{\lambda}) + \boldsymbol{h}(\boldsymbol{x})^{\mathrm{T}} \nabla \boldsymbol{h}(\boldsymbol{x})$$

$$\nabla \boldsymbol{l}(\boldsymbol{x},\boldsymbol{\lambda}) \nabla \boldsymbol{h}(\boldsymbol{x})^{\mathrm{T}} \tag{12.6.9}$$

Thus the inner product of this gradient with the direction vector having components $-\nabla \boldsymbol{l}(\boldsymbol{x},\boldsymbol{\lambda})^{\mathrm{T}}, \boldsymbol{h}(\boldsymbol{x})$ is

$$-\nabla \boldsymbol{l}(\boldsymbol{x},\boldsymbol{\lambda})\boldsymbol{L}(\boldsymbol{x},\boldsymbol{\lambda})\nabla \boldsymbol{l}(\boldsymbol{x},\boldsymbol{\lambda})^{\mathrm{T}} - \boldsymbol{h}(\boldsymbol{x})^{\mathrm{T}} \nabla \boldsymbol{h}(\boldsymbol{x}) \nabla \boldsymbol{l}(\boldsymbol{x},\boldsymbol{\lambda})^{\mathrm{T}} + \nabla \boldsymbol{l}(\boldsymbol{x},\boldsymbol{\lambda}) \nabla \boldsymbol{h}(\boldsymbol{x})^{\mathrm{T}} \boldsymbol{h}(\boldsymbol{x}) = -\nabla \boldsymbol{l}(\boldsymbol{x},\boldsymbol{\lambda})\boldsymbol{L}(\boldsymbol{x},\boldsymbol{\lambda})\nabla \boldsymbol{l}(\boldsymbol{x},\boldsymbol{\lambda})^{\mathrm{T}} \leqslant 0$$

This shows that the search direction is in fact a descent direction for the merit function, unless $\nabla \boldsymbol{l}(\boldsymbol{x},\boldsymbol{\lambda}) = \boldsymbol{0}$. Thus by selecting α_k to minimize the merit function in the search direction at each step, the process will converge to a point where $\nabla \boldsymbol{l}(\boldsymbol{x},\boldsymbol{\lambda}) = \boldsymbol{0}$. However, there is no guarantee that $\boldsymbol{h}(\boldsymbol{x}) = \boldsymbol{0}$ at that point.

We can try to improve the method either by changing the way in which the direction is selected or by changing the merit function. In this case a slight modification of the merit function will work. Let

$$w(\boldsymbol{x},\boldsymbol{\lambda},\gamma) = m(\boldsymbol{x},\boldsymbol{\lambda}) - \gamma[f(\boldsymbol{x}) + \boldsymbol{\lambda}^{\mathrm{T}}\boldsymbol{h}(\boldsymbol{x})]$$

for some $\gamma > 0$. We then calculate that the gradient of w has two components corresponding to $\boldsymbol{x}$ and $\boldsymbol{\lambda}$

$$\nabla \boldsymbol{l}(\boldsymbol{x},\boldsymbol{\lambda})\boldsymbol{L}(\boldsymbol{x},\boldsymbol{\lambda}) + \boldsymbol{h}(\boldsymbol{x})^{\mathrm{T}} \nabla \boldsymbol{h}(\boldsymbol{x}) - \gamma \nabla \boldsymbol{l}(\boldsymbol{x},\boldsymbol{\lambda})$$

$$\nabla \boldsymbol{l}(\boldsymbol{x},\boldsymbol{\lambda}) \nabla \boldsymbol{h}(\boldsymbol{x})^{\mathrm{T}} - \gamma \boldsymbol{h}(\boldsymbol{x})^{\mathrm{T}}$$

and hence the inner product of the gradient with the direction $-\nabla \boldsymbol{l}(\boldsymbol{x},\boldsymbol{\lambda})^{\mathrm{T}}, \boldsymbol{h}(\boldsymbol{x})$ is

$$-\nabla \boldsymbol{l}(\boldsymbol{x},\boldsymbol{\lambda})[\boldsymbol{L}(\boldsymbol{x},\boldsymbol{\lambda}) - \gamma \boldsymbol{I}]\nabla \boldsymbol{l}(\boldsymbol{x},\boldsymbol{\lambda})^{\mathrm{T}} - \gamma|\boldsymbol{h}(\boldsymbol{x})|^2$$

Now since we are assuming that $\boldsymbol{L}(\boldsymbol{x},\boldsymbol{\lambda})$ is positive definite in a compact region of interest, there is a $\gamma > 0$ such that $\boldsymbol{L}(\boldsymbol{x},\boldsymbol{\lambda}) - \gamma \boldsymbol{I}$ is positive definite in this region. Then according to the above calculation, the direction $-\nabla \boldsymbol{l}(\boldsymbol{x},\boldsymbol{\lambda})^{\mathrm{T}}, \boldsymbol{h}(\boldsymbol{x})$ is a descent direction, and the standard descent method will converge to a solution.

Conjugate Directions

Consider the quadratic program

$$\text{minimize} \quad \frac{1}{2}\boldsymbol{x}^{\mathrm{T}}\boldsymbol{Q}\boldsymbol{x} - \boldsymbol{b}^{\mathrm{T}}\boldsymbol{x} \qquad (12.6.10)$$

$$\text{subject to} \quad \boldsymbol{A}\boldsymbol{x} = \boldsymbol{c}$$

The first-order necessary conditions for this problem are

$$\boldsymbol{Q}\boldsymbol{x} + \boldsymbol{A}^{\mathrm{T}}\boldsymbol{\lambda} = \boldsymbol{b}$$

$$\boldsymbol{A}\boldsymbol{x} = \boldsymbol{c} \qquad (12.6.11)$$

As discussed in the previous section, this problem is equivalent to solving a system of linear equations whose coefficient matrix is

$$\boldsymbol{M} = \begin{bmatrix} \boldsymbol{Q} & \boldsymbol{A}^{\mathrm{T}} \\ \boldsymbol{A} & \boldsymbol{0} \end{bmatrix} \qquad (12.6.12)$$

This matrix is symmetric, but it is not positive definite (nor even semidefinite).

Relation to Quadratic Programming

Consider the problem

$$\text{minimize} \quad \boldsymbol{l}_k^{\mathrm{T}}\boldsymbol{d}_k + \frac{1}{2}\boldsymbol{d}_k^{\mathrm{T}}\boldsymbol{L}_k\boldsymbol{d}_k \qquad (12.6.13)$$

$$\text{subject to} \quad \boldsymbol{A}_k\boldsymbol{d}_k + \boldsymbol{h}_k = \boldsymbol{0}$$

The first-order necessary conditions of this problem are exact, where $\boldsymbol{y}_k$ corresponds to the Lagrange multiplier of eq. (12.6.13). Thus, the solution of eq. (12.6.13) produces a Newton step.

Alternatively, we may consider the quadratic program

$$\text{minimize} \quad \nabla f(\boldsymbol{x}_k)\boldsymbol{d}_k + \frac{1}{2}\boldsymbol{d}_k^{\mathrm{T}}\boldsymbol{L}_k\boldsymbol{d}_k \qquad (12.6.14)$$

$$\text{subject to} \quad \boldsymbol{A}_k\boldsymbol{d}_k + \boldsymbol{h}_k = \boldsymbol{0}$$

The necessary conditions of this problem are exact, where $\boldsymbol{\lambda}_{k+1}$ now corresponds to the Lagrange multiplier of eq. (12.6.14). The program eq. (12.6.14) is obtained from eq. (12.6.13) by merely subtracting $\boldsymbol{\lambda}_k^{\mathrm{T}}\boldsymbol{A}_k\boldsymbol{d}_k$ from the objective function; and this change has no influence on $\boldsymbol{d}_k$, since $\boldsymbol{A}_k\boldsymbol{d}_k$ is fixed.

Consider the problem

$$\text{minimize} \quad f(\boldsymbol{x})$$

$$\text{subject to} \quad \boldsymbol{h}(\boldsymbol{x}) = \boldsymbol{0}$$

$$\boldsymbol{g}(\boldsymbol{x}) \leqslant \boldsymbol{0}$$

Given an estimated solution point $\boldsymbol{x}_k$ and estimated Lagrange multipliers $\boldsymbol{\lambda}_k, \boldsymbol{\mu}_k$, one solves the quadratic program

$$\begin{aligned} &\text{minimize} \quad \nabla f(\boldsymbol{x}_k)\boldsymbol{d}_k + \frac{1}{2}\boldsymbol{d}_k^{\mathrm{T}}\boldsymbol{L}_k\boldsymbol{d}_k \\ &\text{subject to} \quad \nabla \boldsymbol{h}(\boldsymbol{x}_k)\boldsymbol{d}_k + \boldsymbol{h}_k = \boldsymbol{0} \\ &\qquad\qquad\quad \nabla \boldsymbol{g}(\boldsymbol{x}_k)\boldsymbol{d}_k + \boldsymbol{g}_k \leqslant \boldsymbol{0} \end{aligned} \tag{12.6.15}$$

where $\boldsymbol{L}_k = \boldsymbol{F}(\boldsymbol{x}_k) + \boldsymbol{\lambda}_k^{\mathrm{T}}\boldsymbol{H}(\boldsymbol{x}_k) + \boldsymbol{\mu}_k^{\mathrm{T}}\boldsymbol{G}(\boldsymbol{x}_k)$, $\boldsymbol{h}_k = \boldsymbol{h}(\boldsymbol{x}_k)$, $\boldsymbol{g}_k = \boldsymbol{g}(\boldsymbol{x}_k)$. The new point is determined by $\boldsymbol{x}_{k+1} = \boldsymbol{x}_k + \boldsymbol{d}_k$, and the new Lagrange multipliers are the Lagrange multipliers of the quadratic program eq. (12.6.15).

12.6.5 Modified Newton Methods

A modified Newton method is based on replacing the actual linearized system by an approximation.

First, we concentrate on the equality constrained optimization problem

$$\begin{aligned} &\text{minimize} \quad f(\boldsymbol{x}) \\ &\text{subject to} \quad \boldsymbol{h}(\boldsymbol{x}) = \boldsymbol{0} \end{aligned} \tag{12.6.16}$$

in order to most clearly describe the relationships between the various approaches.

The basic equations for Newton's method can be written as

$$\begin{bmatrix} \boldsymbol{x}_{k+1} \\ \boldsymbol{\lambda}_{k+1} \end{bmatrix} = \begin{bmatrix} \boldsymbol{x}_k \\ \boldsymbol{\lambda}_k \end{bmatrix} - \alpha_k \begin{bmatrix} \boldsymbol{L}_k & \boldsymbol{A}_k^{\mathrm{T}} \\ \boldsymbol{A}_k & \boldsymbol{0} \end{bmatrix}^{-1} \begin{bmatrix} \boldsymbol{I}_k \\ \boldsymbol{h}_k \end{bmatrix}$$

where as before $\boldsymbol{L}_k$ is the Hessian of the Lagrangian, $\boldsymbol{A}_k = \nabla \boldsymbol{h}(\boldsymbol{x}_k)$, $\boldsymbol{I}_k = [\nabla f(\boldsymbol{x}_k) + \boldsymbol{\lambda}_k^{\mathrm{T}} \nabla \boldsymbol{h}(\boldsymbol{x}_k)]^{\mathrm{T}}$, $\boldsymbol{h}_k = \boldsymbol{h}(\boldsymbol{x}_k)$. A structured modified Newton method is a method of the form

$$\begin{bmatrix} \boldsymbol{x}_{k+1} \\ \boldsymbol{\lambda}_{k+1} \end{bmatrix} = \begin{bmatrix} \boldsymbol{x}_k \\ \boldsymbol{\lambda}_k \end{bmatrix} - \alpha_k \begin{bmatrix} \boldsymbol{B}_k & \boldsymbol{A}_k^{\mathrm{T}} \\ \boldsymbol{A}_k & \boldsymbol{0} \end{bmatrix}^{-1} \begin{bmatrix} \boldsymbol{I}_k \\ \boldsymbol{h}_k \end{bmatrix} \tag{12.6.17}$$

where $\boldsymbol{B}_k$ is an approximation to $\boldsymbol{L}_k$. The term "structured" derives from the fact that only second-order information in the original system of equations is approximated; the first-order information is kept intact.

Of course, the method is implemented by solving the system

$$\begin{aligned} \boldsymbol{B}_k\boldsymbol{d}_k + \boldsymbol{A}_k^{\mathrm{T}}\boldsymbol{y}_k &= -\boldsymbol{I}_k \\ \boldsymbol{A}_k\boldsymbol{d}_k &= -\boldsymbol{h}_k \end{aligned} \tag{12.6.18}$$

for $\boldsymbol{d}_k$ and $\boldsymbol{y}_k$ and then setting $\boldsymbol{x}_{k+1} = \boldsymbol{x}_k + \alpha_k\boldsymbol{d}_k$, $\boldsymbol{\lambda}_{k+1} = \boldsymbol{\lambda}_k + \alpha_k\boldsymbol{y}_k$, for some value of α_k. In this section we will not consider the procedure for selection of α_k, and thus for simplicity we take $\alpha_k = 1$. The simple transformation used earlier can be applied to write eq. (12.6.18) in the form

$$\begin{aligned} \boldsymbol{B}_k\boldsymbol{d}_k + \boldsymbol{A}_k^{\mathrm{T}}\boldsymbol{\lambda}_{k+1} &= -\nabla f(\boldsymbol{x}_k)^{\mathrm{T}} \\ \boldsymbol{A}_k\boldsymbol{d}_k &= -\boldsymbol{h}_k \end{aligned} \tag{12.6.19}$$

Then $\boldsymbol{x}_{k+1} = \boldsymbol{x}_k + \boldsymbol{d}_k$, and $\boldsymbol{\lambda}_{k+1}$ is found directly as a solution to system eq. (12.6.19).

There are, of course, various ways to choose the approximation $\boldsymbol{B}_k$. One is to use a fixed, constant matrix throughout the iterative process. A second is to base $\boldsymbol{B}_k$ on some readily accessible information in $\boldsymbol{L}(\boldsymbol{x}_k, \boldsymbol{\lambda}_k)$, such as setting $\boldsymbol{B}_k$ equal to the diagonal of $\boldsymbol{L}(\boldsymbol{x}_k, \boldsymbol{\lambda}_k)$. Finally, a third possibility is to update $\boldsymbol{B}_k$ using one of the various quasi-Newton formulae.

One important advantage of the structured method is that $\boldsymbol{B}_k$ can be taken to be positive definite even though $\boldsymbol{L}_k$ is not. If this is done, we can write the explicit solution

$$\boldsymbol{y}_k = (\boldsymbol{A}_k\boldsymbol{B}_k^{-1}\boldsymbol{A}_k^{\mathrm{T}})^{-1}[\boldsymbol{h}_k - \boldsymbol{A}_k\boldsymbol{B}_k^{-1}\boldsymbol{l}_k] \tag{12.6.20}$$

$$\boldsymbol{d}_k = -\boldsymbol{B}_k^{-1}[\boldsymbol{I} - \boldsymbol{A}_k^{\mathrm{T}}(\boldsymbol{A}_k\boldsymbol{B}_k^{-1}\boldsymbol{A}_k^{\mathrm{T}})^{-1}\boldsymbol{A}_k\boldsymbol{B}_k^{-1}]\boldsymbol{l}_k - \boldsymbol{B}_k^{-1}\boldsymbol{A}_k^{\mathrm{T}}(\boldsymbol{A}_k\boldsymbol{B}_k^{-1}\boldsymbol{A}_k^{\mathrm{T}})^{-1}\boldsymbol{h}_k \tag{12.6.21}$$

Quadratic Programming

Consider the quadratic program

$$\begin{aligned} \text{minimize} \quad & \nabla f(\boldsymbol{x}_k)\boldsymbol{d}_k + \frac{1}{2}\boldsymbol{d}_k^{\mathrm{T}}\boldsymbol{B}_k\boldsymbol{d}_k \\ \text{subject to} \quad & \boldsymbol{A}_k\boldsymbol{d}_k + \boldsymbol{h}(\boldsymbol{x}_k) = \boldsymbol{0} \end{aligned} \tag{12.6.22}$$

The first-order necessary conditions for this problem are

$$\begin{aligned} \boldsymbol{B}_k\boldsymbol{d}_k + \boldsymbol{A}_k^{\mathrm{T}}\boldsymbol{\lambda}_{k+1} + \nabla f(\boldsymbol{x}_k)^{\mathrm{T}} &= \boldsymbol{0} \\ \boldsymbol{A}_k\boldsymbol{d}_k &= -\boldsymbol{h}(\boldsymbol{x}_k) \end{aligned} \tag{12.6.23}$$

which are again identical to the system of equations of the structured modified Newton method—in this case in the form eq. (12.6.20). The Lagrange multiplier of the quadratic program is $\boldsymbol{\lambda}_{k+1}$. The equivalence of eq. (12.6.22) and eq. (12.6.23) leads to a recursive quadratic programming method, where at each $\boldsymbol{x}_k$ the quadratic program eq. (12.6.22) is solved to determine the direction $\boldsymbol{d}_k$. In this case an arbitrary symmetric matrix $\boldsymbol{B}_k$ is used in place of the Hessian of the Lagrangian. Note that the problem eq. (12.6.22) does not explicitly depend on $\boldsymbol{\lambda}_k$; but $\boldsymbol{B}_k$, often being chosen to approximate the Hessian of the Lagrangian, may depend on $\boldsymbol{\lambda}_k$.

12.6.6 Descent Properties

Absolute-value Penalty Function

Let us consider the constrained minimization problem

$$\begin{aligned}&\text{minimize} \quad f(\boldsymbol{x})\\ &\text{subject to} \quad \boldsymbol{g}(\boldsymbol{x}) \leqslant \boldsymbol{0}\end{aligned} \tag{12.6.24}$$

where $\boldsymbol{g}(\boldsymbol{x})$ is r-dimensional. In accordance with the recursive quadratic programming approach, given a current point $\boldsymbol{x}$, we select the direction of movement $\boldsymbol{d}$ by solving the quadratic programming problem

$$\begin{aligned}&\text{minimize} \quad \frac{1}{2}\boldsymbol{d}^{\mathrm{T}}\boldsymbol{B}\boldsymbol{d} + \nabla f(\boldsymbol{x})\boldsymbol{d}\\ &\text{subject to} \quad \nabla \boldsymbol{g}(\boldsymbol{x})\boldsymbol{d} + \boldsymbol{g}(\boldsymbol{x}) \leqslant \boldsymbol{0}\end{aligned} \tag{12.6.25}$$

where $\boldsymbol{B}$ is positive definite.

The first-order necessary conditions for a solution to this quadratic program are

$$\boldsymbol{B}\boldsymbol{d} + \nabla f(\boldsymbol{x})^{\mathrm{T}} + \nabla \boldsymbol{g}(\boldsymbol{x})^{\mathrm{T}}\boldsymbol{\mu} = \boldsymbol{0} \tag{12.6.26a}$$

$$\nabla \boldsymbol{g}(\boldsymbol{x})\boldsymbol{d} + \boldsymbol{g}(\boldsymbol{x}) \leqslant \boldsymbol{0} \tag{12.6.26b}$$

$$\boldsymbol{\mu}^{\mathrm{T}}[\nabla \boldsymbol{g}(\boldsymbol{x})\boldsymbol{d} + \boldsymbol{g}(\boldsymbol{x})] = 0 \tag{12.6.26c}$$

$$\boldsymbol{\mu} \geqslant \boldsymbol{0} \tag{12.6.26d}$$

Note that if the solution to the quadratic program has $\boldsymbol{d} = \boldsymbol{0}$, then the point $\boldsymbol{x}$, together with $\boldsymbol{\mu}$ from eq. (12.6.26), satisfies the first-order necessary conditions for the original minimization problem eq. (12.6.24).

Proposition 1: Let $\boldsymbol{d}, \boldsymbol{\mu}$ (with $\boldsymbol{d} \neq \boldsymbol{0}$) be a solution of the quadratic program eq. (12.6.25), then if $c \geqslant \max\limits_{j}(\mu_j)$, the vector $\boldsymbol{d}$ is a descent direction for the penalty function

$$\boldsymbol{P}(\boldsymbol{x}) = f(\boldsymbol{x}) + c\sum_{j=1}^{r} g_j(\boldsymbol{x})^{+}.$$

Theorem: Let $\boldsymbol{B}$ be positive definite and assume that throughout some compact region $\subset E^n$, the quadratic program eq. (12.6.25) has a unique solution $\boldsymbol{d}, \boldsymbol{\mu}$ such that at each point the Lagrange multipliers satisfy $\max\limits_{j}(\mu_j) \leqslant c$. Let the sequence $\{\boldsymbol{x}_k\}$ be generated by $\boldsymbol{x}_{k+1} = \boldsymbol{x}_k + \alpha_k \boldsymbol{d}_k$, where $\boldsymbol{d}_k$ is the solution to eq. (12.6.25) at $\boldsymbol{x}_k$ and where k minimizes $\boldsymbol{P}(\boldsymbol{x}_{k+1})$. Assume that each $\boldsymbol{x}_k \in \Omega$. Then every limit point $\bar{\boldsymbol{x}}$ of $\{\boldsymbol{x}_k\}$ satisfies the first-order necessary conditions for the constrained minimization problem eq. (12.6.24).

The Quadratic Penalty Function

Thus we consider the problem

$$\begin{aligned} &\text{minimize} \quad f(\boldsymbol{x}) \\ &\text{subject to} \quad \boldsymbol{h}(\boldsymbol{x}) = \mathbf{0} \end{aligned} \tag{12.6.27}$$

and the standard quadratic penalty objective

$$\boldsymbol{P}(\boldsymbol{x}) = f(\boldsymbol{x}) + \frac{1}{2}c|\boldsymbol{h}(\boldsymbol{x})|^2 \tag{12.6.28}$$

we know that minimization of the objective with a quadratic penalty function will not yield an exact solution to eq. (12.6.27). In fact, the minimum of the penalty function eq. (12.6.28) will have $c\boldsymbol{h}(\boldsymbol{x}) \cong \boldsymbol{\lambda}$, where $\boldsymbol{\lambda}$ is the Lagrange multiplier of eq. (12.6.27). Therefore, it seems appropriate in this case to consider the quadratic programming problem

$$\begin{aligned} &\text{minimize} \quad \frac{1}{2}\boldsymbol{d}^{\mathrm{T}}\boldsymbol{B}\boldsymbol{d} + \nabla f(\boldsymbol{x})\boldsymbol{d} \\ &\text{subject to} \quad \nabla\boldsymbol{h}(\boldsymbol{x})\boldsymbol{d} + \boldsymbol{h}(\boldsymbol{x}) = \hat{\boldsymbol{\lambda}}/c \end{aligned} \tag{12.6.29}$$

where $\hat{\boldsymbol{\lambda}}$ is an estimate of the Lagrange multiplier of the original problem. A particularly good choice is

$$\hat{\boldsymbol{\lambda}} = [(1/c)\boldsymbol{I} + \boldsymbol{Q}]^{-1}[\boldsymbol{h}(\boldsymbol{x}) - \boldsymbol{A}\boldsymbol{B}^{-1}\nabla f(\boldsymbol{x})^{\mathrm{T}}] \tag{12.6.30}$$

where $\boldsymbol{A} = \nabla\boldsymbol{h}(\boldsymbol{x})$, $\boldsymbol{Q} = \boldsymbol{A}\boldsymbol{B}^{-1}\boldsymbol{A}^{\mathrm{T}}$ which is the Lagrange multiplier that would be obtained by the quadratic program with the penalty method.

Proposition 2: For any $c > 0$, let $\boldsymbol{d}, \boldsymbol{\lambda}$ (with $\boldsymbol{d} \neq \mathbf{0}$) be a solution to the quadratic program eq. (12.6.29), then $\boldsymbol{d}$ is a descent direction of the function $\boldsymbol{P}(\boldsymbol{x}) = f(\boldsymbol{x}) + (1/2)c|\boldsymbol{h}(\boldsymbol{x})|^2$.

12.6.7 Interior Point Methods

Consider the inequality constrained problem

$$\begin{aligned} &\text{minimize} \quad f(\boldsymbol{x}) \\ &\text{subject to} \quad \boldsymbol{Ax} = \boldsymbol{b} \\ &\qquad\qquad\quad \boldsymbol{g}(\boldsymbol{x}) \leqslant \mathbf{0} \end{aligned} \tag{12.6.31}$$

In general, a weakness of the active constraint method for such a problem is the combinatorial nature of determining which constraints should be active.

Logarithmic Barrier Method

A method that avoids the necessity to explicitly select a set of active constraints is based on the logarithmic barrier method, which solves a sequence of equality constrained minimization problems. Specifically,

$$\begin{aligned} &\text{minimize} \quad f(\boldsymbol{x}) - \mu \sum_{i=1}^{p} \log(-g_i(\boldsymbol{x})) \\ &\text{subject to} \quad \boldsymbol{A}\boldsymbol{x} = \boldsymbol{b} \end{aligned} \tag{12.6.32}$$

where $\mu = \mu^k > 0, k = 1, \ldots, \mu^k > \mu^{k+1}, \mu^k \to \infty$. μ^k can be pre-determined. Typically, we have $\mu^{k+1} = \gamma\mu^k$ for some constant $0 < \gamma < 1$. Here, we also assume that the original problem has a feasible interior-point $\boldsymbol{x}^0$; that is, $\boldsymbol{A}\boldsymbol{x}^0 = \boldsymbol{b}$ and $\boldsymbol{g}(\boldsymbol{x}^0) < \boldsymbol{0}$, and $\boldsymbol{A}$ has full row rank.

For fixed μ, and using $S_i = \mu/g_i$, the optimality conditions of the barrier problem eq. (12.6.32) are:

$$\begin{aligned} -\boldsymbol{S}\boldsymbol{g}(\boldsymbol{x}) &= \mu 1 \\ \boldsymbol{A}\boldsymbol{x} &= \boldsymbol{b} \\ -\boldsymbol{A}^{\mathrm{T}}\boldsymbol{y} + \nabla f(\boldsymbol{x})^{\mathrm{T}} + \nabla \boldsymbol{g}(\boldsymbol{x})^{\mathrm{T}}\boldsymbol{s} &= \boldsymbol{0} \end{aligned} \tag{12.6.33}$$

where $\boldsymbol{S} = \operatorname{diag}(\boldsymbol{s})$; that is, a diagonal matrix whose diagonal entries are $\boldsymbol{s}$, and $\nabla \boldsymbol{g}(\boldsymbol{x})$ is the Jacobian matrix of $\boldsymbol{g}(\boldsymbol{x})$.

If $f(\boldsymbol{x})$ and $g_i(\boldsymbol{x})$ are convex functions for all i, $f(\boldsymbol{x}) - \mu \sum_i \log(-g_i(\boldsymbol{x}))$ is strictly convex in the interior of the feasible region, and the objective level set is bounded, then there is a unique minimizer for the barrier problem. Let $(\boldsymbol{x}(\mu) > 0, \boldsymbol{y}(\mu), \boldsymbol{s}(\mu) > 0)$ be the (unique) solution of eq. (12.6.33), then, these values form the primal-dual central path of eq. (12.6.31):

$$\ell = \{(\boldsymbol{x}(\mu) > 0, \boldsymbol{y}(\mu), \boldsymbol{s}(\mu) > 0) : 0 < \mu < \infty\}$$

This can be summarized in the following theorem.

Theorem 1: Let $(\boldsymbol{x}(\mu), \boldsymbol{y}(\mu), \boldsymbol{s}(\mu))$ be on the central path.

1) If $f(\boldsymbol{x})$ and $g_i(\boldsymbol{x})$ are convex functions for all i, then $\boldsymbol{s}(\mu)$ is unique.

2) Furthermore, if $f(\boldsymbol{x}) - \mu \sum_i \log(-g_i(\boldsymbol{x}))$ is strictly convex, $(\boldsymbol{x}(\mu), \boldsymbol{y}(\mu), \boldsymbol{s}(\mu))$ are unique, and they are bounded for $0 < \mu \leqslant \mu^0$ for any given $\mu^0 > 0$.

3) For $0 < \mu' < \mu$, $f(\boldsymbol{x}(\mu')) < f(\boldsymbol{x}(\mu))$ *if* $\boldsymbol{x}(\mu') \neq \boldsymbol{x}(\mu)$.

4) $(\boldsymbol{x}(\mu), \boldsymbol{y}(\mu), \boldsymbol{s}(\mu))$ converges to a point satisfying the first-order necessary conditions for a solution of eq. (12.6.31) as $\mu \to 0$.

Once we have an approximate solution point $(\boldsymbol{x}, \boldsymbol{y}, \boldsymbol{s}) = (\boldsymbol{x}_k, \boldsymbol{y}_k, \boldsymbol{s}_k)$ for eq. (12.6.33) for $\mu = \mu^k > 0$, we can again use the primal-dual methods described for linear programming to generate a new approximate solution to eq. (12.6.33) *for* $\mu = \mu^{k+1} < \mu^k$. The Newton direction $(\boldsymbol{d}_x, \boldsymbol{d}_y, \boldsymbol{d}_s)$ is found from the system of linear equations:

$$\begin{aligned} -\boldsymbol{S}\,\nabla \boldsymbol{g}(\boldsymbol{x})\boldsymbol{d}_x - \boldsymbol{G}(\boldsymbol{x})\boldsymbol{d}_s &= \mu\boldsymbol{1} + \boldsymbol{S}\boldsymbol{g}(\boldsymbol{x}) \\ \boldsymbol{A}\boldsymbol{d}_x &= \boldsymbol{b} - \boldsymbol{A}\boldsymbol{x} \\ -\boldsymbol{A}^{\mathrm{T}}\boldsymbol{d}_y + \left(\nabla^2 f(\boldsymbol{x}) + \sum_i s_i \nabla^2 g_i(\boldsymbol{x})\right)\boldsymbol{d}_x &+ \\ \nabla \boldsymbol{g}(\boldsymbol{x})^{\mathrm{T}}\boldsymbol{d}_s &= \boldsymbol{A}^{\mathrm{T}}\boldsymbol{y} - \nabla f(\boldsymbol{x})^{\mathrm{T}} - \nabla \boldsymbol{g}(\boldsymbol{x})^{\mathrm{T}}\boldsymbol{s} \end{aligned} \tag{12.6.34}$$

where $\boldsymbol{G}(\boldsymbol{x}) = \mathrm{diag}(\boldsymbol{g}(\boldsymbol{x}))$.

Recently, this approach has also been used to find points satisfying the first-order conditions for problems when $f(\boldsymbol{x})$ and $g_i(\boldsymbol{x})$ are not generally convex functions.

Quadratic Programming

Let $f(\boldsymbol{x}) = (1/2)\boldsymbol{x}^{\mathrm{T}}\boldsymbol{Q}\boldsymbol{x} + \boldsymbol{c}^{\mathrm{T}}\boldsymbol{x}$ and $g_i(\boldsymbol{x}) = -\boldsymbol{x}_i$ for $i = 1, \ldots, n$, and consider the quadratic program

$$\begin{aligned} &\text{minimize} \quad \frac{1}{2}\boldsymbol{x}^{\mathrm{T}}\boldsymbol{Q}\boldsymbol{x} + \boldsymbol{c}^{\mathrm{T}}\boldsymbol{x} \\ &\text{subject to} \quad \boldsymbol{A}\boldsymbol{x} = \boldsymbol{b} \\ &\qquad\qquad\quad \boldsymbol{x} \geqslant \boldsymbol{0} \end{aligned} \tag{12.6.35}$$

where the given matrix $\boldsymbol{Q} \in E^{n\times n}$ is positive semidefinite (that is, the objective is a convex function), $\boldsymbol{Q} \in E^{n\times m}$, $\boldsymbol{c} \in E^n$ and $\boldsymbol{b} \in E^m$. The problem reduces to finding $\boldsymbol{x} \in E^n$, $\boldsymbol{y} \in E^m$ and $\boldsymbol{s} \in E^n$ satisfying the following optimality conditions:

$$\begin{aligned} \boldsymbol{S}\boldsymbol{x} &= \boldsymbol{0} \\ \boldsymbol{A}\boldsymbol{x} &= \boldsymbol{b} \\ -\boldsymbol{A}^{\mathrm{T}}\boldsymbol{y} + \boldsymbol{Q}\boldsymbol{x} - \boldsymbol{s} &= -\boldsymbol{c} \\ (\boldsymbol{x}, \boldsymbol{s}) &\geqslant \boldsymbol{0} \end{aligned} \tag{12.6.36}$$

The optimality conditions with the logarithmic barrier function with parameter μ are:

$$\begin{aligned} \boldsymbol{S}\boldsymbol{x} &= \mu\boldsymbol{1} \\ \boldsymbol{A}\boldsymbol{x} &= \boldsymbol{b} \\ -\boldsymbol{A}^{\mathrm{T}}\boldsymbol{y} + \boldsymbol{Q}\boldsymbol{x} - \boldsymbol{s} &= -\boldsymbol{c} \end{aligned} \tag{12.6.37}$$

Note that the bottom two sets of constraints are linear equalities.

Thus, once we have an interior feasible point $(\boldsymbol{x}, \boldsymbol{y}, \boldsymbol{s})$ for eq. (12.6.37), with $\mu = \boldsymbol{x}^{\mathrm{T}}\boldsymbol{s}/n$, we can apply Newton's method to compute a new (approximate) iterate $(\boldsymbol{x}^+, \boldsymbol{y}^+, \boldsymbol{s}^+)$ by solving for $(\boldsymbol{d}_x, \boldsymbol{d}_y, \boldsymbol{d}_s)$ from the system of linear equations:

$$Sd_x + Xd_s = \gamma\mu\mathbf{1} - Xs$$
$$Ad_x = \mathbf{0}$$
$$-A^{\mathrm{T}}d_y + Qd_x - d_s = \mathbf{0} \tag{12.6.38}$$

where X and S are two diagonal matrices whose diagonal entries are $x > \mathbf{0}$ and $s > \mathbf{0}$, respectively. Here, γ is a fixed positive constant less than 1, which implies that our targeted μ is reduced by the factor γ at each step.

Potential Function

For any interior feasible point (x, y, s) of eq. (12.6.35) and its dual, a suitable merit function is the potential function introduced in Chapter 5 for linear programming:

$$\psi_{n+\rho}(x, s) = (n + \rho)\log(x^{\mathrm{T}}s) - \sum_{j=1}^{n} \log(x_j s_j)$$

The main result for this is outlined below.

Theorem 2: In solving eq. (12.6.38) for (d_x, d_y, d_s), let $\gamma = n/(n + \rho) < 1$ for fixed $\rho \geqslant \sqrt{n}$ and assign $x^+ = x + \alpha d_x$, $y^+ = y + \alpha d_y$, and $s^+ = s + \alpha d_s$ where $\alpha = \dfrac{\bar{\alpha}\sqrt{\min(Xs)}}{\left|(Xs)^{-1/2}\left(\dfrac{x^{\mathrm{T}}s}{n+\rho}\mathbf{1} - Xs\right)\right|}$, where $\bar{\alpha}$ is any positive constant less than 1. (Again X and S are matrices with components on the diagonal being those of x and s, respectively.) then,

$$\psi_{n+\rho}(x^+, s^+) - \psi_{n+\rho}(x, s) \leqslant -\bar{\alpha}\sqrt{3/4} + \frac{\bar{\alpha}^2}{2(1-\bar{\alpha})}$$

We outline the algorithm here:

Given any interior feasible (x_0, y_0, s_0) of eq. (12.6.35) and its dual. Set $\rho \geqslant \sqrt{n}$ and $k = 0$.

Step 1. Set $(x, s) = (x_k, s_k)$ and $\gamma = n/(n + \rho)$ and compute (d_x, d_y, d_s) from eq. (12.6.38).

Step 2. Let $x_{k+1} = x_k + \bar{\alpha}d_x$, $y_{k+1} = y_k + \bar{\alpha}d_y$, and $s_{k+1} = s_k + \bar{\alpha}d_s$ where $\bar{\alpha} = \arg\min_{\alpha \geqslant 0} \psi_{n+\rho}(x_k + \alpha d_x, s_k + \alpha d_s)$.

Step 3. Let $k = k + 1$. If $s_k^{\mathrm{T}}x_k / s_0^{\mathrm{T}}x_0 \leqslant \varepsilon$, stop. Otherwise, return to Step 1.

Bibliography

Alizadeh, F. ,(1992), Combinatorial Optimization with Interior Point Methods and Semi-definite Matrices, Ph. D. Thesis, University of Minnesota, Minneapolis, Minn.

Andersen, E. D. , and Ye, Y. ,(1999), On a Homogeneous Algorithm for the Monotone Complementarity Problem, *Math. Prog.* 84, 375 -400.

Anstreicher, K. M. , den Hertog, D. , Roos, C. , and Terlaky, T. , (1993), A Long Step Barrier Method for Convex Quadratic Programming, *Algorithmica*, 10, 365 -382.

Bayer, D. A. , and Lagarias, J. C. , (1989), The Nonlinear Geometry of Linear Programming, Part I: Affine and Projective Scaling Trajectories, *Trans. Ame. Math. Soc.* , 314,(2),499 -526.

Bayer, D. A. , and Lagarias, J. C. , (1989), The Nonlinear Geometry of Linear Programming, Part II: Legendre Transform Coordinates, *Trans. Ame. Math. Soc.* 314 (2), 527 -581.

Beale, E. M. L. ,(1967), Numerical Methods, *Nonlinear Programming*, J. Abadie (editor), North-Holland, Amsterdam.

Bertsekas, D. P. ,(1982), Constrained Optimization and Lagrange Multiplier Methods, *Academic Press*, New York.

Bertsekas, D. P. ,(1995), Nonlinear Programming, *Athena Scientific*, Belmont, Mass.

Bertsimas, D. M. , and Tsitsiklis, J. N. ,(1997), Linear Optimization. *Athena Scientific*, Belmont, Mass.

Bixby, R. E. ,(1994), Progress in Linear Programming, *ORSA J. on Compute.* 6, 1,15 -22.

Blum, L. , Cucker, F. , Shub, M. , and Smale, S. ,(1996), Complexity and Real Computation, Springer-Verlag.

Boyd, S. , Ghaoui, L. E. , Feron, E. , Balakrishnan, V. , (1994), Linear Matrix Inequalities in System and Control Science, SIAM Publications. SIAM, Philadelphia.

Boyd, S. , and Vandenberghe, L. ,(2004), Convex Optimization, Cambridge University Press, Cambridge.

Burges, C. J. C. , (1998) , A Tutorial on Support Vector Machines for Pattern Recognition, *Data Mining and Knowledge Discovery* 2, 121 – 167.

Cottle, R. W. , Pang, J. S. & Stone, R. E. , (1992) , The Linear Complementarity Problem, Academic Press, Boston.

Cottle, R. W. , (2002), Linear Programming, Lecture Notes for MS&E 310, Stanford University, Stanford.

Collette Y. , Siarry P. , (2003), Multi-objective Optimization: Principles and Case Studies. Springer, Berlin Heidelberg, New York.

Dantzig, G. B. , and Thapa, M. N. , (1997) , Linear Programming 1: Introduction, Springer-Verlag, New York.

Dantzig, G. B. , and Thapa, M. N. , (2003) , Linear Programming 2: Theory and Extensions, Springer-Verlag, New York.

Dempe S. , (2002) , Foundations of Bilevel Programming, Kluwer, Boston, MA.

Ehrgott M. , (2005) , Multicriteria Optimization. Springer, Berlin Heidelberg, New York.

Eiselt H. A. , Sandblom C. L. (eds & authors), (2004), Decision Analysis, Location Models, and Scheduling Problems. Springer, Berlin Heidelberg New York.

Fang, S. C. , and Puthenpura, S. , (1994) , Linear Optimization and Extensions, Prentice-Hall, Englewood Cliffs. , N. J.

Fiacco, A. V. , and McCormick, G. P. , (1990) , Nonlinear Programming: Sequential Unconstrained Minimization Techniques, Volume 4 of *SIAM Classics in Applied Mathematics*, SIAM Publications, Philadelphia, PA.

Fletcher, R. , (1980) , Practical Methods of Optimization *1*: Unconstrained Optimization, John Wiley, Chichester.

Fletcher, R. , (1981) , Practical Methods of Optimization *2*: Constrained Optimization, John Wiley, Chichester.

Freund, R. M. , (1991) , Polynomial-time Algorithms for Linear Programming Based Only on Primal Scaling and Projected Gradients of a Potential Function, *Math. Prog.* , 51, 203 – 222.

Gass, S. (1985) , Linear Programming Methods and Applications, 5th ed. , Boyd and Fraser, Danvers, Massachusetts.

Gass S. I. , Assad A. A. , (2005) , An Annotated Timeline of Operations

Research: An Informal History. Kluwer, New York, NY.

Golub, G. H. &Van Loan, C. F. ,(1996),Matrix Computations, 3rd ed. , The Johns Hopkins University Press, Baltimore.

Greub, W. H. , (1967), Linear Algebra, 3rd ed. , Springer-Verlag, New York.

Greub, W. H. , (1967),Multilinear Algebra, Springer-Verlag, New York.

Hestenes, M. R. ,(1980), Conjugate-Direction Methods in Optimization, Springer-Verlag, Berlin.

Huang L. ,(1984), Linear Algebra and Its Application in System and Control Theory, *China Science Press*, Beijing (in Chinese).

Jarre, F. ,(1992), Interior-point Methods for Convex Programming, *Applied Mathematics & Optimization*, 26:287 – 311.

Kelley, J. L. , (1965),Introduction to Modern Algebra, Van Nostrand, New York.

Lawler, E. ,(1976), Combinatorial Optimization: Networks and Matroids, Holt, Rinehart, and Winston, New York.

Luenberger, D. G. ,(1984), Linear and Nonlinear Programming, 2nd edtion. Addison-Wesley, Reading.

Mangasarian, O. L. , Musicant, D. R. ,(2000), Robust Linear and Support Vector Regression, IEEE Transactions on Pattern Analysis and Machine Intelligence 22, 950 – 955.

Mangasarian, O. L. , Street, W. N. , Wolberg, W. H. ,(1995), Breast Cancer Diagnosis and Prognosis via Linear Programming, *Operations Research* 43, 570 – 577.

Mehrotra, S. ,(1992), On the Implementation of a Primal-dual Interior Point Method, *SIAM J. Optimization* 2(4), 575 – 601.

Mizuno, S. , Todd, M. J. , and Ye, Y. ,(1993), On Adaptive Step Primal-dual Interior-point Algorithms for Linear Programming, *Math. Oper. Res.* 18, 964 – 981.

Murtagh, B. A. ,(1981), Advanced Linear Programming, McGraw-Hill, New York.

Murty, K. G. ,(1976),Linear and Combinatorial Programming, John Wiley & Sons, New York.

Nash, S. G. , and Sofer, A. ,(1996), Linear and Nonlinear Programming,

New York, McGraw-Hill Companies, Inc.

Nazareth, J. L. ,(1987),Computer Solution of Linear Programs, Oxford University Press, Oxford, UK.

Nemhauser, G. L. ,Wolsey, L. A. ,(1988), Integer and Combinatorial Optimization, John Wiley & Sons, NewYork.

Nesterov, Y. , and Nemirovskii, A. , (1994), Interior Point Polynomial Methods in Convex Programming: Theory and Algorithms, SIAM Publications. SIAM, Philadelphia.

Nesterov, Y. ,(2004), Introductory Lectures on Convex Optimization: A Basic Course, Kluwer, Boston.

Nocedal, J. & Wright, S. J. ,(2006),Numerical Optimization, 2nd ed. , Springer-Verlag, NewYork.

Nomizu, K. , (1966),Fundamentals of Linear Algebra, McGraw-Hill, New York.

Papadimitriou, C. ,Steiglitz, K. , (1982), Combinatorial Optimization Algorithms and Complexity, Prentice-Hall, Englewood Cliffs, N. J.

Powell, M. J. D. ,(1978), Algorithms for Nonlinear Constraints that Use Lagrangian Functions, *Math. Prog.* 14, 224 – 248.

Rudin,W. ,(1974),Real and Complex Analysis, 2nd ed. , McGraw-Hill, Tokyo, Japan.

Renegar, J. , (2001), A Mathematical View of Interior-Point Methods in Convex Optimization, *Society for Industrial and Applied Mathematics*, Philadelphia.

Saigal, R. , (1995), Linear Programming: Modern Integrated Analysis. Kluwer Academic Publisher, Boston.

Schölkopf, B. , Smola, A. ,(2002), Learning with Kernels, MIT Press, Cambridge,Massachusetts.

Shephard, G. C. , (1966),Vector Spaces of Finite Dimensions, *Interscience*, New York.

Strang, G. ,(1993),Introduction to Linear Algebra, Wellesley-Cambridge Press, Wellesley, Massachusetts.

Vanderbei, R. J. , (1997),Linear Programming: Foundations and Extensions, Kluwer Academic Publishers, Boston.

Vavasis, S. A. ,(1991),Nonlinear Optimization: Complexity Issues, Oxford

Science, New York, N. Y.

Winston, W. L., Venkataramanan, M., (2003), Introduction to Mathematical Programming, Vol. 4, *Brooks/Cole-Thomson Learning*, Pacific Grove, California.

Wolsey, L. A., (1998), Integer Programming, Series in Discrete Mathematics and Optimization, JohnWiley & Sons, NewYork.

Wright, S. J., (1997), Primal-dual Interior-point Methods, *SIAM*, Philadelphia References.

Wu Q. H., (2013), Linear Algebra and Its Applications in Automatic Control, Defense Industrial Press, Beijing.

Ye, Y., (1997), Interior Point Algorithms, Wiley, New York.

图书在版编目（CIP）数据

线性代数及其在规划中的应用 = Linear Algebra and Its Applications in Programming：英文/郭树理，韩丽娜主编．—北京：北京理工大学出版社，2017.4

ISBN 978-7-5682-3969-1

Ⅰ.①线… Ⅱ.①郭… ②韩… Ⅲ.①线性代数-高等学校-教材-英文 ②线性规划-高等学校-教材-英文 Ⅳ.①O151.2 ②O221.1

中国版本图书馆 CIP 数据核字（2017）第 084091 号

出版发行／北京理工大学出版社有限责任公司
社　　址／北京市海淀区中关村南大街 5 号
邮　　编／100081
电　　话／（010）68914775（总编室）
（010）82562903（教材售后服务热线）
（010）68948351（其他图书服务热线）
网　　址／http：//www.bitpress.com.cn
经　　销／全国各地新华书店
印　　刷／保定市中画美凯印刷有限公司
开　　本／710 毫米×1000 毫米　1/16
印　　张／18
字　　数／296 千字
版　　次／2017 年 4 月第 1 版　2017 年 4 月第 1 次印刷
定　　价／45.00 元

责任编辑／梁铜华
文案编辑／梁铜华
责任校对／周瑞红
责任印制／王美丽